Lecture Notes in Earth Sciences

93

Editors:
S. Bhattacharji, Brooklyn
G. M. Friedman, Brooklyn and Troy
H. J. Neugebauer, Bonn
A. Seilacher, Tuebingen and Yale

Springer-Verlag Berlin Heidelberg GmbH

Alessandro Montanari   Christian Koeberl

Impact Stratigraphy

The Italian Record

With 174 Figures

 Springer

Authors

Dr. Alessandro Montanari
Osservatorio Geologico
di Coldigioco
62020 Frontale di Apiro, Italy

E-mail: sandro.ogc@fastnet.it

Professor Dr. Christian Koeberl
University of Vienna
Institute of Geochemistry
Althanstrasse 14
1090 Vienna, Austria

E-mail: christian.koeberl@univie.ac.at

Cataloging-in-Publication data applied for

Die Deutsche Bibliothek - CIP-Einheitsaufnahme

Montanari, Alessandro:
Impact stratigraphy: the Italian record/Alessandro Montanari;
Christian Koeberl. ESF IMPACT. - Berlin; Heidelberg; New York;
Barcelona; Hong Kong; London; Milan; Paris; Singapore; Tokyo:
Springer 2000
(Lecture notes in earth sciences; Vol. 93)

Corrected 2nd printing 2002

"For all Lecture Notes in Earth Sciences published till now please see final pages of
the book"

ISSN 0930-0317

ISBN 978-3-540-66368-3   ISBN 978-3-540-48366-3 (eBook)
DOI 10.1007/978-3-540-48366-3

Originally published by Springer-Verlag Berlin Heidelberg New York in 2000.

Typesetting: Camera ready by author
Printed on acid-free paper   SPIN: 10886563   32/3130as-54321

This volume is dedicated to
Walter Alvarez and Jan Smit,
whose studies on the K/T boundary started it all.

Preface

In this volume, we hope to bring together a general introduction to the study of impact structures and ejecta deposits, and a detailed description of how and where to find such impact deposits (for example, in the form of so-called impactoclastic layers) in the field. The marine sedimentary rocks of the Umbria-Marche (U-M) basin in Italy provide a remarkably complete and continuous stratigraphic record from the Late Triassic to the Pleistocene. Several major impact events happened during this time period, and the signatures of these impact events can be recognized in the field in the U-M sequence. It was in this region of Italy, near the town of Gubbio, where, in 1979, a team of researchers from Berkeley discovered extraterrestrial signal in the Cretaceous-Tertiary (K/T) boundary layer. It was this discovery that led to the hypothesis that the major extinction event, which occurred at the end of the Cretaceous, was caused by the impact of an asteroid or comet on Earth. The resulting vigorous scientific debate gave great publicity (not only among geoscientists) to impact research, which had been pursued for several decades by planetary geologists, cosmochemists, and mineralogists/petrologists who were interested in shock metamorphism.

The wider geological community had not been aware of their work, which led to misunderstandings and heated debates that could have been avoided if the data and evidence on impact research would have been more widely known. The last two decades have seen great progress. The findings of shocked minerals at the K/T boundary and the identification of the (buried) Chicxulub impact structure in Mexico as the K/T source crater led to a wide acceptance of the reality of an impact event at the K/T boundary among researchers. Nevertheless, no textbook is currently available that reviews impact processes and their deposits, and how to find them in the field. We hope that the present book will help to fill this gap.

The impetus for writing the present volume came from two different sources. First, an enormous amount of studies has been done over many decades on the U-M sequence (including work by one of us [A.M.], impact-related or not), and nowhere was this work summarized or discussed, especially with respect to the well-preserved impact signatures in this record. Second, in 1998, the European Science Foundation (ESF) launched a new scientific program on "Response of the Earth System to Impact Processes" (IMPACT), with one of us (C.K.) being the chairman of this program, and one of the first activities within the IMPACT program was the organization of a (probably annual) short course on "Impact Stratigraphy" in Coldigioco, to study the impact record of the U-M sequence. The idea of this book grew out of the need to produce course notes for this class, which was held for the first time in May 1999. The present volume is the result of a year-long gestation period following the preparation of this course.

The book is organized in two major parts and several chapters. Chaps. 1–3 set the stage on impact cratering and provide background information on the study of shock metamorphism, impact ejecta, and descriptions of distal ejecta from around the world. Chap. 4 introduces the geology of the Umbria-Marque sequence, and Chap. 5 gives detailed descriptions of the rock units in the U-M sequence, especially of those that have been found to contain impactoclastic layers. Chap. 6 discusses a few pertinent sampling and analysis techniques. Detailed location and outcrop maps and photographs should allow the reader or student to locate the discussed layers in the field. Almost all locations are easily accessible. Thus, the book will hopefully serve as both, an introduction into impact stratigraphy, and a field guidebook.

Coldigioco/Vienna, January 2000

Christian Koeberl
University of Vienna
Vienna
Austria

Alessandro Montanari
Osservatorio Geologico di Coldigioco
Frontale di Apiro
Italy

Acknowledgements

The authors wish to thank a long list of colleagues, friends, and family, for help, comments, data, images, support, and patience. The following are especially acknowledged for providing figures: B.F. Bohor (U.S. Geological Survey, Denver, USA), P. Claeys (Museum of Natural History, Berlin, Germany), V.L. Masaitis (Karpinsky Institute, St. Petersburg, Russia), L. Pesonen (Geological Survey of Finland), C.W. Poag (U.S. Geological Survey, Woods Hole, USA), W.U. Reimold (Wits University, Johannesburg, South Africa), V.L. Sharpton (University of Alaska, Fairbanks), J. Smit (Free University, Amsterdam, The Netherlands). Other figures without acknowledgements are by the authors. Dona Jalufka (Vienna) is much appreciated for drafting many of the line drawings in Chaps. 1–3 and 6. In addition, several colleagues read the various chapters of this book and prevented us from publishing too many errors and misconceptions (which does not mean that none remain): W.U. Reimold (Chap. 1), B.P. Glass (Chaps. 2, 4, 5, and 6), I. Gilmour (Chap. 3), and J. Smit (Chaps. 4 and 5). We appreciate financial assistance from the ESF IMPACT program, from the Austrian Fonds zur Förderung der wissenschaftlichen Forschung (Y58-GEO), and from the Austrian-Italian Scientific and Technical Exchange Program (ÖAD) for work related to, and for the preparation of, this volume. It almost goes without saying how grateful we are to our wives (Paula Metallo and Dona Jalufka, respectively) for their patience with us during the long months of work on this book. Furthermore, we acknowledge the students, who participated in the first "Impact Stratigraphy" short course (A. Abels, Münster; D. Boamah, Vienna; I. Dalwigk, Stockholm; S. Heuschkel, Berlin; H. Huber, Vienna; B. Kettrup, Münster; D. Kettrup, Münster; J.-M. Lange, Dresden; J. Monteiro, Lisbon; A. Rimsa, Vilnius; T. Salge, Berlin; E.G. Serrano, Madrid; S. Suuroja, Tallinn), who helped to test the curriculum. We are also grateful to W. Engel, A. Weber-Knapp, J. Sterrit-Brunner, and A. Bernauer-Budimann, all at Springer Verlag, for their help with, and support of, this project, and, last but not least, to J. Dryden (Open University, U.K.) for her excellent typesetting work of the manuscript.

Table of Contents

1 Introduction: Impact Cratering as a Geological Process................1

1.1   Historical Background..1
1.2   General Characteristics of Impact Craters.............................7
1.3   Recognition of Impact Structures.....................................10
1.4   Formation of Impact Craters..14
1.5   Shock Waves and the Stages of Crater Formation.......................16
1.6   Shock Metamorphic Effects..21
    1.6.1   General...21
    1.6.2   Shatter Cones...25
    1.6.3   Mosaicism...27
    1.6.4   Planar Microstructures..27
    1.6.5   Bulk Optical and other Properties.............................33
    1.6.6   Diaplectic Glass..34
    1.6.7   High-Pressure Polymorphs......................................34
    1.6.8   Mineral and Rock Melts..36
1.7   Classification of Impactites...39
1.8   Meteoritic Components in Impactites..................................42
    1.8.1   Meteoritic Components and Siderophile Element Analyses....43
    1.8.2   Problems with Meteoritic Component Identification..............47
    1.8.3   Osmium Isotopes in Impact Studies.............................49
    1.8.4   Chromium Isotope Systematics..................................50
1.9   Dating of Impact Structures..51

2 Distal Ejecta and Tektites...57

2.1   Characteristics of Proximal and Distal Ejecta........................57
2.2   Distal Ejecta as Stratigraphic Markers...............................60
2.3   Importance of the K/T boundary.......................................63
2.4   Tektites and Microtektites...64
2.4.1   Geographical Distribution: Strewn Fields and Age...............68
    2.4.2   Definition and Classification of Tektites.....................69
    2.4.3   Petrology and Geochemistry of Tektites........................71
    2.4.4   Source Rocks of Tektites......................................79
    2.4.5   Source Craters of Tektites....................................80

2.4.5.1 North American Strewn Field 80
2.4.5.2 Central European Strewn Field 85
2.4.5.3 Ivory Coast Strewn Field 86
2.4.5.4 Australasian Strewn Field 88
2.4.6 Tektite Formation: Facts and Open Questions 89
2.5 Other Impact Glasses: Relation to Source Craters 93
2.5.1 Libyan Desert Glass 95

3 Important Distal Ejecta Layers and Their Source
Craters 101
3.1 General 101
3.2 The K/T Boundary and Chicxulub 101
3.2.1 Evidence for Impact at the K/T boundary 104
3.2.1.1 PGE Enrichments 104
3.2.1.2. PGE Abundance Patterns; K/T boundary
Meteorite? 105
3.2.1.3 Meteoritic Os- and Cr-isotopic signature 106
3.2.1.4 Soot and Organic Chemistry 107
3.2.1.5 Evidence of shock metamorphism 109
3.2.1.6 Impact glass 111
3.2.1.7 Impact-derived diamonds 112
3.2.1.8 Occurrence of spinel 113
3.2.2 Chicxulub Impact Structure: Relation to the K/T Event 114
3.2.3 Chicxulub and the K/T Mass Extinction 123
3.3 Late Eocene Impact Layers 130
3.3.1 Chesapeake Bay 132
3.3.2 Popigai 133
3.4 Manson Impact Structure and Ejecta Layer 136
3.5 Acraman Impact Structure and Ejecta Layer 138
3.6 Morokweng and the Jurassic–Cretaceous Boundary 139
3.7 Other Confirmed and Possible Ejecta Layers 142
3.7.1 South African and Australian Archean Spherule Layers 142
3.7.2 Late Devonian Impact Layer and Alamo Breccia 148
3.7.3 Permian–Triassic Boundary 150
3.7.4 Triassic–Jurassic Boundary 151
3.8 Outlook 152

4 The Umbria–Marche Sequence 157
4.1 Tectonic Setting 157
4.2 Stratigraphic Framework of the U–M Carbonate Sequence 162
4.2.1 Triassic–Lower Cretaceous 162
4.2.2 Upper Cretaceous–Oligocene 163
4.2.3 Miocene 166

5 Impact Stratigraphy in the U–M Sequence 167

5.1   Searching for Impact Signatures in the Stratigraphic Record 167
5.2   The TR/J Record 172
5.3   The Pliensbachian–Toarcian boundary at Valdorbìa 173
5.4   The J/K Record... 177
5.5   The Upper Aptian Event... 181
5.6   The Cenomanian–Turonian Boundary 184
5.7   The Terminal Cretaceous Record.. 190
        5.7.1   The Maastrichtian... 192
        5.7.2   Definition of the K/T Boundary in the U–M Basin 197
        5.7.3   Sedimentary Setting and K/T Boundary "facies" 201
        5.7.4   Impact and Catastrophe Signatures in the U–M K/T Clay..... 203
        5.7.5   High-resolution Impact Stratigraphy of the K/T Event......... 215
        5.7.6   Representative K/T Boundary Outcrops in the U–M Basin... 222
5.8   The Paleogene Impact Record.. 250
        5.8.1   The Paleocene–Eocene Boundary 250
        5.8.2   The Terminal Eocene Events 254
        5.8.3   The Global Stratotype Section and Point for the E/O
                  Boundary at Massignano.. 258
        5.8.4   Impact Signatures at Massignano.................................. 262
        5.8.5   High-resolution Impact stratigraphy at Massignano 268
        5.9   The U–M Miocene Record.. 274

6 Documentation and Laboratory Techniques 279

6.1   Documentation and Sampling in the Field 279
6.2   Sample Preparation.. 282
        6.2.1   Bulk Samples.. 282
        6.2.2   Isolation of Grains and Inclusions................................. 283
6.3   Laboratory Techniques 286
        6.3.1   Standard Bulk and Microprobe Analysis Methods 286
        6.3.2   Iridium Determinations 289
        6.3.3   Platinum Group Element Determinations 291
        6.3.4   Osmium Isotope Measurements 292
        6.3.5   Methods for Shocked Quartz Measurements 295

References 301

Index 353

1 Introduction: Impact Cratering as a Geological Process

Over the last 15 years, impact cratering as a geological process has received a much wider acceptance than before, largely due to research associated with the mass extinction at the end of the Cretaceous, and as a result of extensive planetary studies. Geological, geophysical, petrographical, and geochemical studies during mainly the last 30 years have led to the recognition of about 160 impact structures on Earth. Distal ejecta, which are one of the main focal points of this book, were distributed by some of the impact events associated with the formation of these structures. In some cases, distal ejecta were found without an obvious link to a source crater, and their study led to the discovery of previously unknown impact structures. The most prominent case was the discovery that impact debris is associated with the Cretaceous–Tertiary (K/T) boundary layer.

To allow identification of impact ejecta and impact structures it is necessary to understand fundamental mineralogical and geochemical properties of impact-derived rocks. Very high pressures and temperature affect the target rocks during an impact event. These P–T conditions are significantly different, and higher, than those reached during any internal terrestrial process. The most characteristic changes induced by the impact-generated shock waves include irreversible changes in the crystal structure of rock-forming minerals, such as quartz and feldspar, and the formation of high-pressure modifications of minerals. These shock metamorphic effects are characteristic of impact and do not occur in natural materials formed by any other process. Furthermore, geochemical studies allow the detection of minute traces of the meteoritic projectile in impact melt rocks and glasses. In this first section we will briefly review the characteristics of impact structures (following earlier summaries by, e.g., Koeberl 1996, 1997a, 1998; French 1998), their formation, and the mineralogical, petrographical, and geochemical evidence for an impact origin of crater rocks or ejecta.

1.1 Historical Background

Craters are a fundamental and common topographic form on the surfaces of planets, satellites, and asteroids. On large planetary bodies, of the size of the Moon and larger, craters can form in a variety of processes, including volcanism, impact, subsidence, secondary impact, and collapse. The two most important

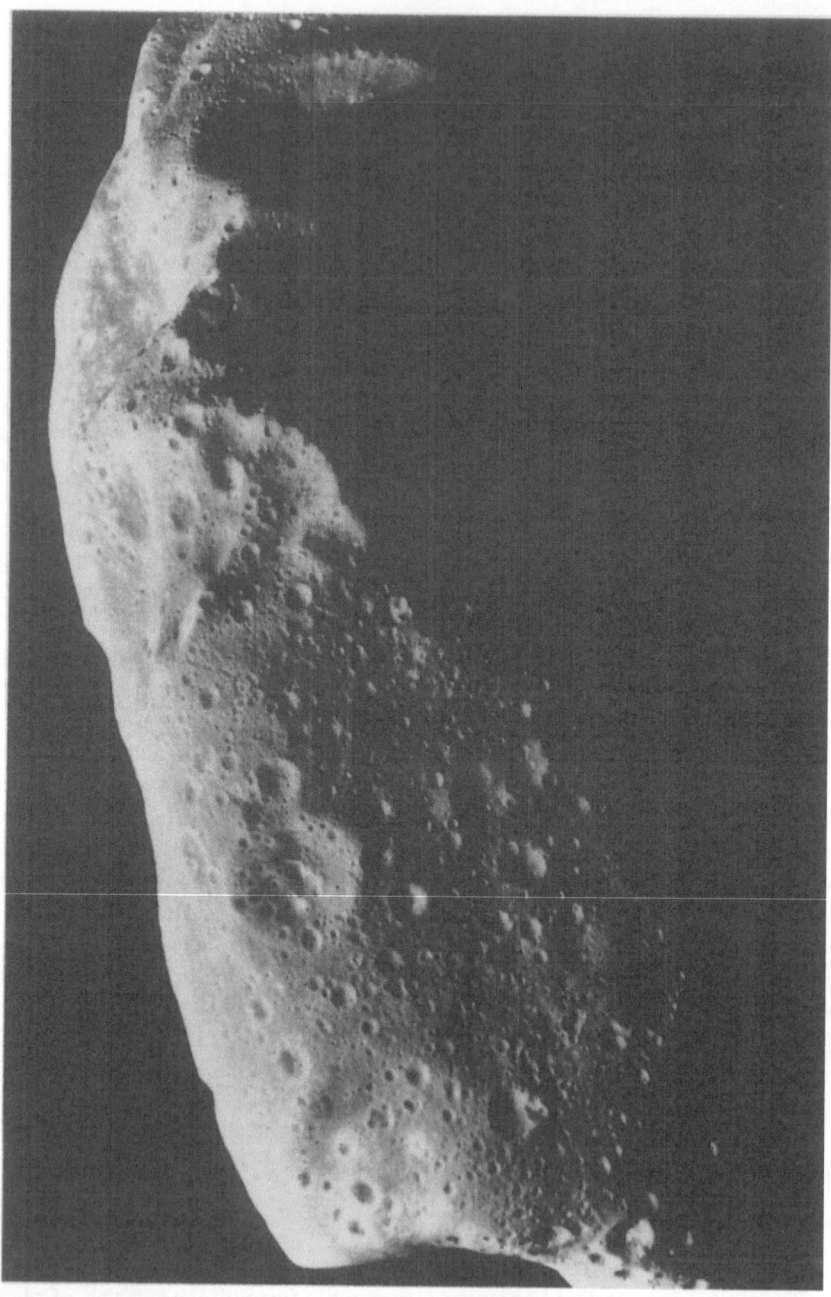

Fig. 1.1.1. Asteroid Gaspra, photographed by the Galileo spacecraft, showing numerous impact craters on its surface.

processes are volcanism and impact. On smaller bodies (e.g., of the size of minor planets), impact may be the only process that can form craters (as shown, for example, by the numerous craters that pockmark the surfaces of asteroids; Fig. 1.1.1). In the explanation of terrestrial crater-like structures, the interpretation as volcanic features and related structures (such as calderas, maars, cinder cones) has traditionally dominated over impact-related interpretations.

During the 1980s and early 1990s, a lively debate was held in the geological community regarding the cause of the mass extinction that marks the end of the Cretaceous Period. Interest in the events at the K/T boundary was renewed by the publication of the paper by Alvarez et al. (1980), who found that the concentrations of some rare siderophile elements are up to 4 orders of magnitude higher in the thin clay layer marking the K/T boundary than normally found in terrestrial crustal rocks. These observations were interpreted by Alvarez et al. (1980) as the result of a large asteroid or comet impact, which caused extreme environmental stress, leading to the mass extinction. At least the first part of this hypothesis was later strongly supported by the finding of shocked minerals in the K/T boundary layer by Bohor et al. (1984, 1987a) (see Sect. 3.2.1). In the absence of detailed knowledge of impact cratering and shock metamorphic processes in the general geological community, many researchers still favored a volcanic explanation for the "K/T event". Similar debates – regarding impact versus internal origin – have been held in discussing the origin of a variety of "unusual" structures around the world.

The history of the study of impact craters begins with Galileo Galilei (1564–1642), when he first pointed a telescope at the Moon in the year 1609. He noticed circular spots and recognized, after studying the moving shadows over a few days time, that these features are depressions on the surface of the Moon. He also noticed that some of these craters have central peaks. The first drawings of lunar craters were published by Galileo in 1610 in his book "Sidereus Nuncius" ("The Starry Messenger"), but the low quality of his early telescope and the distortion of his drawings make it very difficult to identify the craters. Galileo himself did not seem to have any specific opinion regarding the origin of these structures, but most later astronomers preferred the volcanic hypothesis for the formation of the lunar craters. Soon after Galileo's work, much better lunar maps were published. The first map on which lunar formations were named was published by Michel Florent van Langren in 1645. Shortly thereafter (1647), the German astronomer Johannes Hevelius published his famous "Selenographia".

The first speculations about the origin of lunar craters seem to date back to the British physicist Robert Hooke (1635–1703), who compared lunar craters to pits that remained on the surface of boiled alabaster. He also considered an impact theory and made a few crude experiments by dropping objects into mud. However, because the interplanetary space was at that time considered to be empty, he did not follow up on this hypothesis and rather thought that lunar craters may have formed from some kind of internal volcanic activity. The next few hundred years were dominated by the volcanic theory. The astronomer Friedrich Wilhem Herschel (1738–1822) (the discoverer of Uranus) allegedly saw in 1787 a volcanic eruption on the Moon. Most experts had accepted the volcanic hypothesis, although a few voices remained in favor of an impact origin, such as, in 1829, the German astronomer Franz von Paula Gruithuisen. However, it did not help the impact hypothesis that Gruithuisen had announced a few years earlier that he had seen inhabited cities on the Moon, complete with cows grazing on lunar meadows.

During these times, the study of lunar craters was an affair mainly for astronomers (for a review, see Schultz 1998). The few geologists that were concerned about lunar craters supported the volcanic hypothesis. The first serious study of the impact hypothesis of lunar craters dates back to Grove Karl Gilbert (1843–1918), whose positions included Chief Geologist of the U.S. Geological Survey and President of the Philosophical Society of Washington. In 1892, after detailed studies of lunar craters and after performing impact experiments in his hotel room during a lecture tour, he concluded that only the impact hypothesis was able to explain the formation of the lunar craters. However, he also recognized that the circularity of basically all lunar craters present one of the major problems of the impact hypothesis (as formulated at that time): due to the variation in impact angle relative to the surface (from vertical, or 90°, to near 0°) the resulting craters should mostly be elliptical in shape and not circular. Only decades later was the solution to this problem found.

Another scientist, who was ahead of his times in other topics as well, studied lunar craters shortly after Gilbert and concluded that the craters on the Moon were of meteorite impact origin: Alfred Wegener (1880–1930), famous for his work on continental drift, published a little-known booklet (Wegener 1921), in which he discussed his impact experiments and conclusions regarding lunar and some terrestrial craters (cf. Greene 1998). Similar to his ideas on continental drift (which were only accepted many decades later), his views on lunar craters were largely ignored.

Not only during the 19th, but also during most of the 20th century, the concept of impact cratering on Earth has not been much accentuated in classical geological studies. The ideas of "modern" geology, as expressed by, for example, James Hutton (1726–1797) and Charles Lyell (1797–1875), proposed that slow, endogenic processes lead to gradual changes in our geological record. In this uniformitarian view, internal forces are preferred over seemingly more exotic processes to explain geological phenomena that often give the impression of occurring over very long periods of time. In contrast, impact appears as an exogenic, relatively rare, violent, and unpredictable event, which violates every tenet of uniformitarianism (cf. Marvin, 1990). This may explain why many geologists over much of this century have opposed the explanation of craters on the Moon or on Earth as being of impact origin. Taken at face value, the Lyellian view of geology excludes any catastrophic event that is not directly observed. In contrast, the geological history of our planet is marked by rare catastrophes that only occur a few times in a million years (e.g., huge volcanic eruptions, gigantic floods).

The history of study and acceptance of impact cratering over this century is somewhat similar to the record of the acceptance of plate tectonics (see, e.g., Mark 1987; Hoyt 1987; Marvin 1990; Glen 1994; D'Hondt 1998; for historical accounts of impact crater studies). One of the reasons that continental drift (later called plate tectonics) found little acceptance early on was the absence of any known mechanism for moving continents. Only later, after irrefutable geophysical evidence had been found that the continents have indeed moved, were geologists forced to find a mechanism to explain these observations. Impact cratering had a similar fate: no mechanism was known to produce large circular craters by impact. Here, the situation is even more complicated, because even after some craters on Earth had clearly been shown to be the result of impact events, and after a physically plausible mechanism for their formation had been found, the process of impact cratering was still not widely accepted among geologists. Part of the reason was that impact mechanics were explained by astronomers and physicists whose goal it was to explain the formation of lunar craters. In contrast, the existence of internal mechanisms on Earth that result in craters of various forms and sizes made it unnecessary for geologists to resort to extraterrestrial explanations.

One of the first studies that demonstrated that, due to very high velocities with which a body hits the Moon or the Earth, impacts are similar to explosions, and, therefore, resulting craters are always circular, was formulated by the Estonian astronomer Ernest Öpik in 1916. Unfortunately he published his findings in Russian, with a French abstract, in a little-known Estonian journal, which made sure that almost nobody (certainly no geologists) paid any attention (Öpik 1916). Later work by the New Zealand astronomer Algernon Charles Gifford, which was published in 1924 and 1930 in English in a more widely read journal (Gifford 1924), finally caused at least astronomers and physicists to take notice.

These developments happened around the time that the mining engineer Daniel Moreau Barringer (1860–1929) was actively involved in studying the "Coon Butte" or "Crater Mountain" structure in central Arizona. Iron meteorite fragments had been found around this 1.2 km diameter crater. Despite the opinion of several leading geologists (including Gilbert) that this structure was of volcanic origin and the presence of the meteorite fragments was only a coincidence, Barringer was convinced that this was an impact crater. His subsequent exploration of the crater, from about 1903 until his death in 1929, was mainly driven by the desire to find large metallic (meteoritic) deposits underneath the crater floor, which could be exploited for rare metals. No such large iron meteorite mass was ever found leading many geologists to be more convinced than ever that impact had nothing to do with the formation of this structure. Nevertheless, physicists and astronomers, by now familiar with Öpik's and Gifford's arguments, had no problem in explaining both the impact origin of the structure (now called "Meteor Crater") and the absence of large amounts of meteoritic material: only a body of 30 to 50 m in diameter is required to cause a crater with a diameter of more than 1 km.

The early work by Barringer and others on Meteor Crater laid the foundations for a detailed understanding of impact cratering. In the second half of our century it was mainly planetary exploration and extensive lunar research that led to the conclusion that essentially all craters visible on the Moon (and many on Mercury, Venus, and Mars) are of impact origin. From these observations it was concluded that, over its history, the Earth had been subjected to a significant number of impact events. Part of the reason why this conclusion was not widely accepted among geologists was the obvious absence of large numbers of craters on the Earth's surface. The reason is of course that terrestrial processes (weathering, plate tectonics, etc.) effectively work to obliterate the surface expression of these structures on Earth.

The science historian William Glen found, in an interesting historical and sociological evaluation of the K/T boundary debates, that "resistance to the [impact] hypothesis seemed inverse to familiarity with impacting studies." (Glen 1994, p. 52). Planetary scientists, astronomers, and meteoriticists, who deal on a daily basis with topics including asteroids and meteorites (products of collisions between asteroids), are used to view "large-body impact as a normal geological phenomenon – something to be expected throughout Earth history – but another group, the paleontologists, is confounded by what appears to be an ad hoc theory about a nonexistent phenomenon" (D. Raup in Glen 1994, p. 147). Thus, it seems that what is uniformitarian for a member of one scientific field (in this case a planetary scientist), would be considered by the representative of another field (e.g., paleontology) as an unnecessary *deus ex machina*.

However, it is also important to take the time scales into account that are involved in this discussion. What geologists have called "uniformitarianism" is the result of a large number of individual catastrophes of various magnitudes over a sufficiently long time span. Earthquakes, volcanic eruptions, landslides, etc., are locally devastating during periods on the order of about 20 to 100 years, but if the whole world and longer time spans are taken into account, these (local) "catastrophes" become part of the (global) "uniformitarian" process of explosive volcanism, earthquake history, or erosion. The bias in what is considered uniformitarian is related to the life span of humans and the human civilization. As large meteorite impacts have not been observed during the last few millennia (with rare exceptions, such as the Tunguska event, which occurred in a remote Siberian location in 1908, but even this event was too small to produce a crater), such events tend to be neglected when constructing the "uniformitarian" history of the Earth. In contrast, small meteorites have been observed to fall from the sky quite frequently, but scientists have failed to make the connection to impact events by applying the same principle that is being used for extrapolating the frequency of volcanic eruptions and earthquakes: large and devastating ones occur less often than small events. There is no real conflict between uniformitarianism and meteorite impact. Over long periods of time, impact is a common process on the Earth.

About 160 impact structures are currently (in the year 2000) known on Earth (e.g., Grieve and Shoemaker 1994; Grieve et al. 1995; updates are currently available on the internet, see web page of the Canadian Geological Survey at: http://gdcinfo.agg.emr.ca/toc.html?/crater/world_craters.html). A few of these structures were only recognized because layers of impact ejecta were found and eventually traced back to their source crater (e.g., Chicxulub, from studies of the K/T boundary impact layer). Considering that some impact events demonstrably affected the geological and biological evolution on Earth, and that even small impacts can disrupt the biosphere and lead to local and regional devastation (Chapman and Morrison 1994), the understanding of impact structures and the processes by which they form should be of interest not only to earth scientists, but also to society in general.

1.2
General Characteristics of Impact Craters

As no record seems to exist that documents the observation of a large impact event by humans over the last several thousand years (which is, of course, not a geologically long period of time), impact experiments and the detailed study of impact craters on Earth are indispensable for our understanding of these features. During an impact event, the target area is changed in a characteristic way, which can be used to help distinguish volcanic structures from meteorite impact craters. Meteorite impact craters are surficial features without deep roots, whereas in volcanic craters or calderas the disturbances continue to (or, rather, emerge from) great depth. Impact craters are almost always circular. On the Moon, with its surface covered by impact craters, only a small number of non-circular features are known. On Earth there are also only few structures that result either from highly oblique impacts (see, e.g., the Rio Cuarto crater field in Argentina – Schultz et al. 1994), or from post-formational distortion due to, for example, tectonism or erosion (e.g., the Sudbury structure in Canada – e.g., Stöffler et al. 1994; Deutsch et al. 1995; or the Vredefort structure in South Africa – e.g., Reimold and Gibson 1996; Gibson and Reimold 2000).

When discussing morphological aspects, one should note the distinction between an impact *crater*, i.e., the feature that results from the impact, and an impact *structure*, which is what we observe today, i.e., long after formation and modification of the crater. Impact craters (before post-impact modification by erosion and other processes) occur on Earth in two distinctly different morphological forms. These are known as *simple* craters (small bowl-shaped craters) with diameters of up to ≤2 to 4 km, and *complex* craters, which are larger and have diameters of ≥2 to 4 km (the exact change-over diameter between simple and complex crater depends on the composition of the target).

The striking morphological difference between the simple and complex craters is best illustrated by looking at lunar craters, where post-formational modification is minimal compared to crater structures on the Earth (e.g., Melosh 1989). Figs. 1.2.1 and 1.2.2 show a simple and a complex lunar crater, respectively.

Complex craters are characterized by a central uplift. Craters of both types have an outer rim and are filled by a mixture of fallback ejecta and material slumped in from the walls and crater rim during the early phases of formation. Such crater infill may include brecciated and/or fractured rocks, and impact melt rocks. Fresh simple craters have an apparent depth (crater rim to present-day crater floor) that is about one third of the crater diameter (Fig. 1.2.3). For complex craters, this value is closer to one fifth or one sixth. The central structural uplift in complex craters consists of a central peak or of one or more peak ring(s) and exposes rocks that are usually uplifted from considerable depth. On average, the actual stratigraphic uplift amounts to about 0.1 of the crater diameter (e.g., Melosh 1989).

Fig. 1.2.1. Simple (bowl-shaped) impact crater on the Moon. Crater Isidorus D has a diameter of 15 km. (NASA Apollo 16 photo.)

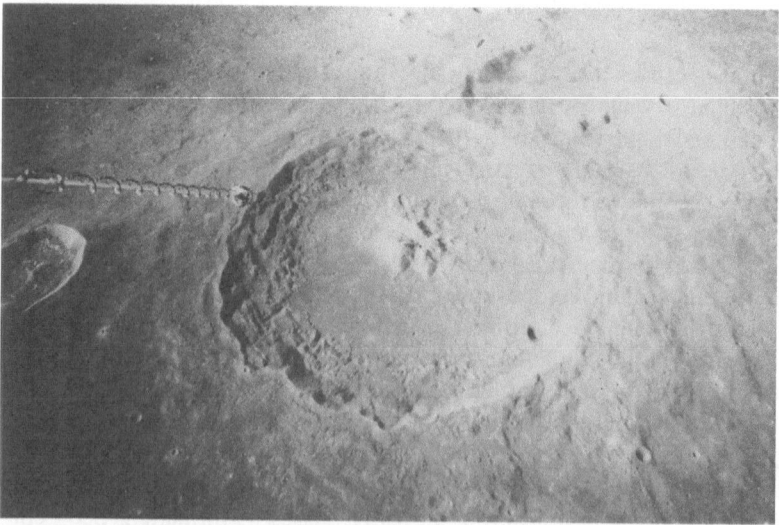

Fig. 1.2.2. Complex impact crater on the Moon. Crater Theophilus has a diameter of about 100 km. The feature on the left is the antenna of the Apollo spacecraft. (NASA Apollo 16 photo.)

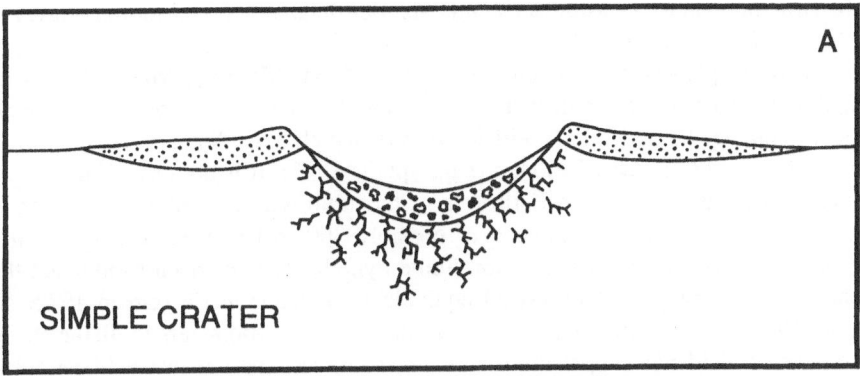

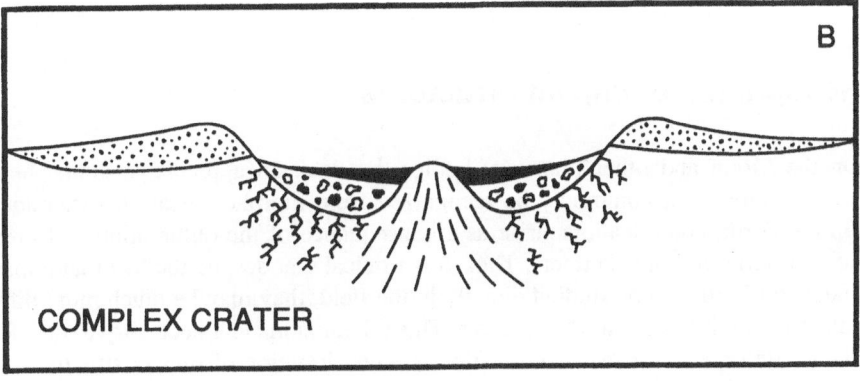

Fig. 1.2.3. Schematic cross-section through a simple (a) and a complex (b) impact crater. Ejecta around the crater rim are indicated by a stipple pattern, whereas brecciated, shocked, and melted crater fill rocks (including fallback ejecta) are indicated by larger open and filled irregular symbols. Fracturing beneath the crater floor is also indicated, as are the central stratigraphic uplift and a possible melt sheet in the complex crater case.

The variation in diameter of impact craters on Earth is quite different from that on (e.g.) the Moon, because on Earth basically all of the small craters are relatively young. This size relationship is biased and results from typical terrestrial processes. The erosional processes that obliterate small (0.5–10 km diameter) craters after a few million years result in a severe deficit of such small craters, compared to the crater count that would be expected from the (observed) number of larger craters and from astronomical considerations (Grieve and Shoemaker 1994). This effect also explains why most small craters known on Earth are young. Older craters of larger initial diameter also suffer erosional degradation leading to the destruction of the original topographical expression, or to burial of the structures under post-impact sediments. For details on crater morphology, see, e.g., Melosh (1989).

On other planets and satellites (e.g., the Moon, Mercury, Mars, Ganymede, Callisto) larger structures than the simple and complex craters have been identified. These are the so-called *multiring basins*, which have diameters ranging from a few hundred to at least 2000 km (e.g., Spudis 1993). It is not clear if the largest impact structures known on Earth (e.g., Chicxulub with ca. 200 km diameter, or Vredefort with an initial diameter of possibly 300 km) may represent terrestrial examples of multiring basins or not. Some hypotheses have been formulated that call for the existence of terrestrial equivalents of lunar maria (Green 1972; Alt et al. 1988), which would be hundreds to thousands of kilometers in diameter, but no physical evidence supporting such far-reaching assumptions has yet been forthcoming.

1.3
Recognition of Impact Structures

On the Moon and other planetary bodies that lack an appreciable atmosphere, impact craters can commonly be recognized from morphological characteristics. On the Earth, complications arise as a consequence of the obliteration, deformation, or burial of impact craters. Thus, it is ironical that despite the fact that impact craters on Earth can be studied directly in the field, they may be much more difficult to recognize than on other planets. This dilemma made it necessary to develop diagnostic criteria for the identification and confirmation of impact structures on Earth (see also French 1998). The most important of these characteristics are: a) crater morphology, b) geophysical anomalies, c) evidence for shock metamorphism, and d) the presence of meteorites or geochemical evidence for traces of the meteoritic projectile.

Of these characteristics, only the presence of diagnostic shock metamorphic effects and, in some cases, the discovery of meteorites, or traces thereof, are generally accepted to provide unambiguous evidence for an impact origin. Shock deformation can be expressed in macroscopic form (shatter cones) or in microscopic form (see Sect. 1.6). The same two criteria apply to distal impact ejecta layers and allow to confirm that material found in such layers originated in an impact event at a possibly still unknown location. So far (2000), the presence of such evidence led to the confirmation of about 160 impact structures (some exposed on the surface, others are subsurface structures) on Earth (Fig. 1.3.1).

The distribution of impact structures, as shown in Fig. 1.3.1, is not uniform over the planet. Even though about two thirds of the planet are covered by water, no impact structures have so far been found on the ocean floor. Only two structures are currently underwater, but are located on the continental shelf. There are several reasons for this: first, small projectiles make a crater in the ocean without penetrating through up to several kilometers of water and, therefore, do no leave a permanent imprint on the ocean floor. A good example is the Eltanin event, where an asteroidal body impacted the Southern Ocean during the Late Pliocene and only (partly melted) meteoritic debris was recovered from up to about 4 km water depth, but no crater was formed (e.g., Kyte and Brownlee 1985; Kyte et al. 1988; Margolis et al. 1991; Gersonde et al. 1997). Only large bodies would leave a crater, but those are much rarer than small ones. Second, due to plate tectonics, the ocean floor is relatively young (less than about 200 Ma), and records only recent impact events. Third, the ocean floor has never been searched in any detail for evidence of impact events, and even if indications are found, structures on the deep ocean floor are very difficult to reach and explore. The distribution of impact craters on land is largely governed by two factors: age of target areas and efficiency of crater search programs. Among the young craters there is a predominance of relatively recent, small structures (which are mostly simple craters), because erosion either removes or rapidly covers other craters (Figs. 1.3.2 and 1.3.3).

Whereas petrographical and geochemical data can provide confirming evidence for impact, morphological and geophysical observations are important in providing supplementary (or initial) information. Geological structures with a circular outline that are located in places with no other obvious mechanism for producing near-circular features may be of impact origin and at least deserve further attention. Geophysical methods are also useful in identifying candidate sites for further studies, especially for subsurface features. In complex craters the central uplift usually consists of dense basement rocks and usually contain severely shocked material. This uplift is often more resistant to erosion than the rest of the crater, and, thus, in old eroded structures the central uplift may be the only remnant of the crater that can be identified.

Geophysical characteristics of impact craters include gravity, magnetic properties, reflection and refraction seismics, electrical resistivity, and others (see Pilkington and Grieve 1992, and Grieve and Pilkington 1996, for reviews). In general, simple craters have negative gravity anomalies due to the lower density of the brecciated rocks compared to the unbrecciated target rocks, whereas complex craters often have a positive gravity anomaly associated with the central uplift that is surrounded by an annular negative anomaly. Magnetic anomalies can be more varied than gravity anomalies. Seismic data often show the loss of seismic coherence due to structural disturbance, slumping, and brecciation. Such geophysical surveys are important for the recognition of anomalous subsurface structural features, which may be deeply eroded craters or impact structures covered by post-impact sediments (e.g., Chicxulub – see Hildebrand et al. 1991; or in the U.S.: Ames, Avak, Chesapeake Bay, Manson, Newporte, Red Wing Creek – see Koeberl and Anderson 1996a, b, for reviews).

No book or review publication currently exists that gives detailed descriptions of all impact structures known on the Earth, with the possible exception of Hodge (1994), who gives a short, but unfortunately rather uneven and incomplete, overview of all impact structures. There are some more detailed reviews of impact structures in some geographical areas, e.g., Africa (Koeberl 1994a), Australia (Glikson 1996; Shoemaker and Shoemaker 1996), northern Europe (Pesonen

1996; Pesonen and Henkel 1992), Russia (Masaitis 1999), and the United States of America (Koeberl and Anderson 1996b).

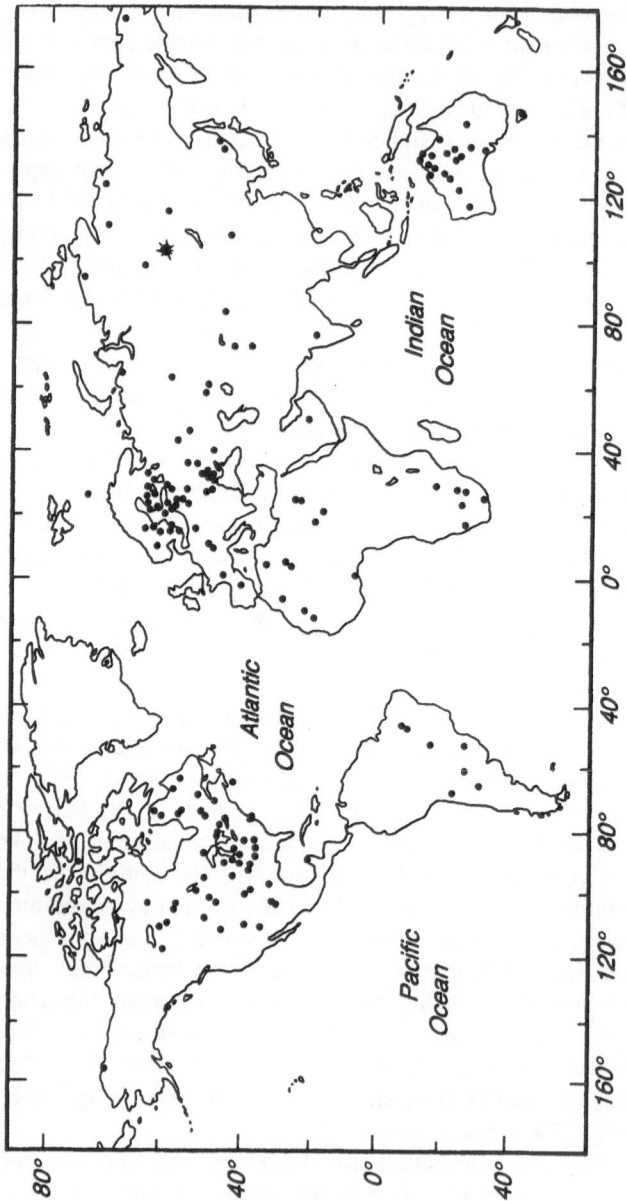

Fig. 1.3.1. Distribution of currently (2000) known impact structures on Earth. The star in central Siberia denotes the location of the 1908 Tunguska airblast.

Fig. 1.3.2. The Roter Kamm crater in Namibia, an example of a relatively young (3.7 Ma) simple impact crater (2.5 km diameter), which is already suffering from erosion and is mostly covered by aeolian sediments.

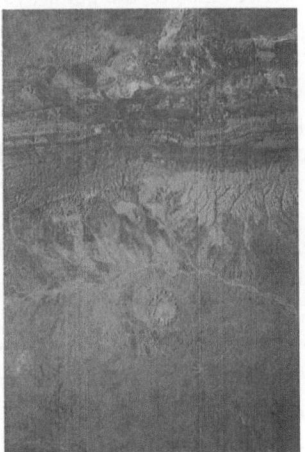

Fig. 1.3.3. Gosses Bluff impact structure, Australia. This complex impact structure is 142.5 million years old and severely eroded. The inner ring has a diameter of about 5 to 6 km and is the remnant of the central uplift. A faint circular feature, notable from drainage patterns, with a diameter of 22 km, outlines the original diameter of the crater. The original crater rim itself has long since been removed by erosion, leaving only the prominent central uplift behind. Space Shuttle photograph S84–14–41–028; north is to the left.

1.4
Formation of Impact Craters

Impact craters form when a small object (in the solar system) collides with a large one. For a planet like the Earth, the impacting small bodies are derived either from asteroidal or cometary sources. From observational programs that study the orbits of asteroids and comets, astronomers have a relatively good understanding of the rate at which these objects strike the Earth (e.g., Shoemaker et al. 1990; Weissman 1990). For example, minor objects in the solar system with diameters of about 1 km (mainly asteroids) collide with the Earth at a frequency of about 4.3 impacts per million years (Shoemaker et al. 1990). Depending on the impact velocity and angle, each such impact is capable of forming a crater with a diameter of at least 10 km. Impactors about 2 km in diameter collide with the Earth about every 1–$2 \cdot 10^6$ years. Collisions of Earth-orbit crossing asteroids with the Earth dominate the formation of craters that are smaller than about 30 km in diameter, whereas cometary impacts are probably responsible for the majority of craters with diameters of more than about 50 km (Shoemaker et al. 1990). However, the orbits of asteroids are better known than those of comets, because many of the latter have such long periodicities that no appearance has yet been observed during the time of human civilization. Thus, the long-term orbital evolution of asteroids is better known than that of comets, and collisions with asteroids could be calculated longer in advance than those with comets (which may be known only a few months in advance if a comet approaches from the outer reaches of the solar system).

The formation of a crater by hypervelocity impact is – not only in geological terms – a very rapid process that is usually divided into three stages: 1) contact/compression stage, 2) excavation stage, and 3) post-impact crater modification stage. Crater formation processes have been studied for many decades, but space limitations and the focus of this book on impact ejecta require that the reader is referred to the literature (see, e.g., Gault et al. 1968; Roddy et al. 1977; Melosh 1989; and references therein) for more detailed discussions of the physical principles of impact crater formation. Here, we mention only some basic principles.

During the impact of a large meteorite, asteroid, or comet, large amounts of kinetic energy (equal to $\frac{1}{2}mv^2$, m = mass, v = velocity) are released. Earlier in the century, the amount of energy was largely underestimated, because the velocities with which extraterrestrial bodies hit the Earth had not been known or assessed properly. This is shown very well by the historical development of Meteor Crater studies (see Hoyt 1987; Mark 1987), where the mass of the body that created the crater was originally severely overestimated. Any extraterrestrial body that is not slowed down by the atmosphere will hit the Earth with a velocity between about 11 and 72 km/s. These velocities are determined by celestial mechanics: 11 km/s is the escape velocity of the Earth–Moon system, to which a body with zero velocity difference to this system will be accelerated simply by gravitational attraction; 72 km/s is the sum of the escape velocity of the solar system at the distance of 1 Astronomical Unit (A.U.; approximately equal to the average distance of the Earth from the Sun), 42 km/s, and the average orbital velocity of the Earth, 30 km/s.

Thus, every body that collides with the Earth, will do so at cosmic velocity. The Earth accretes about 100 tons of extraterrestrial material every day, but most of that is in the form of dust. Larger dust grains may, when they hit the Earth's atmosphere, burn up and cause a glowing channel of ionized air: a shooting star or *meteor* (before entering the Earth's atmosphere, a small interplanetary body is called a *meteoroid*). Such bodies will be decelerated in the atmosphere until they reach free-fall velocity; during this deceleration they may lose a large fraction of their mass. If a remnant of a small extraterrestrial body reaches the surface of the Earth intact, it is called a *meteorite*. These may hit the surface with appreciable speed (but much less than cosmic velocity) and cause penetration craters or penetration funnels. The largest single-mass meteorite currently known on Earth, which survived passage through the atmosphere and the "hard landing", is the Hoba iron meteorite in Namibia with about 60 tons. In contrast, if a body is large enough to not be decelerated significantly in the Earth atmosphere and hits the surface with cosmic velocities, it will produce a *hypervelocity impact crater*. For the purposes of this book the term *impact* or *impact crater* is always meant to be the result of a hypervelocity impact.

The physical processes that govern the formation of an impact crater are the result of the extremely high amounts of energy that are liberated almost instantaneously when the projectile (also termed impactor, bolide, meteorite, asteroid, comet [nucleus]) hits the ground. Even a small meteoritic body, about 6 m in diameter, would release as much energy as an atomic bomb when it hits the surface of the Earth at about 20 km/s (about 20 kilotons of TNT, or about $8.3 \cdot 10^{13}$ joules [J]). A meteorite with a diameter of 250 m has a kinetic energy that is roughly equivalent to about 1000 megatons of TNT, which would lead to the formation of a crater about 5 km in diameter. The relatively small Meteor (or Barringer) Crater in Arizona (1.2 km diameter) was produced by an iron meteorite of about 30–50 m in diameter and had an energy content of more than 10 megatons of TNT (see Schnabel et al. 1999).

There is a difference between the behavior of a stony impactor and an iron one. Due to the difference in mechanical strength, smaller iron meteorites can reach the ground intact, in contrast to stony meteorites, which may undergo catastrophic disintegration in the atmosphere. In 1908, a stony bolide exploded over the Tunguska region in central Siberia (see Fig. 1.3.1) and devastated more than 2000 km^2 of forest (see, e.g., Turco et al. 1982, and Martino et al. 1998, for reviews). It has been estimated that a stony body with a diameter of about 20 to 40 m exploded (due to stress during deceleration) at an altitude of about 10 km, liberating an amount of energy equivalent to about 10 to 30 megatons of TNT; the body was totally disintegrated (see, e.g., Chyba et al. 1993; Svetsov 1996). Events of such magnitude may happen every few hundred years somewhere over the Earth (e.g., Chapman and Morrison 1994).

The impact energy can be compared to that of "normal" terrestrial processes, such as volcanic eruptions or earthquakes. During a small impact event, which may lead to craters of 5–10 km in diameter, about 10^{24-25} ergs (10^{17-18} J) are released, whereas during formation of larger impact craters (<50 km in diameter) about 10^{28-30} ergs (10^{21-23} J) are liberated (e.g., French 1968, 1998; Kring 1993). On the other hand, about $6 \cdot 10^{23}$ ergs ($6 \cdot 10^{16}$ J) were released over several months during the 1980 eruption of Mount St. Helens, and 10^{24} ergs (10^{17} J) in the big San Francisco earthquake in 1906. It may also be surprising that the total annual energy release from the Earth (including heat flow, which is by far the largest component, besides contributions from volcanism and earthquakes) is about $1.3 \cdot 10^{28}$ ergs ($1.3 \cdot 10^{21}$ J/y) (French 1998; Sclater et al. 1980; Morgan 1989). This amount of energy is comparable to the energy that is released within a very short time interval during large impact events. It is also important to realize that the energy that is liberated during an impact is concentrated in a point on the Earth's surface, leading to an enormous local energy increase. The collision of a medium-sized asteroid or cometary nucleus would, therefore, release the same amount of energy within seconds that the whole Earth releases within hundreds or thousands of years.

1.5
Shock Waves and the Stages of Crater Formation

Structural modifications and phase changes in the target rocks occur during the compression stage, whereas the morphology of a crater is defined in the second and third stages of impact cratering. For a more detailed description of crater formation, see, e.g., Grieve (1987, 1991), Melosh (1989), Boslough and Asay (1993), Melosh and Ivanov (1999), and references therein. During this early impact phase, the impacting body is stopped at a depth equivalent to about two projectile radii and the kinetic energy ($\frac{1}{2}mv^2$) is transformed into heat and shock waves that penetrate into the projectile and target. The most important phenomenon, which is characteristic of impact, is the generation of a supersonic shock wave that is propagated into the target rock. The effects of shock waves on matter are well understood from decades of experimental study. The following discussion is based mainly on information from Melosh (1989). Matter is being accelerated very rapidly and, as a consequence of the decrease of compressibility with increasing pressure, the resulting stress wave will become a shock wave moving at supersonic speed (up to about 2/3 of the impact velocity). Shock waves are inherently nonlinear and shock fronts are abrupt. They can be mathematically represented as a discontinuous increase of pressure, density, particle velocity, and internal energy. In reality, shock waves have a finite thickness, which is, however, very limited. For example, the widths of shock waves in gas are limited to about 10 μm, which is roughly equal to one molecular mean free path, but shock waves in solids are wider, up to a few meters in rocks, depending on their porosities.

The shock wave leads to compression of the target rocks at pressures far above a material property called the Hugoniot elastic limit. The Hugoniot elastic limit (HEL) is usually described as the maximum stress in an elastic wave that a material can be subjected to without permanent deformation (Melosh 1989). Above this limit, plastic, or irreversible, distortions occur in the solid medium through which the compressive wave travels (see, e.g., compilations by Roddy et al. 1977; Melosh 1989; and references therein). The value of the HEL is about 5–10 GPa for most minerals and whole rocks. For example, single crystals of quartz have HELs ranging from 4.5 to 14.5 GPa (depending on the orientation of the crystals), for feldspar the HEL is at 3 GPa, and for olivine at 9 GPa. For rocks, the HEL of dolomite is 0.3 GPa, for granite 3 GPa, and for granodiorite, 4.5 GPa. The only known process that produces shock pressures exceeding the HELs of most crustal rocks and minerals in nature is impact cratering. The pressures reached in volcanic processes are not known to exceed 0.5 to 1 GPa (see, e.g., Gratz et al. 1992b). In addition to structural changes, phase changes may occur in minerals once the HEL is exceeded.

For a thermodynamic description of shock fronts travelling through matter, the so-called Hugoniot equations are used (see Melosh 1989). These equations link the pressure P, internal energy E, and density ρ ahead of a shock wave (uncompressed: P_0, E_0, ρ_0) to values behind the shock front (compressed: P, E, ρ). The density is also expressed as the specific volumes $V = 1/\rho$ and $V_0 = 1/\rho_0$ for the compressed and uncompressed cases, respectively. Initial pressure, energy, and density ahead of the shock are known values, whereas the respective values behind the shock are unknown quantities, as are the shock velocity U and particle velocity u_p behind the shock front. The Hugoniot equations are then written as:

$$\rho(U - u_p) = \rho_0 U$$

$$P - P_0 = \rho_0 u_p U$$

$$E - E_0 = (P + P_0)(V_0 - V)/2$$

These equations express the conservation of mass, momentum, and energy across the shock front to reduce the number of unknown variables from five to two. For a derivation of the Hugoniot equations, see Appendix 1 in Melosh (1989), as well as Stöffler (1972) and Boslough and Asay (1993). In the uncompressed material, the initial particle velocity should be zero, and the initial pressure P_0 can be neglected, yielding the approximation $E - E_0 = u_p^2/2$. In addition to the three equations mentioned above, a fourth one, the so-called equation of state, is necessary to specify conditions on either side of the shock front. This equation links pressure, specific volume (density), and internal energy: $P = (V,E)$. Equations of state have been determined experimentally for a large number of different materials (e.g., Marsh 1980).

The shock wave equation of state data can be plotted in pressure versus specific volume or shock velocity versus particle velocity diagrams (Fig. 1.5.1). The curves in these diagrams are not equivalent to conventional equilibria in thermodynamical pressure versus volume (P,V) diagrams, but represent the loci of several individual shock events, i.e., each point on a curve is the result of one particular shock wave compression event. The HEL appears as a kink in the shock curve, indicating yielding at the maximum stress of the elastic wave.

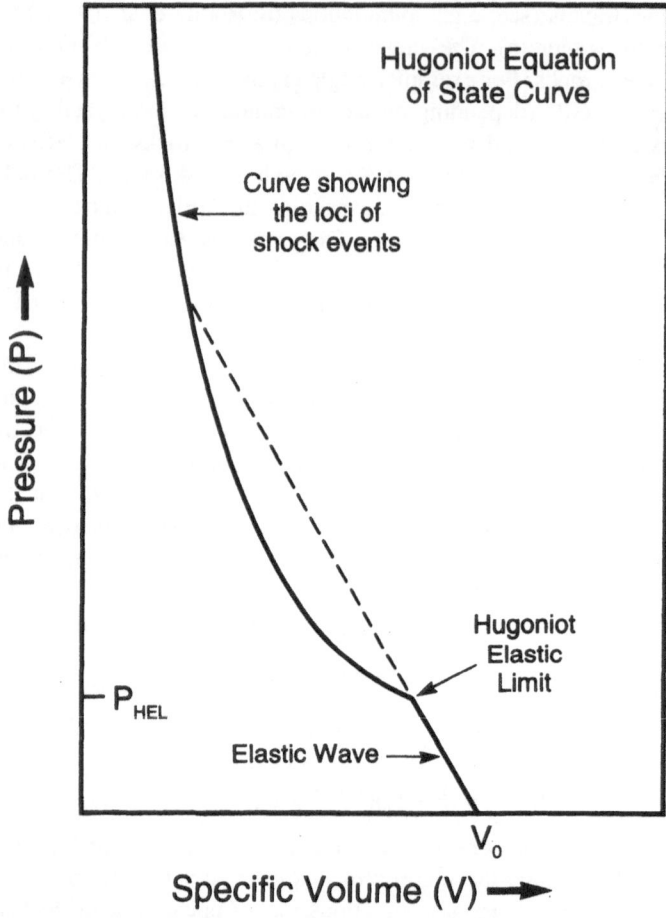

Fig. 1.5.1. Hugoniot equation of state curve. The Hugoniot curve does not represent a continuum of states as in thermodynamical diagrams, but the loci of individual shock compression events. The yielding of the material at the Hugoniot Elastic Limit is indicated (diagram after Koeberl 1997a).

After the passage of the shock wave, the high pressure is released by a so-called rarefaction wave (also called release wave), which follows the shock front. The rarefaction wave is a pressure wave, not a shock wave, and travels at the speed of sound in the shocked material. The release wave moves faster than the shock wave and, therefore, gradually overtakes the shock front and causes a decrease in pressure with increasing distance of propagation. Whereas the pressure behind a rarefaction wave may drop to near zero, the residual particle velocity actually accelerates material, leading to impact crater excavation. In addition, the rarefaction wave not only conserves mass, energy, and momentum (as the shock wave does), but also entropy. Thus, rarefaction is a thermodynamically reversible adiabatic process, whereas shock compression is thermodynamically irreversible. During shock compression, a large amount of energy is being introduced into a rock. Upon decompression, the material follows a release adiabat in a pressure versus specific volume diagram. The release adiabat is located close to the Hugoniot curve, but usually at somewhat higher P and V values, leading to excess heat appearing in the decompressed material, which may result in phase changes (e.g., melting or vaporization). Because the pressure in the shock waves are so high, the release of the pressure (decompression) results in almost instantaneous melting and vaporization of the projectile – and of large amounts of target rocks. All the effects of the phenomena described above can be observed in various forms in shocked minerals and rocks, and form during the contact (or compression) stage, which only lasts up to a few seconds even for large impacts.

The contact stage is followed by the excavation stage, in which the actual crater is excavated. Complex interactions between the shock wave(s) and the target, as well as the release wave(s), lead to an excavation flow. In the upper levels of the target, material moves mainly upwards and out, whereas in lower levels material moves mainly down and outwards, which results in a bowl-shaped depression, the *transient cavity*. This cavity grows in size as long as the shock and release waves are energetic enough to excavate materials from the impact location. Afterwards, gravity and rock-mechanical effects lead to a collapse of the steep and unstable rims of the transient cavity, widening and filling of the crater. Compared to the contact stage, the excavation stage takes longer, but still only up to a minute or two even in large craters of more than 200 km diameter.

Thus, the formation of an impact crater can be summarized schematically as shown in Fig. 1.5.2. First, a relatively small extraterrestrial body hits the surface (A) with cosmic velocity and marks the beginning of the contact and compression stage. Next (B), jetting ejects material from the impact site with velocities that can approach about one half of the impact velocity. A shock wave propagates hemispherically into the ground (C), closely followed by a rarefaction wave, which releases the shock pressure from the compressed rocks and leads to the creation of a mass flow that opens up the crater (D), marking the beginning of the excavation phase. The maximum excavation is reached in (E), showing the extent of the transient cavity. The post-impact modification stage is characterized by enlargement of the transient crater through slumping of the crater walls and infill from fall-back ejecta (F). More details can be found in French (1998) and Melosh (1989). Most ejecta end up close to the crater (<5 crater radii), but a small fraction may travel much greater distances (see Sect. 2.1).

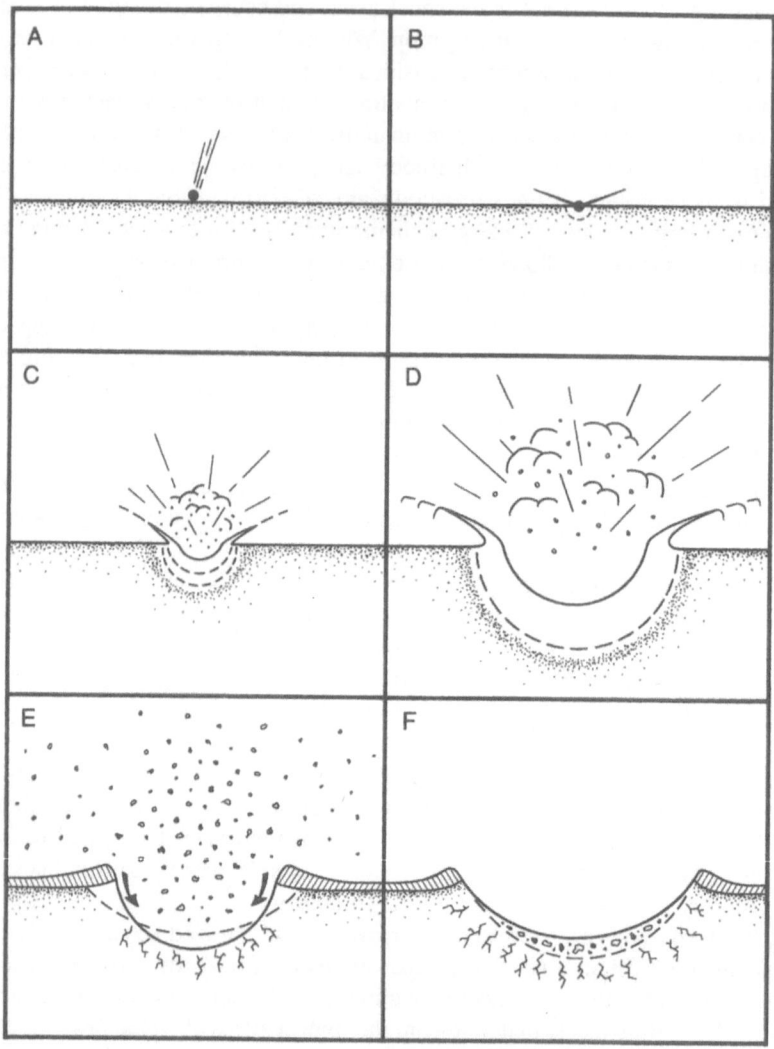

Fig. 1.5.2. Schematic diagram of the stages of the formation of a simple impact crater.

Only small craters on Earth (up to about 2 to 4 km in diameter, depending on the target rock type) are simple bowl-shaped craters. These have a depth to diameter ratio of about 1:5. Larger impact events lead to the formation of complex craters, which have central peaks, wall terraces, and (in the largest structures) internal rings. The formation of the central peak (or central uplift) is not well understood. Theoretical studies (e.g., Melosh 1989; Melosh and Ivanov 1999) indicate that the central uplift forms within minutes by rebound and a series of complicated rock movements and deformations (during the collapse of the transient cavity) that may involve a process called acoustic fluidization (cf. Melosh and Ivanov 1999). This process leads to the stratigraphic uplift of rock strata well above their normal stratigraphic positions. Thus, the formation of large (more than 100 km in diameter) impact structures leads to the (vertical) stratigraphic uplift of material at the center of the structure by about 10 km. The result is a often a geological and/or metamorphic anomaly, which draws attention to the structure (e.g., Gibson and Reimold 2000). The central uplift is usually a geologically stable feature that remains, in some cases (e.g., Vredefort, Gosses Bluff), as the only remnant of the impact crater after long periods of erosion. Moreover, the rocks of the central uplift have been subjected to high shock pressures. This allows the use of rocks from the (eroded) central uplift in the identification of a structure as being of impact origin.

1.6
Shock Metamorphic Effects

1.6.1
General

A large meteorite impact will produce shock pressures of ≥100 GPa and temperatures ≥3000°C in large volumes of the target rock. These conditions are in sharp contrast to conditions for endogenic metamorphism of crustal rocks, with maximum temperatures of 1200°C and pressures of usually <2 GPa (except static pressure affecting some deep-seated rocks, e.g., eclogites) (Fig. 1.6.1.1). As mentioned above, shock compression is not a thermodynamically reversible process, and most of the structural and phase changes in minerals and rocks are uniquely characteristic of the high pressures (5–>50 GPa) and extreme strain rates (10^6–10^8 s^{-1}) associated with impact. Some assemblages of high-pressure and high-temperature mineral phases are preserved together with glass in shocked rocks due to disequilibrium caused by transient high pressures followed by quenching.

As some recent debates indicate (e.g., Lyons et al. 1993 and Luczaj 1998, and the counter-arguments by Reimold 1994 and Koeberl and Reimold 1999), there seems still to be some incomplete understanding in the geological community about the precise nature of shock metamorphism (for a discussion, see, e.g., French 1990a; Sharpton and Grieve 1990). In contrast to some assertions, the existence of definite shock metamorphic features in volcanic rocks has never been substantiated (see, e.g., de Silva et al. 1990; Gratz et al. 1992b). Static compression (see, e.g., Kingma et al. 1993), as well as volcanic or tectonic processes, yield products different from those of shock metamorphism, because of lower peak pressures and especially strain rates that are different by more than 11 orders of magnitude. It must be reaffirmed that the study of the response of materials to shock is not a recent development, but has been the subject of thorough investiga-

22

tions over much of the 20th century, in part stimulated by military research. Numerous shock recovery experiments (i.e., controlled shock wave experiments, which allow the collection of shocked samples for further studies), using various techniques, have been performed. These experiments have led to a good understanding of the conditions for formation of shock metamorphic products and a pressure-temperature calibration of the effects of shock pressures up to about 100 GPa (see, e.g., Hörz 1968; French and Short 1968; Stöffler 1972, 1974; Gratz et al. 1992a, b; Huffman et al. 1993; Stöffler and Langenhorst 1994; Huffman and Reimold 1996; Langenhorst and Deutsch 1998; and references therein).

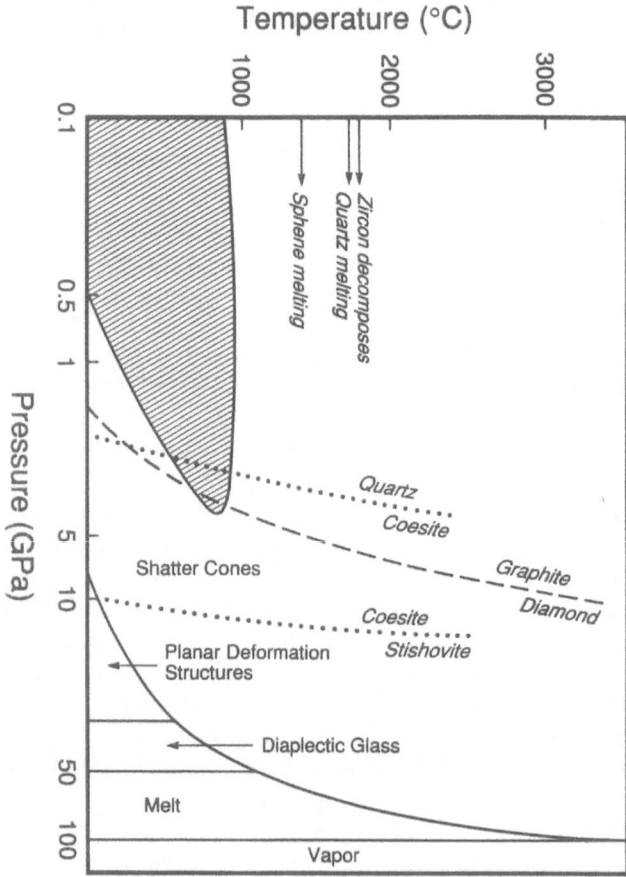

Fig. 1.6.1.1. Comparison of pressure-temperature fields of endogenic metamorphism and shock metamorphism. Also indicated are the onset pressures of various irreversible structural changes in the rocks due to shock metamorphism. The curve on the right side of the diagram shows the relation between pressure and post-shock temperature for shock metamorphism of granitic rocks (after Grieve 1987, and B.M. French, personal communication, 1995).

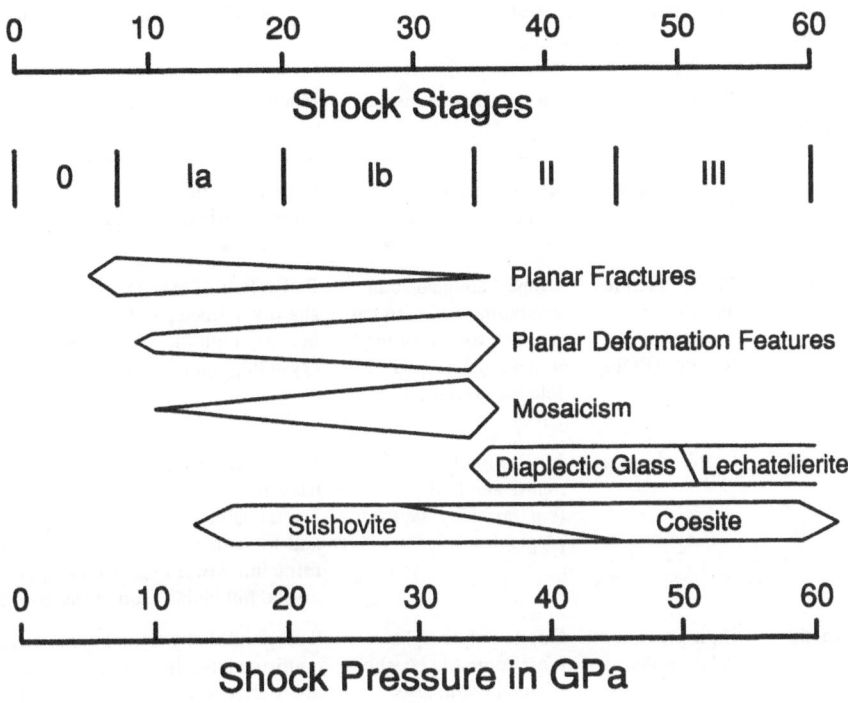

Fig. 1.6.1.2. Pressure dependency of various characteristic shock indicators in quartz, and relation to shock stages (after Stöffler and Langenhorst 1994).

A wide variety of shock metamorphic effects has been identified (Fig. 1.6.1.2). The best diagnostic indicators for shock metamorphism are features that can be studied easily by using the polarizing microscope. They include planar microdeformation features, optical mosaicism, changes in refractive index, birefringence, and optical axis angle, isotropization (e.g., formation of diaplectic glasses), and phase changes (high pressure phases; melting). Kink bands (mainly in micas) have also been described as a result of shock metamorphism (e.g., Hörz and Ahrens 1969), but can also be the result of normal tectonic deformation. Although there seem to be some differences between the characteristics of the kink bands formed by shock and those formed in "normal" deformations (Stöffler 1972), these have not been quantified. Table 1.6.1.1. lists the most characteristic products of shock metamorphism, as well as the associated diagnostic features.

Table 1.6.1.1. Characteristics and formation pressures of various shock deformation features.

Pressure (GPa)	Features	Target Characteristics	Feature Characteristics
2–45	Shatter cones	Best developed in homogeneous fine-grained, massive rocks.	Conical fracture surfaces with subordinate striations radiating from a focal point.
5–45	Planar fractures and Planar deformation features (PDFs)	Highest abundance in crystalline rocks; found in many rock-forming minerals; e.g., quartz, feldspar, olivine and zircon.	PDFs: Sets of extremely straight, sharply defined parallel lamellae; may occur in multiple sets with specific crystallographic orientations.
30–40	Diaplectic glass	Most important in quartz and feldspar (e.g., maskelynite from plagioclase).	Isotropization through solid-state transformation under preservation of crystal habit as well as primary defects and sometimes planar features. Index of refraction lower than in corresponding crystal but higher than in fusion glass.
15–50	High-pressure polymorphs	Quartz polymorphs most common: coesite, stishovite; but also ringwoodite from olivine, and others.	Recognizable by crystal parameters, confirmed usually with XRD or NMR; abundance influenced by post-shock temperature and shock duration; stishovite is temperature-labile.
>15	Impact diamond	From carbon (graphite) present in target rocks; rare.	Cubic (hexagonal?) form; usually very small but occasionally up to mm-size; inherits graphite crystal shape.
45–>70	Mineral melts	Rock-forming minerals (e.g., lechatelierite from quartz).	Impact melts are either glassy (fusion glasses) or crystalline; of macroscopically homogeneous, but microscopically often heterogeneous composition.

Data from: Alexopoulos et al (1988), French and Short (1968), Sharpton and Grieve (1990), Stöffler (1972, 1974), Koeberl (1994), Koeberl et al. (1995a).

Before discussing the various shock metamorphic features, the type and location of impact lithologies should be briefly mentioned (however, see Sect. 1.7 for details). In an impact crater, shocked minerals, impact melts, and impact glasses are commonly found in various impact-derived breccias (e.g., Fig. 1.6.1.3). Well-preserved ejecta at the crater rim may display a stratigraphic sequence that is inverted compared to the normal stratigraphy in the area.

The impact process leads to the formation of various breccia types, which are found within and around the resulting crater (see also Stöffler and Grieve 1994a, b, and Koeberl et al. 1996a). The three main breccia types include monomict or polymict breccias consisting of 1) cataclastic (fragmental), 2) suevitic (fragmental with a melt fragment component), or 3) impact melt (melt breccia – i.e., melt in the matrix – with a clastic component) breccias. The breccias can be allochthonous or autochthonous. In addition, dikes of injected or locally formed fragmental or pseudotachylitic breccias (Reimold 1995, 1998), which contain evidence of melting, can be found in the basement rocks. Whether these various breccia types are indeed present and/or preserved in a crater depends on factors including the size of the crater, the composition, and the porosity of the target area (e.g., Kieffer 1971; Kieffer and Simonds 1980), and the level of erosion (see, e.g., Roddy et al. 1977; Hörz 1982; Hörz et al. 1983; Grieve 1987; and references therein).

Fig. 1.6.1.3. Macroscopic view of a granitic fragmental breccia from the Newporte impact structure, North Dakota; sample D9463.0 from the Duerre 43-5 drill core (from 2884 m depth), showing angular granitic fragments in a dark, fine-grained, clast-rich matrix.

1.6.2
Shatter Cones

The occurrence of shatter cones has long been discussed as the only good macroscopic indicator of impact-generated deformation, and a variety of structures were proposed to be of impact origin on the basis of shatter cone occurrences (e.g., Dietz 1968; Milton 1977). Such cones have also been formed in (chemical) explosion crater experiments (see, e.g., Milton 1977). Their formation is dependent on the type of target rock (i.e., they are better developed in certain lithologies than in others), and has been estimated to take place at pressures in the range of 2 to 45 GPa (e.g., Gash 1971; Milton 1977). In general, shatter cones are cones with regular thin grooves (striae) that radiate from the top (the apex) of a cone. They can range in size from less than one centimeter to more than one meter (Fig. 1.6.2.1).

Fig. 1.6.2.1. Shatter cone from the Haughton Dome (Devon Island, Canada); size of sample about 20 cm long dimension.

Shatter cones occur mostly in the central uplift and in outer and lower parts of a crater and may be preserved even if a structure is deeply eroded. Unfortunately, conclusive criteria for the recognition of "true" shatter cones have not yet been defined. It is possible to confuse impact-produced cones with concussion features, pressure-solution features (cone-in-cone structure), or abraded or otherwise striated features with shatter cones (Reimold and Minnitt 1996). It would be important to arrive at some generally accepted criteria for the correct identification of shatter cones, as some impact craters have been identified almost exclusively by the occurrence of shatter cones (for example, the Red Wing Creek structure, USA; see Koeberl et al. 1996b and the compilation by Koeberl and Anderson 1996b). Shatter cones form in a variety of target rocks, including crystalline igneous and metamorphic rocks, sandstones, shales, and carbonates. The best developed cones are found in fine-grained rocks, especially carbonates (e.g., limestone).

The formation mechanism of shatter cones is poorly understood. It seems that although the shatter cones themselves form at fairly low shock pressures, localized melting on shatter cone surfaces requires locally higher pressures and/or temperatures. The apices of shatter cones have also been used, after graphical restoration to their original horizontal preimpact position, to reconstruct the source of the shock wave (e.g., Dietz 1968; see discussion in French 1998, pp 38–40). However, there can be some confusion about the orientation of shatter cones, with multiple apex directions even within one unit, as shown in Fig. 1.6.2.1, making any reconstruction of shock wave sources impossible. Nevertheless, shatter cones are important as potential macroscopic shock indicators, as they are developed in large volumes of rock, and are useful as a guide for the possible presence of more definitive shock indicators, such as shocked minerals (see below).

1.6.3
Mosaicism

Mosaicism is a microscopic expression of shock metamorphism observed in a number of rock-forming minerals (e.g., Hörz and Quaide 1972) and appears as an irregular mottled optical extinction pattern, which is distinctly different from undulatory extinction that, for example, occurs in tectonically deformed quartz. Mosaicism can be measured in the optical microscope by determining the scatter of optical axes in different regions of crystals showing mosaicism. Mosaicism can be semiquantitatively defined by X–ray diffraction study of the asterism of single crystal grains, where it shows up as a characteristic increase (with increasing shock pressure) of the width of individual lattice diffraction spots in diffraction patterns. Highly shocked quartz crystals show a diffraction pattern that becomes similar to a powder pattern, because of shock-induced polycrystallinity. Many shocked quartz grains that show planar microstructures also show mosaicism. In addition, it should be noted that the crystal lattice of shocked quartz shows lattice expansion above shock pressures of 25 GPa, leading to an expansion of the cell volume by $\leq 3\%$ (Langenhorst 1994).

1.6.4
Planar Microstructures

Two types of planar microstructures can be apparent in shocked minerals: planar fractures (PFs) and planar deformation features (PDFs). Their characteristics are summarized for the example "quartz" in Table 1.6.4.1. PDFs in rock-forming minerals (e.g., quartz, feldspar, or olivine) are generally accepted to provide diagnostic evidence for shock deformation (see, e.g., French and Short 1968; Stöffler

1972, 1974; Alexopoulos et al. 1988; Sharpton and Grieve 1990; Stöffler and Langenhorst 1994; Huffman and Reimold 1996; Grieve et al. 1996). PFs, in contrast to irregular, non-planar fractures, are thin fissures, spaced about 20 μm or more apart, which are parallel to rational crystallographic planes with low Miller indices, such as (0001) or {10$\bar{1}$1} (e.g., Engelhardt and Bertsch 1969). PFs form at lower pressures than PDFs, and may not provide conclusive evidence of shock metamorphism, but can direct attention to other, more characteristic, shock deformation effects.

Table 1.6.4.1. Characteristics of planar fractures and planar deformation features in quartz.

Nomenclature	1. Planar fractures (PF)
	2. Planar deformation features (PDF)
	2.1 Non-decorated PDFs
	2.2 Decorated
Crystallographic orientation	1. PFs: usually parallel to (0001) and {10$\bar{1}$1}
	2. PDFs: usually p to {10$\bar{1}$3}, {10$\bar{1}$2}, {10$\bar{1}$1}, (0001), {11$\bar{2}$2}, {11$\bar{2}$1}, {10$\bar{1}$0}, {11$\bar{2}$0}, {21$\bar{3}$1}, {51$\bar{6}$1}, etc.
Optical microscope properties	Multiple sets of PFs or PDFs (up to 15 orientations) per grain
	Thickness of PDFs: <2–3 μm
	Spacing: >15 μm (PFs), 2–10 μm (PDFs)
TEM properties (PDFs)	Two types of primary lamellae are observed:
	1. Amorphous lamellae with a thickness of ca. 30 nm (at pressures <25 GPa) and ca. 200 nm (at pressure >25 GPa)
	2. Brazil twin lamellae parallel to (0001)

Data after Stöffler and Langenhorst (1994).

PDFs, together with the somewhat less definitive planar fractures (PFs), are well developed in quartz (Stöffler and Langenhorst 1994). PDFs are parallel zones with a thickness of ≤1–3 μm and are spaced about 2–10 μm apart (see Figs. 1.6.4.1a, b). The degree of planarity of the individual sets of PDFs is an important parameter for the correct identification of *bona fide* PDFs and allow their distinction from (sub-)planar features that are produced at lower strain rates, e.g., in tectonically deformed quartz. Tectonic deformation lamellae (Böhm lamellae) usually consist of bands that are >10 μm wide and are spaced at distances of >10 μm. They are usually not completely straight, do not extend through the whole grain, and occur only in one set. As such they can easily be, with some experience, distinguished from PDFs.

It was demonstrated from Transmission Electron Microscopy (TEM) studies (see, e.g., Goltrant et al. 1991) that PDFs consist of amorphous silica. The structural state of the glassy lamellae is, however, slightly different from that of regular silica glass (Goltrant et al. 1991, 1992). The fact that the PDF lamellae are filled by glass allows them to be preferentially etched by, e.g., hydrofluoric acid, emphasizing the planar deformation features (see Fig. 1.6.4.2). Such an etching method has been described by Gratz et al. (1996) and can be used, in combination with SEM studies, to distinguish tectonically deformed quartz from shocked quartz.

Engelhardt and Bertsch (1969) have classified PDFs into four groups: (a) non-decorated PDFs (extremely fine lamellae, cannot be resolved in the optical microscope), (b) decorated PDFs (the lamellae are lined by, or replaced with, small spherical or elliptical bubbles, often representing fluid inclusions), (c) homogeneous lamellae (thicker lamellae that can be resolved in the microscope), and (d) filled PDFs (where the lamellae are filled with very small fine-grained crystals). Types (a) and (b) are the most common ones.

In addition, TEM studies have shown that there is a second type of PDF, which consists of very thin multiple lamellae of Brazil twins. Brazil twins have been observed in hydrothermally grown quartz, but always parallel to the $\{10\bar{1}1\}$ plane, while the impact-derived Brazil twins form at pressures >8 GPa, are of mechanical origin, and are exclusively formed parallel to the (0001) plane (Goltrant et al. 1991, 1992; Leroux et al. 1994). The latter authors have documented that such Brazil twins are recognizable from the Vredefort impact structure in South Africa because they are more resistant to annealing than other shock metamorphic indicators.

Most rock-forming minerals, as well as accessory minerals, such as zircon, develop PDFs (e.g., Stöffler 1972, 1974; Dworak 1969). The occurrence of diagnostic shock features is by far the most important criterion for evaluating the impact origin of a crater, particularly when several of the features that are typical of progressive shock metamorphism, as listed in Table 1.6.1.1, have been found. The occurrence of PDFs and PFs can be used, together with other shock effects, to determine the maximum shock pressure in impactites (Table 1.6.4.2). Most commonly, quartz is used to study these shock effects, as it is the simplest, best studied, and most widely distributed rock-forming mineral that develops PDFs well.

Planar deformation features occur in planes corresponding to specific rational crystallographic orientations. In quartz, the most abundant mineral that develops distinctive PDFs, the (0001) or c (basal), $\{10\bar{1}3\}$ or ω, and $\{10\bar{1}2\}$ or π orientations are the most common ones. In addition, PDFs often occur in more than one crystallographic orientation per grain. With increasing shock pressure, the distances between the planes decrease, and the PDFs become more closely spaced and more homogeneously distributed over the grain, until at about ≥ 35 GPa complete isotropization has been achieved. Depending on the peak pressure, PDFs are observed in 2 to 10 (maximum 18) orientations per grain. To properly characterize PDFs, it is necessary to measure their crystallographic orientations by using either a universal stage (Reinhard 1931; Emmons 1943) or a spindle stage (Medenbach 1985), or by transmission electron microscopy (see, e.g., Goltrant et al. 1991; Gratz et al. 1992a; Leroux et al. 1994).

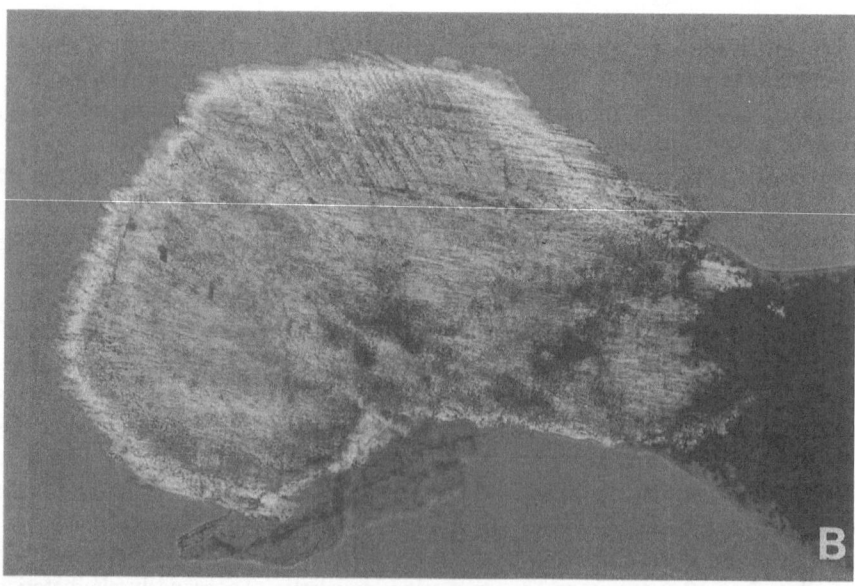

Fig. 1.6.4.1. (a) PDFs in quartz from the Ilyinets impact structure, Ukraine (photo courtesy W.U. Reimold). (b) PDFs in feldspar from the K/T boundary, Raton Basin (photo courtesy B.F. Bohor, U.S. Geological Survey).

Fig. 1.6.4.2. SEM image of quartz grain from the K/T boundary layer at DSDP Site 596 (Southwest Pacific), after brief etching with HF, showing three different sets of PDFs (courtesy B.F. Bohor, U.S. Geological Survey).

The relative frequencies of the crystallographic orientations of PDFs can be used to calibrate shock pressure regimes, as given in Table 1.6.4.2 (see, e.g., Robertson et al. 1968; Hörz 1968; Stöffler and Langenhorst 1994). For example, at 5 to 10 GPa, PDFs with (0001) and {1011} orientations are formed, whereas PDFs with {1013} orientations start to form between about 10 and 12 GPa. Such studies are done by measuring the angles between the c-axis and a set of PDFs in individual quartz grains with a universal stage. In a stereographic projection (see Engelhardt and Bertsch 1969), the optical axis (c-axis) is rotated into the center of projection, the locations of the poles of PDFs are plotted, and then those positions are compared with the stereographic projection of the rational crystallographic planes in quartz (see Sect. 6.3.5). The measured angles that fall within 5° of the theoretical polar angle of the plane are considered valid and can be indexed. The results of this procedure can be shown in the form of a histographic plot of indexed PDFs. Such plots are used to identify the relative frequencies, in which the various shock-characteristic crystallographic orientations occur.

Table 1.6.4.2. Relation between shock stage and crystallographic orientation (indices) of planar microstructures in quartz.

Shock stage	Main orientations	Additional orientations	Optical properties
1. Very weakly shocked	PFs: (0001)	PFs: rarely {10$\bar{1}$1}	normal
2. Weakly shocked	PDFs: {10$\bar{1}$3}	PFs: {10$\bar{1}$1}, (0001) PDFs: rare	normal
3. Moderately shocked	PDFs: {10$\bar{1}$3}	PFs: {10$\bar{1}$1}, (0001) rare PDFs: {11$\bar{2}$2}, {11$\bar{2}$1}, (0001), {10$\bar{1}$0} + {11$\bar{2}$1}, {10$\bar{1}$1}, {21$\bar{3}$1}, {51$\bar{6}$1}	normal or slightly reduced refractive indices
4. Strongly shocked	PDFs: {10$\bar{1}$2} {10$\bar{1}$3}	PFs: rare or absent PDFs: {11$\bar{2}$2}, {11$\bar{2}$1}, (0001), {10$\bar{1}$0} + {11$\bar{2}$1}, {10$\bar{1}$1}, {21$\bar{3}$1}, {51$\bar{6}$1}	reduced refractive indices 1.546–1.48
5. Very strongly shocked	PDFs: {10$\bar{1}$2} {10$\bar{1}$3}	None	reduced refractive indices (<1.48)

PFs: planar fractures; PDFs: planar deformation features; after Stöffler and Langenhorst (1994).

The pre-shock temperature of a target rock also influences the formation and distribution of PDFs (Langenhorst et al. 1992). Reimold (1988), Huffman et al. (1993), and Huffman and Reimold (1996) presented the results of shock experiments with quartzite and granite at room temperature (25 °C), and preheated to 450 °C and 750 °C. They noticed a slight difference in the relative distribution of the {1013} and {1012} orientations and a large difference in the number of PDF sets per grain (Fig. 1.6.4.3). Langenhorst (1993) and Langenhorst and Deutsch (1994) compared PDF orientations in shocked quartz single crystals preheated to a higher temperature than Huffman et al. (1993) and found a distinct change in the relative frequencies of the {1013} and {1012} orientations.

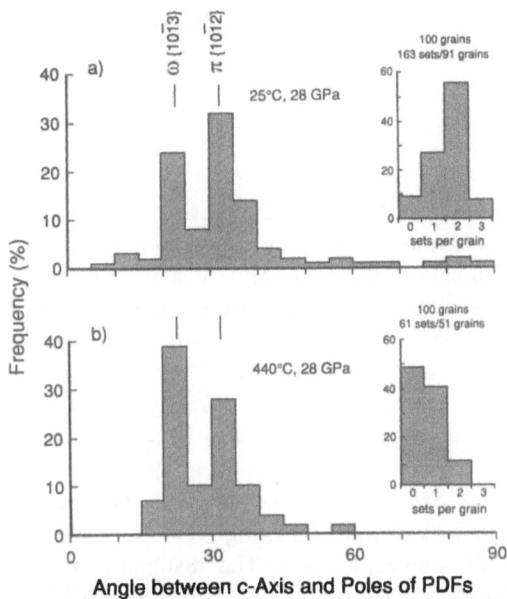

Fig. 1.6.4.3. Histogram with crystallographic orientation of PDFs in quartz from Hospital Hill quartzite from the Witwatersrand Basin, showing the dependency of the orientations on the pre-shock temperature (after Huffman et al. 1993). a) Pre-shock temperature 25 °C, shock pressure 28 GPa. b) Pre-shock temperature 440 °C, shock pressure 28 GPa. The main difference between the sets is that about half of the quartz grains in the high-temperature experiment remain un-shocked, whereas in the low-temperature experiment, almost all quartz grains are shocked.

1.6.5
Bulk Optical and other Properties

It has been shown that there is a decrease of the density of shocked quartz with increasing shock pressure (Langenhorst 1993; cf. Stöffler and Langenhorst 1994). At shock pressures up to about 25 GPa, only a slight decrease is noticeable, followed by a significant drop in density between 25 and 35 GPa, depending on the direction of the shock wave relative to the c-axis of the quartz crystal, and the pre-shock temperature. Optical properties, such as the birefringence of quartz and the refractive index, show also an inverse relationship with shock pressure in the 25 to 35 GPa range. At 35 GPa, isotropization (formation of diaplectic quartz glass) occurs. The data also indicate that with increasing shock pressure the birefringence ($n_\omega - n_\varepsilon$) decreases. Still other properties of shocked minerals can be used to either confirm a shock history or calibrate shock pressures. For example, intensity and wavelength of infrared absorption bands, the electron paramagnetic resonance, and peak width in a ^{29}Si magic angle spinning nuclear magnetic resonance

(NMR) spectrum all depend in a quantitative way on the shock pressure (e.g., Boslough et al. 1995; references in Stöffler and Langenhorst 1994).

1.6.6
Diaplectic Glass

At shock pressures in excess of about 35 GPa, diaplectic glass is formed (Table 1.6.1.1), which has been found at numerous impact craters. It is an isotropic phase that preserves the crystal habit, original crystal defects, and, in some cases, planar features. It forms without melting by solid-state transformation and has been described as a phase "intermediate between crystalline and normal glassy phases" (Stöffler and Hornemann 1972). For example, a phase called maskelynite forms from feldspar (Bunch et al. 1967). Diaplectic glass has a refractive index that is slightly lower, and a density that is slightly higher, than that of synthetic quartz glass. At pressures that exceed about 50 GPa, lechatelierite, a "normal" mineral melt, forms by fusion of quartz (see Sect. 1.6.8). Other minerals also undergo melting (fusion) at similar pressures. This complete melting is not the same process that results in the formation of diaplectic glass. The distinction between diaplectic glass and lechatelierite (both after quartz) was described by Stöffler and Hornemann (1972) and Stöffler and Langenhorst (1994).

1.6.7
High-Pressure Polymorphs

A very important form of shock deformation are phase transitions to high-pressure polymorphs of minerals in a solid state transformation process. Such transformation can be predicted from Hugoniot data. Many minerals form metastable high-pressure phases (Stöffler 1972). These include (density in g/cm^3 is given in parentheses): stishovite (4.23) and coesite (2.93) from quartz (2.65); jadeite (3.24) from plagioclase (2.63–2.76), majorite (3.67) from pyroxene (3.20–3.52); and ringwoodite (3.90) from olivine (3.22–4.34). In contrast to expectations from the equilibrium phase diagram of quartz, which can be used to predict phase transitions under static conditions, stishovite forms at lower shock pressures than coesite, probably because stishovite forms directly during shock compression, whereas coesite crystallizes during pressure release.

The first time that coesite, the high-pressure modification of quartz, was detected in nature was at impact structures, and the presence of this phase has been taken as confirming evidence of an impact event (e.g., Chao et al. 1961). In addition, stishovite was for the first time discovered in nature at Meteor Crater, Arizona (Chao et al. 1962), and, to date, has only been observed in impact-deformed rocks. The formation probabilities and conditions for these phases are strongly dependent on the porosity of the target rocks. While stishovite has never been found in any natural non-impact related rocks, there are rare findings of coesite in metamorphic rocks of ultra-high pressure (UHP) origin or in kimberlites. However, coesite from UHP rocks can easily be distinguished from that in impact-derived rocks. Coesite within UHP metamorphic rocks occurs as large single crystals within, or associated with, high-pressure minerals of metamorphic or volcanic origin, but never associated with quartz. On the other hand, impact-derived coesite occurs as fine-grained, colorless to brownish, polycrystalline aggregates of up to 200 μm in size, which are usually embedded in diaplectic quartz, or, rarely, in nearly isotropic shocked quartz or their recrystallization products. In addition to morphological differences, shock-produced coesite occurs in a disequilibrium assemblage of quartz-coesite-stishovite-glass (see also Grieve et al. 1996). Recently, a rare "post-stishovite" phase has been identified by Sharp et al. (1999), indicating very high shock pressures during its formation.

In addition to high-pressure phases of rock-forming minerals, impact-derived diamonds (the high pressure modification of carbon) have also been found at various craters. These diamonds form from carbon in the target rocks, mainly in graphite-bearing (e.g., graphitic gneiss) or coal-bearing rocks (Koeberl et al. 1997a). Impact diamonds commonly preserve the crystal habit of their precursor material, which is mostly hexagonal graphite (some are coal-derived). For example, the gneissic target rocks at the Popigai impact structure in Siberia contain graphite flakes, crystals, and aggregates with sizes of 20 μm to about 10 mm. Well-shaped tabular graphite crystals are usually 0.1–1 mm in size and are elongated parallel to the gneissic fabric. Diamonds, which are also found in situ in shocked crystalline rocks, commonly show tabular shapes, indicating that they preserve the crystal habit of the precursor graphite crystals (Fig. 1.6.7.1). The diamonds that formed after graphite have been called "apographitic" diamonds.

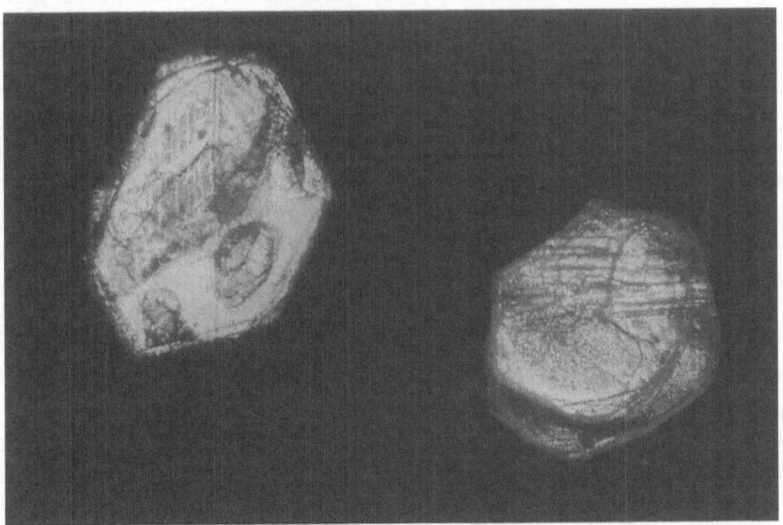

Fig. 1.6.7.1. Shock-produced diamonds from the Popigai impact structure, Russia (grain size about 0.5 mm) (photo courtesy V.I. Masaitis).

All impact diamonds are polycrystalline, with sizes of individual diamond microcrystals being on the order of 1 μm or less. Many of them were thought, from X–ray diffraction work, to contain up to several volume percent lonsdaleite, the rare hexagonal diamond modification. However, TEM investigations show that the diamonds have an internal layered texture with thicknesses of a few micrometers, which may be inherited from the precursor graphite. TEM also shows that locally the cubic diamond phase is intergrown with a lamellar phase or defect. The lamellae could either represent stacking faults or microtwins. If multiple stacking faults occur, the lamellae could be interpreted to be lonsdaleite, the hexagonal diamond polymorph. More recent TEM work also failed to identify the presence of lonsdaleite and rather seems to suggest that stacking faults in the diamond microcrystals mimic the X–ray pattern of the hexagonal phase (Koeberl et al. 1997a; Hough et al. 1998; Langenhorst et al. 1999). Impact-derived diamonds have also been found to be intergrown with silicon carbide (SiC) in suevites from the Ries crater in Germany (Hough et al. 1995), and at the K/T boundary (see Sect. 3.2.1).

1.6.8
Mineral and Rock Melts

At pressures in excess of about 60 GPa, rocks undergo complete (bulk) melting to form impact melts (see Table 1.6.1.1). The melts can reach very high temperatures due to the passage of shock waves that generate temperatures far beyond those commonly encountered in normal crustal processes or in volcanic eruptions. The high temperatures are demonstrated by the presence of inclusions of high-temperature minerals, such as lechatelierite, which is the monomineralic quartz melt and forms from pure quartz at temperatures >1700 °C (see above), or baddeleyite, which is the thermal decomposition product of zircon, forming at a temperature of about 1900 °C. Lechatelierite is not found in any other natural rock, except in fulgurites, which form by fusion of soil or sand when lightning hits the ground. Lechatelierite does not occur in any volcanic igneous rocks.

If higher temperatures persist, not only individual minerals undergo melting, but whole rocks and rock units can melt. Impact melts may also go through a phase of superheating (i.e., staying liquid even though the vaporization temperature has been exceeded) at temperatures of 10,000 °C or higher (e.g., Jakes et al. 1992; Melosh 1998). Depending on the initial temperature, the location within the crater, the composition of the melt, and the speed of cooling, impact melts either form impact glasses (if they cool fast enough), or, more commonly, (mostly) fine-grained impact melt rocks (if they cool slower). Impact melt rocks are also found as inclusions in suevitic breccias in the form of glass fragments or melt clasts (Fig. 1.6.8.1, and see Sect. 1.7). Impact melt rocks contain clasts of shocked minerals or lithic clasts. Recently, carbonate melts have been identified at the Ries crater in Germany (Graup 1999).

As glasses are metastable supercooled liquids, impact glasses slowly recrystallize (if dissolution is not acting faster), at a rate that depends on the composition of the glass and post-impact environmental conditions. Therefore, impact glasses are more commonly found at young impact craters than at old impact structures. Very fine-grained recrystallization textures are often characteristic for devitrified impact glasses. Impact glasses have chemical and isotopic compositions that are very similar to those of individual target rocks or mixtures of several rock types. For example, it is possible to use the rare earth element (REE) distribution patterns, or the Rb-Sr isotopic composition, which are identical to those of the (often sedimentary or metasedimentary) target rocks, to distinguish the impact melt rocks from intrusive or volcanic rocks (e.g., Blum and Chamberlain 1992; Blum et al. 1993). Furthermore, impact glasses have much lower water contents (about 0.001–0.05 wt.%) than volcanic or other natural glasses (e.g., Koeberl 1992b). Detailed descriptions of impact melts and glasses and their characteristics and compositions are discussed by, for example, El Goresy et al. (1968), Dence (1971), Stöffler (1984), Koeberl (1986, 1992a, b), Vishnevsky and Montanari (1999), and references therein. See also Sects. 2.4 and 2.5.

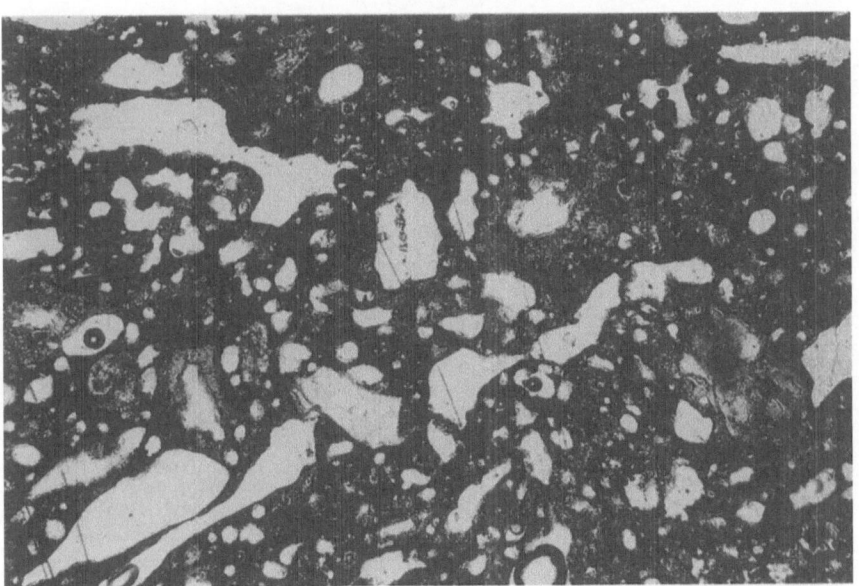

Fig. 1.6.8.1. Aphanitic impact melt breccia from suevite deposits on the north side of the Bosumtwi crater, Ghana. A small quartz clast near the center of the image shows ballen structure. Width of image 3.4 mm, parallel polars (photo courtesy W.U. Reimold).

Impact melt rocks are true igneous rocks that have formed by cooling and crystallization of high-temperature silicate melts. They often have textures and mineral compositions that are similar to those of volcanic igneous rocks, which may make it difficult to judge the impact origin from an isolated sample. Field evidence should be used to determine if large-scale melt rock volumes are associated with other, more distinctive, shocked rocks (that did not undergo complete melting). Occasionally, in individual samples, evidence for an impact origin can be obtained from careful petrographic and geochemical studies. Impact melt rocks often preserve inclusions of not completely melted rock and mineral fragments that may show evidence for shock metamorphism in the form of PDFs in rock-forming minerals. The presence of lechatelierite may be characteristic, as well as the so-called "Ballen" texture, where lechatelierite forms a distinctive "crackled" pattern of curved fractures that may result from (partial) devitrification of the quartz melt (see Carstens 1975 and French 1998, pp 94–95). Chemical studies may also provide evidence for an impact origin of a melt rock. For example, Sr isotope analyses are able to distinguish between the composition of volcanic rocks and that of locally melted crustal rocks. The presence of a meteoritic component in such rocks also provides clear evidence of an impact origin (see Sect. 1.8). Impact melts and glasses (or minerals that have recrystallized from the melt; e.g., Krogh et al. 1993a, b; Izett et al. 1994) have another important use, as they often are the most suitable material for the dating of an impact structure (see Sect. 1.9).

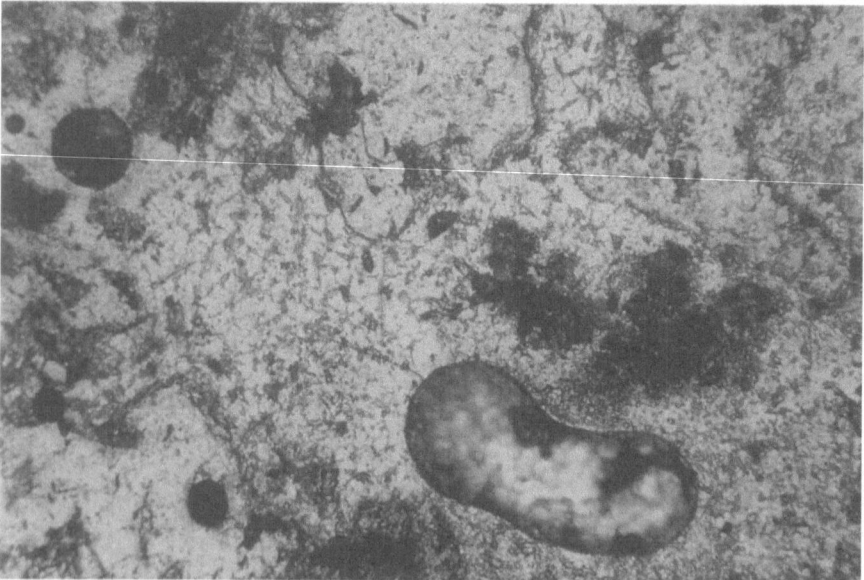

Fig. 1.6.8.2. Ballen structure in quartz from loose suevitic gravel of the Saltpan (Tswaing) impact crater, South Africa. Width of image 1.1 mm, parallel polarizers (photo courtesy W.U. Reimold).

1.7
Classification of Impactites

At this point a short review of the nomenclature of impactites might be useful, as different terminologies for the rock types at a crater have been used by various authors. In general, it is best to follow well-established and widely accepted classification criteria of impactites discussed by, e.g., Stöffler et al. (1979), Stöffler and Grieve (1994a, b), and French (1998). In some aspects crater-specific terminologies might have some advantages, because they refer to the local rock types; on the other hand, local terms make it very difficult to compare rock units from different impact structures. A possible shortcoming of the general impactite terminology may be the lack of genetic information, for example, with regard to the possibility that more than one type of suevitic or fragmental impact breccia may have been deposited, perhaps by different processes, such as atmospheric fall-out, or base surge deposition, within the crater (e.g., Vishnevsky and Montanari 1999).

Following Stöffler and Grieve (1994b) and French (1998), the following terms and definitions of common impact formations can be used. An *impactite* is a collective term for all rocks affected by (an) impact(s) resulting from collision(s) of planetary bodies. Their classification scheme (for single impacts) uses two sets of criteria for a) a combination of lithological components, texture, and degree of shock metamorphism, and b) mode of occurrence (in- or outside the crater). In terms of location, a fundamental distinction is made between parautochthonous rocks beneath and allochthonous (or allogenic) rocks that fill the crater (crater-fill units, e.g., breccias and melt rocks) and also occur as ejecta around the crater. French (1998) distinguishes four locations in and around an impact structure, which are, with their associated rock units: a) sub-crater (parautochthonous rocks, cross-cutting allogenic units, pseudotachylite); b) crater interior (allogenic crater-fill deposits: lithic [fragmental] breccias, suevitic breccias, impact melt breccias); c) crater rim region (proximal ejecta deposits), and d) distal ejecta (for a discussion of the latter, see Sect. 2.1).

Parautochthonous rocks may include target rocks that were subjected to shock metamorphism but remained in place, as well as impactites (e.g., monomict fragmental impact breccia that remained in situ, but was internally brecciated, and where breccia clasts were subjected to small-scale movements or rotation). Another breccia type, *pseudotachylite*, has been reported from sub-crater basement, in the form of small veins and dikes hardly ever exceeding a few centimeters width. Only in the large Sudbury and Vredefort structures have abundant volumes of such breccia been observed. As pseudotachylite in general usage is a term reserved for friction melt rock – not only in impact but also in tectonic settings – and because in impact crater floors a whole range of different dike breccias (including impact melt breccia, fragmental, and suevitic breccia injections, or clastic breccias, such as cataclasites, or even pre- or post-impact tectonically produced pseudotachylites) may occur, it has been recommended (Reimold 1995, 1998) to use the term pseudotachylite with caution. All the breccia types mentioned may macroscopically appear like true pseudotachylite, and diligent micro-analysis is required to resolve the true nature of these breccias in question. Particularly, it should be reserved for occurrences of true friction melts only. Instead, the term "pseudotachylitic breccia" could be used as a field term and in such instances where the true nature of such material has not been confirmed (yet). It has been suggested that some pseudotachylites formed during late-stage crater developments and modification (e.g., Spray 1995, 1997; Spray and Thompson 1995).

The crater fill contains a variety of breccia types. *Fragmental impact breccia* is a "monomict or polymict impact breccia with clastic matrix containing shocked and unshocked mineral and lithic clasts, but lacking cogenetic impact melt particles" (Stöffler and Grieve 1994b). This type has also been called a *lithic breccia* (French 1998). *Impact melt breccia* is also a common type, which is defined by Stöffler and Grieve (1994b) as an "impact melt rock containing lithic and mineral clasts displaying variable degrees of shock metamorphism in a crystalline, semi-hyaline or hyaline matrix (crystalline or glassy impact melt breccias)" (with an *impact melt rock* being a "crystalline, semihyaline or hyaline rock solidified from impact melt"). *Suevite* (or suevitic breccia; Figs. 1.7.1–1.7.3) is defined as a "polymict breccia with clastic matrix containing lithic and mineral clasts in various stages of shock metamorphism including cogenetic impact melt particles which are in a glassy or crystallized state". The distribution of the rock types is a function of their formation and the order in which they formed. For example, lithic breccias can occur not only inside, but also outside a crater. At the Ries crater in Germany, a distinctively colored polymict lithic breccia, the so-called "Bunte Breccia", occurs beneath melt-bearing suevites both inside and outside the crater, with a sharp contact between the two units (see, e.g., Hörz et al. 1983).

For the identification of meteorite impact structures, suevites and impact melt breccias (or impact melt rocks) are the most commonly studied units. Despite the detailed definition given above, an easy way to distinguish between the two types is the following: suevites are polymict breccias that contain inclusions of melt rock (or impact glass), i.e., they are clast-dominated ("melt fragment breccias"), and impact melt breccias have a melt matrix with a variable amount of (often shocked) rock fragment inclusions (they are matrix-dominated breccias that also have been termed "melt-matrix breccias").

The rocks in the crater rim zone are usually only subjected to relatively low shock pressures (commonly <2 GPa), leading mostly to fracturing and brecciation, and often do not show shock-characteristic deformation. Thus, even at larger craters *in situ* crater rim rocks rarely show evidence for shock deformation (Reimold et al. 1998a; Koeberl et al. 1998a). The best indication for impact in these locations is the occurrence of shocked ejecta, such as suevites and melt rocks (or impact glasses), or the presence of so-called dike breccias (e.g., Lambert 1981). In well-preserved impact structures the area immediately outside the crater rim is covered by a sequence of different impactite deposits (see, e.g., Hörz et al. 1983; French 1998), which often allow the identification of these structures as being of impact origin.

Fig. 1.7.1. Large blocks of suevitic breccia are found within less than one crater radius outside the rim of the Bosumtwi crater, Ghana (with one of the authors, C.K.).

Fig. 1.7.2. Cut slab of suevitic breccia from the Bosumtwi crater, Ghana, showing the variety of clasts within the fine-grained matrix. Dark frothy areas are glass.

Distal impact deposits are difficult to describe in any standard nomenclature prompting a suggestion of the use of stratigraphic terminology by King and Petruny (1999).

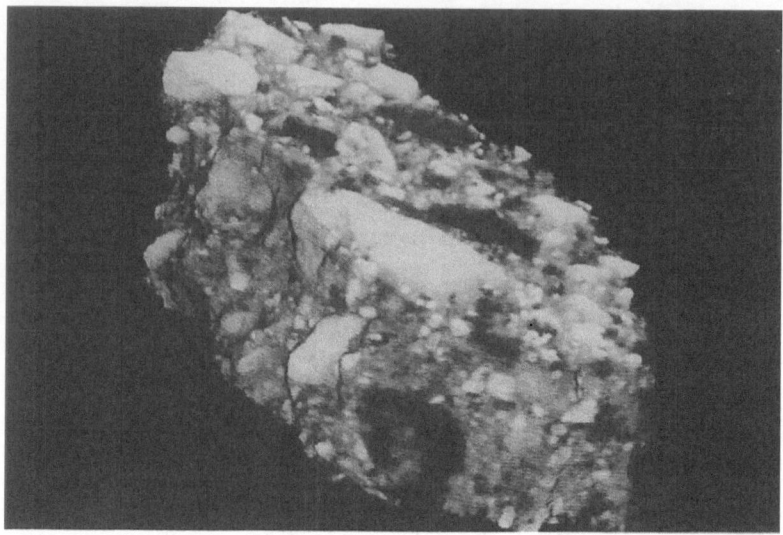

Fig. 1.7.3. High-resolution X–ray computed tomography image of a 5-cm-sized suevite from the Bosumtwi crater, Ghana. The matrix has been removed by computer processing, showing the clast population (different clast densities are indicated by variations in brightness). From a study by Koeberl et al., in preparation (see also Koeberl et al. 1998c).

1.8
Meteoritic Components in Impactites

The verification of an extraterrestrial component in impact-derived melt rocks or breccias can be of diagnostic value to provide confirming evidence for an impact origin of a geological structure. Geochemical methods are used to determine the presence of the traces of such a component. The presence of actual meteorite fragments at a crater structure is rare. This is a consequence of the physics of an impact event. A shock wave, similar to the one that penetrates through the target, also passes through the meteoritic impactor and, within fractions of a second, vaporizes most, or all, of the projectile. Only during the impact of small objects (less than about 40 m in diameter, depending on impact angle and velocity), because of spallation during entry into the atmosphere or due to lower impact velocity resulting from atmospheric drag, a small fraction of the initial mass of the meteorite may survive. The cut-off diameter of impact craters at which some fraction of meteoritic material may be preserved is about 1–1.5 km.

A good example is Meteor Crater, Arizona, where only about 2% of the estimated total mass of the iron meteorite impactor have been found around the crater. In addition, a large number of small iron-rich spheroids were found around the crater, which did undergo melting during the cratering event; these amount to about 5% of the original impactor mass. Recent theoretical modeling calculations, as well as [59]Ni isotopic measurements on the spherules and meteorites show that most of the projectile had melted and only little was vaporized, and that a part of the outer shell of the projectile (amounting to a maximum of 16% of its mass; see Schnabel et al. 1999) remained solid on the rear surface. Thus, even under optimistic conditions, meteoritic fragments are only preserved at very young and small craters. The absence of meteorite fragments can, therefore, not be used as evidence against an impact origin of a crater structure.

In the absence of actual meteorite fragments, it is necessary to search for traces of meteoritic material that is mixed in with the target rocks in breccias and melt rocks. This can be done by studying the concentrations and interelement ratios of siderophile elements, especially the platinum group elements, which are several orders of magnitude more abundant in meteorites than in terrestrial upper crustal rocks. The usage of platinum group element abundances and ratios avoids some of the ambiguities that may result if only common siderophile elements (e.g., Cr, Co, Ni) are considered. However, problems may arise if the target rocks have high abundances of siderophile elements, or if the siderophile element concentrations in the impactites are very low. In such cases, the Re–Os isotopic system (Sect. 1.8.3) and the Cr isotope system (Sect. 1.8.4) have recently been used for establishing the presence of a meteoritic component in a number of impact melt rocks and breccias. Meteoritic components have been identified for about 41 impact structures (out of more than 160 known on Earth; see Koeberl 1998 for a listing), which reflects also the detail in which these structures were studied. The following discussion follows mostly the review by Koeberl (1998).

1.8.1
Meteoritic Components and Siderophile Element Analyses

During impact, a small amount of the finely dispersed meteoritic melt (droplets) or vapor is mixed in with a much larger quantity of target rock vapor and melt, which later forms impact melt rocks, melt breccias, or glass. In most cases, the contribution of meteoritic matter to these impactite lithologies is very small (commonly <<1% by weight), leading to only slight chemical changes in the resulting impactites. The detection of such small amounts of meteoritic matter within the normal upper crustal compositional signature of the target rocks is extremely difficult. Only elements that have high abundances in meteorites, but low abundances in terrestrial crustal rocks, such as the siderophile elements, can be used to detect such a meteoritic component. Another complication is the existence of a variety of meteorite groups and types (the three main groups are stony meteorites, iron meteorites, and stony-iron meteorites, in order of decreasing abundance), which have widely varying siderophile element compositions. Distinctly higher siderophile element contents in impact melts, compared to target rock abundances, can be indicative of the presence of either a chondritic or an iron meteoritic component. Achondritic projectiles are much more difficult to discern, because they have significantly lower abundances of the key siderophile elements.

Siderophile (and related) elements that have often been used are Ni, Co, Cr. In addition, the interelement ratios of these elements are an effective discriminator. If the meteoritic contribution is exceeding 0.1%, it is possible to distinguish between stony meteorites (chondrites) and iron meteorites, because chondrites have high abundances of Cr (typically about 0.26 wt.%), whereas iron meteorites have Cr abundances that are much more variable, but typically about 100 times lower than those of chondrites, and low Ni/Cr or Co/Cr ratios. For this reason, several authors (e.g., Palme et al. 1978; Evans et al. 1993) have used the Cr abundance, together with the Ni/Cr or Co/Cr ratios in impact melts, to distinguish between iron and chondritic projectiles. Unfortunately, the abundances of Co, Cr, and Ni are not particularly low in most terrestrial rocks. Thus, there may be a relatively high (and variable) indigenous component that is derived from the target rocks. If any conclusions are to be drawn from the abundances of elements such as Cr, Ni, and Co, a detailed study of their abundances in the target rocks needs to be done (cf. Palme 1980). From mixing calculations it is possible to determine the relative proportions of the various target rocks types that are contributing to breccias or melt rocks. Thus, a bulk composition of a breccia (or melt rock) can be modeled, and the siderophile element composition of this model calculation compared with the actually observed siderophile element abundances in the impactites.

The first studies on meteoritic components were done not on terrestrial impactites, but on lunar rocks. These have shown that the platinum group elements (PGEs: Ru, Rh, Pd, Os, Ir, Pt) and Au are better suited for identifying a meteoritic component than Cr, Co, and Ni (e.g., Morgan et al. 1975; Palme et al. 1978, 1979; Evans et al. 1993; Schmidt and Pernicka 1994). The abundances of the PGEs in chondrites and most iron meteorites are several orders of magnitude higher than those in terrestrial crustal rocks, as illustrated for Ir in Fig. 1.8.1.1. The range of Ir and Os abundances in chondrites is about 400 to 800 ppb, while iron meteorites show a much wider range. In contrast, continental crustal rocks contain on the order of 0.02 ppb Ir or Os (e.g., Taylor and McLennan 1985). Thus, the signal to background ratio is very high for the PGEs (i.e., low indigenous concentrations; high concentrations in the "contaminating" meteorite).

The admixture of minor quantities of meteoritic material (chondrite, iron meteorite) to crustal target rocks yields significantly elevated abundances of the PGEs in the resulting impact melt rocks or breccias. For example, if only 1 wt.% of a chondritic meteorite is mixed into terrestrial crustal target rocks, the resulting breccia or melt will contain about 4 ppb of Ir. Even 0.1 wt.% of a chondritic component still results in 0.4 ppb of Ir (or Os) in the breccia, which is at least one order of magnitude higher than average crustal background values. Achondrites have a much wider range of Ir and Os contents, and, in general, lower abundances than for chondrites (e.g., Morgan et al. 1979; Schmidt and Pernicka 1994). Iron meteorites (Buchwald 1975) show widely varying PGE abundances, depending on the meteorite type. Using siderophile element abundances and interelement ratios, the projectile type was determined for a number of terrestrial craters.

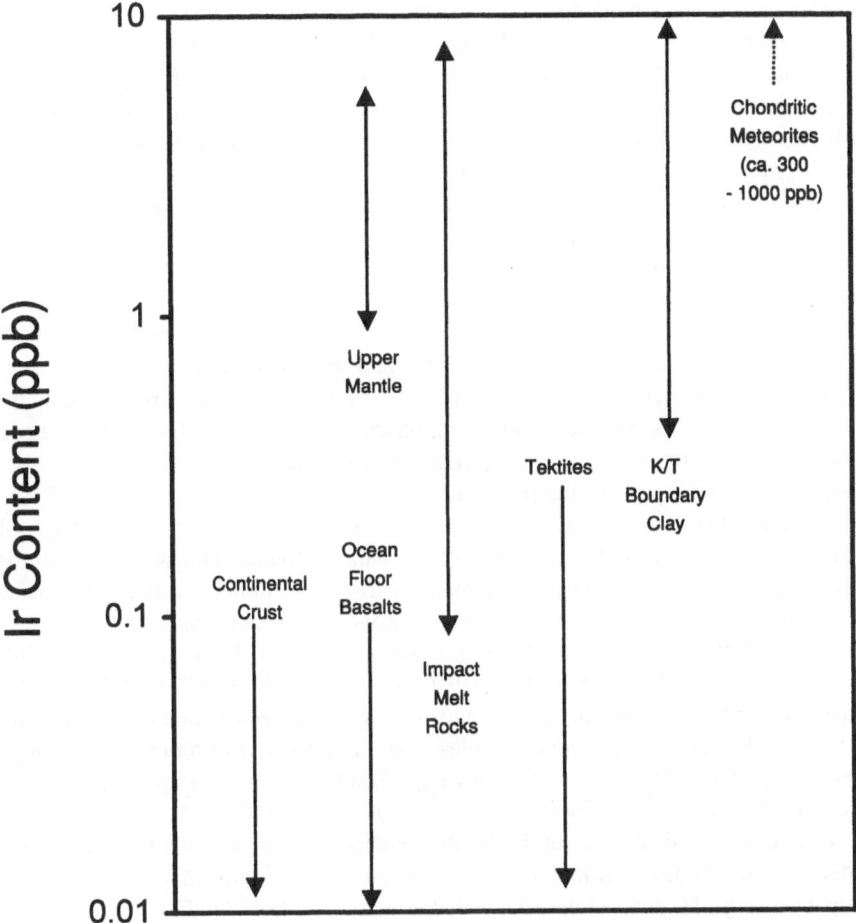

Fig. 1.8.1.1. Range of Ir contents in typical terrestrial and extraterrestrial rocks (diagram after Koeberl 1998).

As a general rule, the larger the crater, the more likely it is that the projectile has been melted or vaporized, although the impact angle also plays an important role (e.g., Pierazzo and Melosh 2000). Most small impact craters (those with less than about 1.5 km diameter) were formed by iron projectiles of known composition (because meteorite fragments were recovered, e.g., at Wolf Creek or Meteor Crater), with the exception of the Saltpan (Tswaing) crater (South Africa), which was formed by the impact of a chondritic body (see Koeberl et al. 1994a). Some of the smaller structures are in fact crater fields consisting of several more or less circular structures, such as Wabar, Henbury, and Odessa (at all of which iron meteorite fragments have been recovered).

All craters that are associated with meteorite fragments are relatively young (less than about 0.1 Ma), whereas at older craters (e.g., the 3.1 Ma Aouelloul crater, Mauritania) no remnants of the actual impactor were recovered. While the Saltpan crater is only 0.22 Ma old, no remnants of the impacting chondrite were found (most likely do to erosion of any remaining fragments). An unusual case is the Rio Cuarto crater field in Argentina, because it consists of at least ten rimmed, oblong structures ranging in size between about 0.1 x 0.3 and 4.5 x 1.1 km, which formed by a very low angle (grazing) impact of a chondritic body, with the top of the impactor being decapitated and producing the down-range crater structures (e.g., Schultz et al. 1994). In contrast to the Saltpan crater, a small mass of the chondritic impactor (of H4/5 type) that formed the much larger Rio Cuarto crater field is still preserved. This could be due to the young age of the structure (probably as low as 2000 to 4000 years, based on preliminary fission track ages), and because of the oblique impact (and resulting impactor decapitation), which helped to preserve parts of the projectile.

In most impact structures, impact melt rocks (or breccias) are often the only material in which a meteoritic component can be found. The results of studies using PGE and other siderophile element distributions indicated that in almost all cases the amount of the extraterrestrial component in such impact-derived rocks is less than 1% (e.g., Morgan et al. 1975; Palme et al. 1978, 1979; Göbel et al. 1980; Koeberl et al. 1994a; French et al. 1997; see also Koeberl 1998). Only in a few cases are higher amounts found, such as at East Clearwater (Canada), where some – but not all – impact melt rock samples contain up to 10% of a meteoritic component (Palme et al. 1981), or at Morokweng (South Africa) with up to about 4% of a chondritic component (Koeberl et al. 1997c).

In a number of studies of PGE abundances and interelement ratios, it was attempted to use these values to determine the type or class of meteorite for the impactor (e.g., Morgan et al. 1975; Palme et al. 1978, 1979; Evans et al. 1993), but these attempts were not always successful. Suggestions regarding a distinction between asteroidal and cometary sources are relatively useless because so far no unambiguously cometary material has ever been retrieved and analyzed for its trace element composition. Not only is it difficult to decide between chondrite types, but it should also be pointed out that conflicting identifications have been made for a number of impact structures. Many type identifications are highly uncertain. For example, both iron and chondritic projectiles were proposed for the Bosumtwi (Ghana), Zhamanshin (Kazakhstan), and Rochechouart (France) impact structures (see Koeberl 1998 for details). Clearly, the identification of a meteoritic component in impactites is not a trivial problem.

1.8.2
Problems with Meteoritic Component Identification

Apart from the seemingly straightforward cases in which meteorite fragments are found at a crater, inferences about the projectile type have to be made from the abundances and interelement ratios of various siderophile elements in impactites. However, several complications are introduced by complex fractionation processes that seem to take place during the formation of impact glasses and melts.

In recent studies of impact glasses from some small craters for which the meteorite has been partly preserved (e.g., Meteor Crater, Wabar, and some Australian craters), it was found that the siderophile elements show strong (and highly variable!) fractionation in the interelement ratios compared to the initial ratios in the impacting meteorite. The changes do not seem to be correlated in a straightforward way with any physical or chemical properties of the elements (Hörz et al. 1989; Attrep et al. 1991a, Mittlefehldt et al. 1992a, b). Mittlefehldt et al. (1992b) argued against simple vapor fractionation of the chemical elements as well as post-depositional fractionation. Instead, they proposed that the siderophile elements may have been fractionated from each other during the early phases of the impact, while the projectile was undergoing decompression and before mixing with the target materials, although they were unable to explain the fractionation with a specific model.

In general, though, there is a correlation between the amount of melt and vapor with increasing impact velocity (e.g., Pierazzo et al. 1997), which may, thus, be the most important factor controlling the incorporation of a meteoritic component (the amount of which is known to vary widely between craters of similar size, and also within impact breccias and melt rocks from a single crater). However, this energy scaling relationship would not explain the fractionation within a single crater. This observation does not bode well for any attempts to directly infer projectile types of small craters from siderophile element ratios in impactites. And, indeed, the element ratios at, for example, Aouelloul or Brent do not readily conform to those of any known meteorite types, but can be interpreted in various ways.

To allow the identification of a meteoritic component in impactites, the abundances of PGEs and other siderophile elements in the impactites have to be compared with those in the target rocks. Ideally, these indigenous concentrations should be determined and subtracted from the abundances found in the impact melt rocks, and yield the "pure" meteoritic abundance ratios (e.g., Morgan and Wandless 1983). Thus, it is necessary to analyze all rock types that are known or suspected to be present in the pre-impact target area (in some cases, clasts within breccias can be used). Mixing calculations (see, e.g., French et al. 1997) can be used to establish the relative proportions of the target rocks types that were mixed to form impact melt rocks or breccias, which then allows to subtract a proper indigenous component from the melt or breccia composition. This procedure may yield good results only if the target stratigraphy is simple (i.e., not involving too many rock types, which should also be distinct in composition). In reality, though, the exact target rocks that were involved in forming impact melt rocks or breccia are not always well known (e.g., due to later erosion of the target stratigraphy), or because of very low or highly variable indigenous PGE concentrations (see, e.g., Schmidt and Pernicka 1994; Pernicka et al. 1996).

Palme (1982) used the Au/Ir ratio to distinguish between cosmic and terrestrial signatures. However, Au is much more mobile than Ir under a number of terrestrial conditions (e.g., weathering, diagenesis, metamorphism), which may lead to non-chondritic Ir/Au ratios even if a meteoritic component is present in impact-derived rocks. Specifically, Au often shows high indigenous abundances in some terrestrial rocks, yielding low (non-chondritic) Ir/Au ratios. Additional complications arise because PGEs are enriched in some rock types (e.g., mafic and ultramafic rocks and mineralized rocks). The PGE patterns of the mantle and in some mantle-derived rocks may be similar to those of chondrites, and the PGE abundances in mantle rocks are also higher than those in the crust by factors of about 10^2 to 10^3, making a distinction between an exposed mantle section or a component in the impactites derived from a mafic to ultramafic precursor rock and a meteoritic component quite difficult, if not impossible.

Siderophile element fractionation may also occur in the impact melt while it is still molten. This effect may be significant in larger craters, because there the melt could stay hot for many thousand years. Different mineral phases, such as sulfides or oxides (e.g., magnetite, chromite), may take up various proportions of the PGEs or other siderophile elements, leading to an irregular distribution of these elements and possibly fractionated interelement ratios and patterns. Such irregular distribution of siderophiles is known from, for example, the East and West Clearwater impact structures (Palme et al. 1979), or the Chicxulub impact structure (Koeberl et al. 1994c; see also below). Hydrothermal processes associated with the hot impact melt may also change PGE abundances. In contrast to a widely held opinion, actual data show that meteoritic components are often inhomogeneously distributed in impact melt rocks and breccias (e.g., data in Palme et al. 1979; Koeberl et al. 1994a–c; Schuraytz et al. 1994).

A variety of fractionation effects have also been documented for distal ejecta from the K/T boundary impact at various localities around the world (e.g., Evans et al. 1993). For example, high PGE abundances were discovered in impact ejecta from the Acraman structure in Australia (see Sect. 3.4), but show deviations from chondritic patterns due to low-temperature hydrothermal alteration (e.g., Gostin et al. 1989; Wallace et al. 1990b; cf. also Colodner et al. 1992).

The limitations and complications discussed above make the proper identification of a meteoritic component difficult. The determination of a specific projectile type is even more difficult and can yield ambiguous results. Thus, the use of some more sensitive and selective methods for meteoritic component detection and identification would be desirable.

1.8.3
Osmium Isotopes in Impact Studies

The Os isotopic system is such a more sensitive method and has some advantages over the use of PGE elemental abundances. The method is superior with respect to detection limit and selectivity, as discussed by Koeberl and Shirey (1993, 1997) and Koeberl et al. (1994a–c). In principle, the abundances of Re and Os and the $^{188}Os/^{187}Os$ isotopic ratios, which are measured by very sensitive mass spectrometric techniques, allow to distinguish the isotopic signatures of meteoritic and terrestrial Os; this was for the first time shown by Fehn et al. (1986), whose technique, however, lacked the sensitivity of modern analytical methods. Meteorites (and the terrestrial mantle) have much higher (by factors of 10^2–10^5) PGE contents than terrestrial crustal rocks. In addition, meteorites have relatively low Re and high Os abundances, resulting in Re/Os ratios less or equal to 0.1, while the Re/Os ratio of terrestrial crustal rocks is usually no less than 10. More importantly even, the $^{188}Os/^{187}Os$ isotopic ratios for meteorites and terrestrial crustal rocks are significantly different.

^{187}Os is formed from the ß-decay of ^{187}Re (with a half-life of 42.3±1.3 Ga). Thus, due to the high Re and low Os concentrations in old crustal rocks, their $^{187}Os/^{188}Os$ ratio increases rapidly with time. The present day $^{187}Os/^{188}Os$ ratio of mantle rocks is about 0.13. Meteorites also have low $^{187}Os/^{188}Os$ ratios of about 0.11 to 0.18. Osmium is much more abundant in meteorites than Re, leading to only small changes in the meteoritic $^{187}Os/^{188}Os$ ratio with time. Because of the high Os abundances in meteorites, the addition of a minute meteoritical contribution to the crustal target rocks leads to an almost complete change of the Os isotopic signature of the resulting impact melt or breccia (see Fig. 1.8.3.1 for an example). The Os isotopic method does not, however, allow the determination of the projectile type.

It is necessary, however, to distinguish between a possible mantle signal and an extraterrestrial signal, as the Os isotopic ratios for mantle and meteoritic material are very similar. Fortunately, however, there is a significant difference in the total abundance of Os between mantle rocks (which have about 1–4 ppb Os) and meteorites (typical chondrites have about 400–800 ppb Os). Thus, about 100 times more mantle material than meteoritic material would need to be added to normal crustal rocks (e.g., in a breccia) to result in the same Os isotopic ratio of the bulk rock. A detailed geological field investigation, and a petrographic study of the rocks in question, if necessary combined with trace element and Rb–Sr and/or Sm–Nd isotopic analyses, will easily show if significant amounts of ultramafic materials are present. For details about this method, see Koeberl and Shirey (1993, 1996, 1997). Case studies for a variety of impact structures are presented by (e.g.) Walker et al. (1991), Dickin et al. (1992), Koeberl et al. (1994a–c, 1996d, 1998a) and French et al. (1997). For the use of the method in investigations of the K/T boundary, see Sect. 3.2.1. Similar to studies of shock metamorphism, Os isotopic measurements of target rocks and impactites may be used as evidence in favor of an impact origin of a new or doubtful structure.

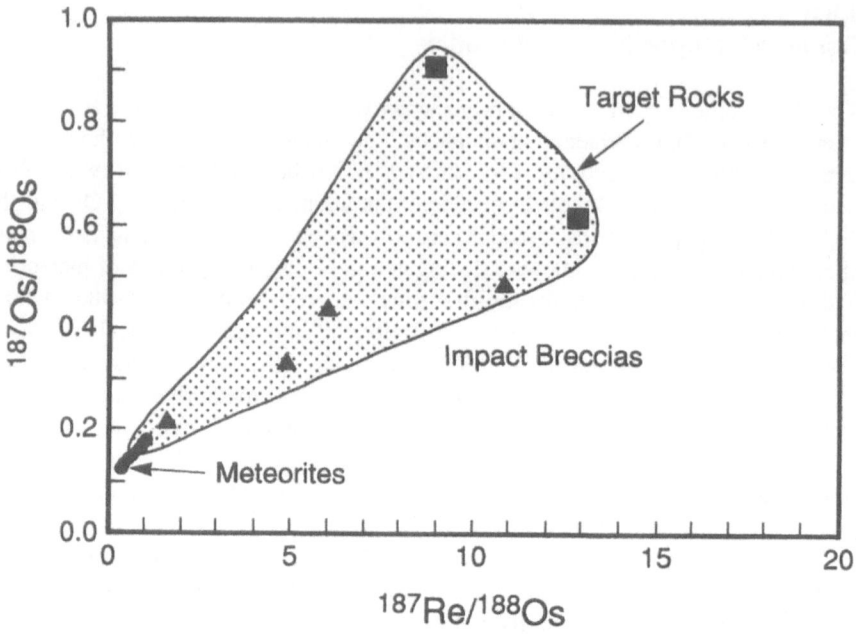

Fig. 1.8.3.1. Ratios of $^{187}Os/^{188}Os$ versus $^{187}Re/^{188}Os$ for target rocks (shale and sandstone) of the Kalkkop impact crater, South Africa (see Reimold et al. 1998b), in comparison with data for four impact breccias (solid triangles) and the data array for chondritic and iron meteorites (small solid dots). The sample depths in the drill core are given as well (after Koeberl et al. 1994b). The dotted area marks the mixing field between target rocks and meteorites. All impact breccias fall within this mixing field; breccia-3 plots very close to the meteorite data array, indicating that this sample contains a larger meteoritic contributions than analyzed for any other breccia sample, which is in good agreement with the high Os content of this sample (about 0.2 ppb). In addition to distinct differences in the Os isotopic composition, the Os content of the target rocks, at about 0.02 ppb, is ca. 10 times lower than the Os content of the breccias.

1.8.4
Chromium Isotope Systematics

Recently, the usage of another isotopic system has been explored for impact studies. It has been shown that the isotopic composition of the element chromium, in terms of the isotopes ^{52}Cr and ^{53}Cr, differs between individual meteorite groups (e.g., Shukolyukov and Lugmair 1998). For example, ordinary chondrites (types H and L) have isotopic characteristics that are slightly different from those of enstatite chondrites and are, even more significantly, different from those of carbonaceous chondrites. In contrast, all terrestrial materials analyzed so far (see Shukolyukov and Lugmair 1998) show no or only insignificant variation and are indistinguishable from the zero value $^{53}Cr/^{52}Cr \equiv \varepsilon 0$. Thus, assuming that a certain (fairly significant) percentage of the chromium in an impact melt or ejecta layer is of extraterrestrial origin, it should be possible by measuring the Cr isotopic composition to determine not only that an extraterrestrial component is present, but also which meteorite type might have been involved.

Shukolyukov and Lugmair (1998) used this method to determine a carbonaceous chondritic component at the K/T boundary and Shukolyukov et al. (1999) showed that the meteoritic component in impact melt from the Morokweng impact structure in South Africa (see Sect. 3.5) is of ordinary chondritic affinity. However, this powerful and potentially very selective method is not without problems either. The differences in isotopic composition are very small, requiring extremely precise and time-consuming measurements, which severely limit the number of samples that can be measured and, therefore, invalidate the use of this technique as a screening tool for structures of unknown origin. This is far from a routine method that is only done in very few laboratories. Furthermore, a substantial amount of the chromium has to be of extraterrestrial origin to show an effect in the Cr isotopic composition that is larger than the precision of the measurement. Thus, while being more selective than the Os isotopic method, the Cr isotopic method is much less sensitive. Once again it becomes clear that no silver bullet exists in impact geochemistry.

1.9
Dating of Impact Structures

Of the about 160 impact structures known today, precise ages have been determined only for a small fraction. In principle, it should be possible to precisely date impact craters, as they mark geologically instantaneous events. Precise ages (with a precision better than 1 or 2 rel.%) are necessary to a) understand the frequency with which impact events of a given size occur, b) to compare this value with astronomical considerations for impact frequency, c) to link impact craters and distal ejecta horizons, d) to compare the ages of large impact structures with those of mass extinction horizons in the geological record, and e) to test the hypothesis that impacts occur with a certain periodicity. Some of these relationships require the comparison of radiometric age determination techniques and biostratigraphic methods. The methods most commonly used for dating of impact melt rocks or glasses include the K–Ar, ^{40}Ar–^{39}Ar, fission track, Rb–Sr, Sm–Nd, or U–Th–Pb isotope methods. However, dating of impact craters is complicated and tedious and, if not done with utmost care, can easily lead to erroneous results. Bottomley et al. (1990), and, especially, Deutsch and Schärer (1994) provided excellent and detailed reviews of impact crater dating; thus, a short summary will be given here.

There are two different approaches to the dating of impact structures. The first approach uses impact-derived rocks directly in or around the crater, usually for radiometric dating of melt rocks, and the second approach uses (distal) impact ejecta and tries to obtain a date for them based on (bio)stratigraphic data, and then uses this age to date an associated crater. First, we will discuss some aspects and methods used in dating crater rocks (breccias, melt rocks). A wide variety of geochronological methods has been used for the dating of impact craters, as their ages span the whole geological history of the Earth. The oldest currently known impact structure is Vredefort in South Africa at 2.023 Ga (e.g., Reimold and Gibson 1996), and the youngest craters formed only within the last few thousand years (e.g., Wabar in Saudi Arabia is only about 6000 years old; Storzer and Wagner 1977).

This text cannot provide descriptions of the background of each dating method. The reader has to be referred to a standard textbook that includes geochronological methods (e.g., Faure 1986). Thus, we assume that the reader is familiar with the basics and methodology of these techniques. A very useful method in impactite dating is fission track dating. Most glassy materials, both within a crater and in distal ejecta (e.g., tektites, see Sect. 2.4.1), can be dated by counting fossil tracks and then comparing the number of fossil tracks with the number of tracks induced by irradiation in a reactor (for the determination of the uranium content). It is also possible to measure tracks in shocked minerals, such as apatite or zircon. Track fading and annealing may cause systematic errors that can be corrected by extensive experimental work (see, e.g., Storzer and Wagner 1982 for a review).

Direct dating of impactites is often difficult due to inhomogeneity of the rocks, alteration, and limitations of the methods. For many methods it will be necessary to extract mineral separates and perform separate isotopic analyses. For example, the Rb–Sr isochron method (see Faure 1986) requires separation of several mineral phases, such as biotite, K-feldspar, and plagioclase to obtain the spread in isotopic ratios and elemental abundances that is necessary to define an isochron. If all minerals grew in the impact melt rock upon recrystallization, use of the isochron method will yield the age of the impact event. Problems can arise from post-impact hydrothermal or metamorphic processes and from devitrification of glassy or melt rocks. Another method involves U–Th–Pb dating of zircons that have either been reset by the impact event (e.g., Kamo and Krogh 1995; Krogh et al. 1993a, b) or have grown in the hot melt (e.g., Kamo et al. 1996; Koeberl et al. 1997c). The small size of the zircons makes it necessary to use high precision methods (e.g., single zircon dating after air abrasion; Krogh et al. 1993a, b) or ion beam methods with high spatial resolution (Koeberl et al. 1997c).

A common and precise method is the Ar–Ar technique, which uses the same decay system as the K–Ar method (see Faure 1986), but allows the use of very small sample sizes and also to perform the K abundance determination and the Ar isotopic measurement in the same sample. Glassy or mineral fragments need to be isolated from the impactites for dating. After irradiation the sample chips are heated either in an oven (stepwise heating) or with a laser to release the Ar, from which the age can then be calculated. Details of the use of this method for impact crater dating are given by Bottomley et al. (1990). Sample selection is also a crucial step – different samples from the same crater may give a variety of different results (e.g., Koeberl et al. 1990). While this is a potentially very precise method, it is still possible that some post-impact regional event reset the Ar isotopic system and ages obtained by this method, while still precise, do not date the actual impact event. An example is provided by the Gardnos crater in Norway, for which an age of about 550 to 600 Ma is preferred based on stratigraphic relationships, but Ar–Ar dating of several samples gave a consistent age of 385 ± 5 Ma (Grier et al. 1999). Clearly, abandoning local stratigraphic results in favor of an Ar–Ar age is an unpopular option.

Distal impact ejecta are often discovered accidentally and their ages are determined from biostratigraphic relationships and paleomagnetic considerations. Such ejecta layers (in the impactite terminology they are called *impactoclastic layers*) may subsequently lead to the discovery of a source crater. For example, the impact ejecta layer at the K/T boundary led to the discovery of the Chicxulub structure, and the findings of impact ejecta in Australia led to the discovery of the Acraman structure (e.g., Gostin et al. 1986; Wallace et al. 1990a, b). These topics, and the relationship between impact ejecta and geological boundaries, will be discussed in detail in Chap. 3. The reverse problem, namely to search for the ejecta of a specific and known impact structure in a well-preserved stratigraphic sequence (e.g., at the Umbria–Marche sequence), is much more difficult, because of the lack of precise radiometric ages for many of the rock units involved, and because of correlation problems between radiometric and biostratigraphic (and paleomagnetic) age determinations. This question is discussed in detail for all of the geological boundaries and possible impact events recorded in the Umbria–Marche sequence in Chaps. 4 and 5. Table 1.9.1 lists the ages of impact structures that coincide with the stratigraphic record that is preserved in the Umbria-Marche sequence.

Impact crater ages are important to determine the frequency of such events, and to discuss any possible periodicities in the crater record. Following the intriguing suggestion that the mass extinctions in the geological history are not randomly distributed in time, but follow a certain periodicity (Raup and Sepkoski 1984; Sepkoski 1990, 1992, 1996), Alvarez and Muller (1984) proposed that the age of impact structures on Earth record a similar periodicity on the order of 26 to 30 million years. However, their analysis was based on an insufficient number of craters (some of which were imprecisely dated), and subsequent, statistically more extensive, studies by Grieve et al. (1985, 1988) failed to detect any periodicity. Thus, at present this hypothesis cannot be confirmed, mainly because of the insufficient number of (large) impact structures that have been dated with the necessary precision, and because most well-dated craters would only cover the past two or three cycles of an about 30 million years periodicity. It is hoped that in the future a major emphasis of crater studies will be to obtain precise dates for a large number of impact structures.

Table 1.9.1. List of impact structures on Earth younger than 250 Ma and larger than 5 km in diameter.

Structure name and country	Diameter (km)	Age (Ma)	Error (Ma)	Dating method	Reference
1 Bosumtwi (Ghana)	10.5	1.07	0.05	^{40}Ar/^{39}Ar	Koeberl et al. 1997b
2 Zhamanshin (Kazakhstan)	13.5	1.09	0.05	fission track	Koeberl and Storzer 1988
3 Kara-Kul (Tajikistan)	52	2.5	2.5	stratigraphic?	Grieve et al. 1995
4 El'gygytgyn (Russia)	18	3.5	0.5	K/Ar	Gurov and Gurova 1980
5 Karla (Russia)	12	5	1	stratigraphic	Masaitis 1999
6 Bigach (Kazakhstan)	7	5	3	stratigraphic	Kilesev and Korotushenko 1986; Masaitis 1999
7 Ries (Germany)	24	15.1	1.0	^{40}Ar/^{39}Ar	Staudacher et al. 1982
8 Haughton (Canada)	24	23.4	1.0	^{40}Ar/^{39}Ar	Jessberger 1988
9 Chesapeake (U.S.A.)	90	35.3	0.2	stratigraphic	Poag and Aubry 1995
10 Popigai (Russia)	100	35.7	0.8	^{40}Ar/^{39}Ar	Bottomley et al. 1997
11 Wanapitei (Canada)	7.5	37	2	K/Ar	Winzer et al. 1976
12 Mistastin (Canada)	28	38	4	^{40}Ar/^{39}Ar	Mak et al. 1976
13 Logoisk (Russia)*	17	40	5	stratigraphic	Masaitis et al. 1980; Grieve et al. 1995
14 Chiyly (Kazakhstan)	5.5	46	7	?	Grieve et al. 1995
15 Kamensk (Russia)	25	49.2	0.2	^{40}Ar/^{39}Ar	Izett et al. 1994
16 Montagnais (Canada)	45	50.5	0.8	^{40}Ar/^{39}Ar	Bottomley and York 1988
17 Ragozinka (Russia)*	9	55	5	stratigraphic	Vishnevsky and Lagutenko 1986
18 Marquez (U.S.A.)	13	58.3	2	fission track; stratrigraphic	McHone and Sorkhabi 1994; Sharpton and Gibson 1990
19 Chicxulub (Mexico)	180	64.98	0.05	^{40}Ar/^{39}Ar	Swisher et al. 1992
20 Kara (Russia)	65	70.3	2.2	^{40}Ar/^{39}Ar	Trieloff et al. 1998
21 Ust Kara (Russia)	25	70.3	2.2	^{40}Ar/^{39}Ar	Trieloff et al. 1998
22 Manson (U.S.A.)	37	74.1	0.1	^{40}Ar/^{39}Ar	Izett et al. 1998
23 Lappajärvi (Finland)	23	77.3	0.4	^{40}Ar/^{39}Ar	Jessberger and Reimold 1980
24 Boltysh (Ukraine)*	24	88	3	K/Ar	Boiko et al. 1985
25 Dellen (Sweden)	15	89.0	2.7	Rb/Sr	Deutsch et al. 1992
26 Steen River (Canada)	25	95	7	K/Ar	Carrigy 1968
27 Avak (U.S.A.)	12	100	5	stratigraphic	Kirschner et al. 1992

28 Carswell (Canada)	39	115	10	$^{40}Ar/^{39}Ar$	Bottomley et al. 1989
29 Mien (Sweden)	9	121.0	2.3	$^{40}Ar/^{39}Ar$	Bottomley et al. 1989
30 Tookoonooka (Australia)	55	128	5	stratigraphic	Gorter et al. 1989
31 Gosses Bluff (Australia)	22	142.5	0.8	$^{40}Ar/^{39}Ar$	Milton and Sutter 1987
32 Mjølnir (Barents Sea)	40	143.1	0.8	stratigraphic	Dypvik et al. 1996; H. Dypvik, pers.comm. 1999
33 Morokweng (South Africa)	≥80	144.7	1.9	$^{208}Pb/^{232}Th$	Koeberl et al. 1997c
34 Puchezh-Katunki (Russia)	80	167	3	stratigraphic	Masaitis 1999
35 Obolon (Ukraine)	15	169	7	$^{40}Ar/^{39}Ar$	Masaitis 1999
36 Red Wing (U.S.A.)	9	200	25	stratigraphic	Brenan et al. 1975; Koeberl et al. 1996b
37 Manicoaugan (Canada)	100	214	1	U/Pb	Hodych and Dunning 1992
38 Saint Martin (Canada)	40	219	32	Rb/Sr	Reimold et al. 1990
39 Araguainha (Brazil)	40	245	3	$^{40}Ar/^{39}Ar$	Hammerschmidt and Engelhardt 1995
40 Kursk (Russia)	5.5	250	80	stratigraphic	Masaitis 1999

Note: * Masaitis (1999) lists the following ages for these structures, but without giving any source for these values: Logoisk, 30 ± 5 Ma; Ragozinka, 46 ± 3 Ma; Boltysh, 95 ± 10 Ma

2 Distal Ejecta and Tektites

Distal ejecta are, by definition, those ejecta that occur at considerable distances from the source crater (>5 crater radii from the crater rim; see, e.g., Melosh 1989). They either consist of (usually fine-grained) rock and mineral fragments or are of glassy consistency. It is often not immediately possible to recognize that they are directly connected to a specific impact structure. However, distal ejecta can act as a guide to major impact events and even lead to the discovery of large impact structures (Sect. 2.3). In this chapter we will review mainly glassy distal ejecta (with tektites being the main representative of this class), whereas Chap. 3 deals with the characteristics and importance of lithic distal ejecta.

2.1
Characteristics of Proximal and Distal Ejecta

During crater formation, at the end of the excavation stage, about 90% of all material ejected (excavated) from the crater will be deposited relatively close to the crater (less than about 5 crater radii) and is called *proximal ejecta*. The mass of the ejecta decreases outwards: about half of the ejected material can be found within about one crater radius from the rim. At this proximity ejecta blankets can be up to several hundred meters thick, depending on the size of the crater. Further away the ejecta blankets become increasingly thinner and discontinuous. Comparison of data from impact experiments with those of chemical and nuclear explosion shows that at least 90% of all material ejected from a crater is deposited at a distance of 5 crater radii from the center of the structure (Fig. 2.1.1.). Most material is ejected from the crater in ballistic trajectories to a distance (range) R_b, which, for short distances, follows the simple formula:

$$R_b = (v_e^2/g)\sin(2\Phi)$$

where v_e is the ejection velocity, g is the surface acceleration of gravity, and Φ is the ejection angle. If the range becomes a significant part of the radius of the planet on which the event occurs, the relations become more complicated because of the curvature of the planet (Melosh 1989, p. 87).

There is a significant difference in the ejection and distribution mechanisms for craters formed with and without an atmosphere. On an airless planet, ejecta will be emplaced ballistically. Many studies on ejecta behavior were done on lunar craters (see review by Oberbeck 1975). Ejection of large blocks of target material may lead to the formation of proximal secondary craters, which are – in contrast to the main crater – the result of low-velocity impacts. Calculations and experiments show that (as might be expected) not only the total ejecta mass decreases with increasing distance from the crater, but also the average particle size decreases.

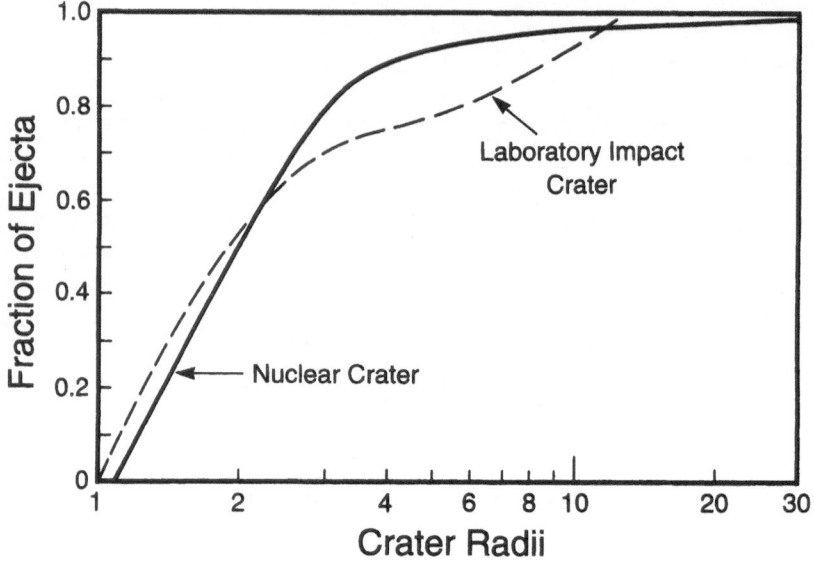

Fig. 2.1.1. Proportion of ejecta mass shown in relation to distance from crater center, as deduced from laboratory impact as well as nuclear explosion crater experiments (data after Oberbeck 1975).

Ejection of material from a crater is a complicated, multistage process. Some material is ejected early on during the contact and compression stage at very high velocities (which can reach more than one third of the impact velocity; Melosh 1989), but most material is ejected during the excavation phase under much lower conditions of stress. The material seems to be ejected at relatively low angles, with very little material being thrown out at angles of more than 45° (Oberbeck 1975). On Earth (or another planet with an atmosphere), ejecta near the crater can also be deposited in a base surge, which is a gravity-driven density current (composed of air and dust entrained by the fireball) that flows down (and outwards) from the expanding mushroom-shaped cloud. All ejecta are subject to subsequent erosion (e.g., Grant and Schultz 1993), unless they are rapidly covered by sediments.

An important conclusion from models (Oberbeck 1975) is that material launched early in the crater formation will still be in flight after the crater is completely excavated and after the more massive local ejecta have been deposited to form the continuous deposits around the crater. Thus, material ejected late in the crater formation (when the shock wave has already decayed somewhat) is ejected at lower velocities and will be deposited first and close to the crater rim, whereas material ejected early and at higher velocities will be deposited later and at greater distances. The low velocity ejecta deposited at the crater rim often preserve the initial stratigraphic relationship, which leads to the inverted stratigraphy (the overturned flap) at the crater rim (because material from greater depth at the target is deposited last on the rim). This also means that material that is at greater stratigraphic depth at a crater location will, in general, be deposited close to the crater, whereas distal ejecta more commonly consist of the uppermost stratigraphic layers of the target. The higher energies available early in the crater formation sequence make it also more likely that the earliest ejecta are molten or at least strongly shocked, while the later ejecta may only be slightly shocked or not at all. On Earth the interaction with the atmosphere complicates matters somewhat. Early ejecta may also be entrained in the rapidly expanding fireball and, if the event is large enough, they will be ejected outside of the atmosphere, leading to a global distribution. Size sorting will result in larger particles settling out first and finer dust being deposited later.

There are several different empirically derived formulas that describe the thickness of ejecta blankets with distance from the crater (McGetchin et al. 1973; Stöffler et al. 1975; see Melosh 1989, p. 90). One of the most commonly used ones, after McGetchin et al. (1973), gives a power law for the ejecta blanket thickness t as a function of distance r from the crater center:

$$t = 0.14R^{0.74}(r/R)^{-3.0} \text{ for } r \bullet R$$

where R is the radius of the transient cavity and dimensions are in meters. The exact values for the function of R are debated, and Melosh (1989) proposed a more general version in which the term $0.14R^{0.74}$ is replaced by f(R), a poorly defined function that depends on a variety of parameters and may be different for different planets. From the study of the distribution of microtektites in the Australasian strewn field, Glass and Pizzuto (1994) arrived at a modified relation:

$$t = 0.02R(r/R)^{-4.4 \pm 0.3}$$

which can be rewritten as a function of the distance alone:

$$t = 10^{24.0 \pm 0.4} r^{-4.4 \pm 0.3}$$

Whereas the original relations seem to have been deduced from proximal ejecta, it seems that this formula, with somewhat different coefficients, also describes the distribution of distal ejecta.

A glance at the full Moon through binoculars shows an interesting feature around some craters: the very conspicuous so-called crater rays, which are bright radially-oriented streaks along approximately great circles, originating from some craters. For example, the crater rays from the lunar crater Tycho extend almost across the whole visible hemisphere. Only the younger and fresher craters seem to have rays (Melosh 1989). The maximum radius R_r of an average ray pattern on the Moon has been described as a power function of crater diameter R, where

$R_r = 10.5R^{1.25}$, but individual ray systems may be larger or smaller than the formula would indicate. The nature and source of the rays is poorly understood. It is not clear if the material that forms the rays comes from deeper zones of the target or from near-surface layers. Explosion experiments have shown that jets of high-velocity gas seem to be forming patterns similar to those of crater rays, which could explain a role of the expanding vapor cloud in their formation. Crater rays have only been observed on airless bodies (e.g., Moon, Mercury), either because the atmosphere prevents their emplacement or because they are eroded soon after their formation, but it is tantalizing to speculate if some distal ejecta on Earth may not be the result of a ray-like formation process (see Sects. 2.4.1. and 2.4.6.). Clearly the atmosphere influences the distribution of proximal ejecta (e.g., Kring 1997; Barnouin-Jha and Schultz 1998), so possibly there is also an interaction during distal ejecta emplacement.

2.2
Distal Ejecta as Stratigraphic Markers

Distal ejecta can serve as markers for impact events in the stratigraphic record. "Impact markers" can be described as all chemical, isotopic, and mineralogical species derived from the encounter of cosmic bodies (such as cometary nuclei or asteroids) with the Earth. They include material derived directly from the extraterrestrial bodies (siderophile elements, rare gases, meteoritic minerals) and species resulting from mechanical and physical interaction between the impactor and the target (e.g., shock-metamorphosed minerals and rocks). Such markers are of prime importance to detect and study accretionary events in the sedimentary record, to identify their origin, and to evaluate their possible role in global change and on the Earth's biotic and climatic evolution throughout geological time (see Sect. 5.1). Distal ejecta layers are better suited to study a possible relationship between biotic changes and impact events, as the direct correlation can be confirmed in the stratigraphic record, whereas a correlation with radiometrically determined crater ages always suffers from uncertainties in age determinations.

There is a variety of the different types of stratigraphic impact markers. In principle, the concept behind interpreting them is similar as for volcanic markers in the stratigraphic record, such as biotite and other typical volcanogenic minerals (or shards of volcanic glass). Such volcanic markers have various functions. They often provide an excellent means for obtaining radiometric ages for a stratigraphic sequence that consists of, e.g., carbonate rocks that are otherwise difficult to date directly (see Chap. 5 for several examples). This allows then to tie in the biostratigraphic record with direct radiometric ages. By studying the composition of mineral and glass fragments in such volcanogenic layers it is also often possible to determine from which volcano these materials had been erupted. Impactoclastic layers may be studied in similar ways.

Impactoclastic layers are composed of distal ejecta. For example, tektites and microtektites are glassy ejecta that are described in detail in Sect. 2.4. The occurrence of microtektites in well-defined and thin layers in deep-sea sediments provides an excellent time marker and means for correlation the stratigraphy between different locations (e.g., Zhou and Shackleton 1999). The study of ejecta at the K/T boundary or in southern Australia eventually led to the discovery of previously unknown large impact structures (Chicxulub and Acraman, respectively). In other cases, comparative analyses of the composition and age of tektites and/or microtektites and rocks at specific crater sites led to the determination of tektite source craters (e.g., Ivory Coast tektites and Bosumtwi crater – see Sect. 2.4.5.3).

Another example for the study of potential impact markers involves the search for material resulting from the 1908 airblast at Tunguska, Siberia, where most likely a stony body (if of chondritic/asteroidal or cometary provenance is not yet clear) exploded in the air (see Sect. 1.4), distributing cosmic dust over a wide area. At the airblast site, minor amounts of extraterrestrial dust were found to be mixed in with local soil (e.g., Rasmussen et al. 1999), but it was debated if any such signature could also be detected elsewhere around the world. Ganapathy (1983) claimed to have found Tunguska-derived extraterrestrial material in ice near the South Pole, whereas other studies (e.g., Rocchia et al. 1990; Rasmussen et al. 1995) found no such evidence.

The studies of the K/T ejecta layers led to improved detection sensitivities for impact markers, allowing identification of smaller events and the study of their effects. A very promising example has been detected in terminal Eocene sediments, as described in Sect. 5.8.4. The most commonly used impact markers are elevated contents of siderophile elements (especially the Ir and other PGEs; e.g., Asaro et al. 1982, 1988; Orth et al 1988, 1990; Alvarez et al. 1990; see Sect. 1.8.1) and the presence of shocked minerals, especially quartz. However, more recently a number of other markers have been identified that seem to be characteristic of impact. For example, Ni-rich spinels (Fig. 2.2.1) have been found in several impactoclastic layers (e.g., Kyte and Smit 1986; Robin et al. 1992; Gayraud et al. 1996; Rocchia et al. 1996b; Pierrard et al. 1998). Robin et al. (1999) report on finding several spinel-bearing horizons in a core from the Central Pacific Ocean, which may be related to several different impact events. Impact-derived diamonds have been found at different K/T boundary locations around the world (e.g., Carlisle and Braman 1991; Hough et al. 1997, 1998), but the problem with using diamonds as impact markers in distal ejecta is that, so far, they have only been found in terrestrial or relatively shallow-water and proximal K/T boundary sites (however, they have been found within proximal ejecta at several impact structures; see Sect. 1.6.7).

Another geochemical marker possibly associated with impact events is the presence of fullerenes (Becker et al. 1994), which have been found in K/T boundary samples (Heymann et al. 1994, 1996), although their origin (if extraterrestrial or produced in the impact fireball) is disputed (Becker et al. 1996; Heymann et al. 1996). Recently, Chijiwa et al. (1999) reported on the discovery of fullerenes in a Permo–Triassic boundary section (see Sect. 3.6.3).

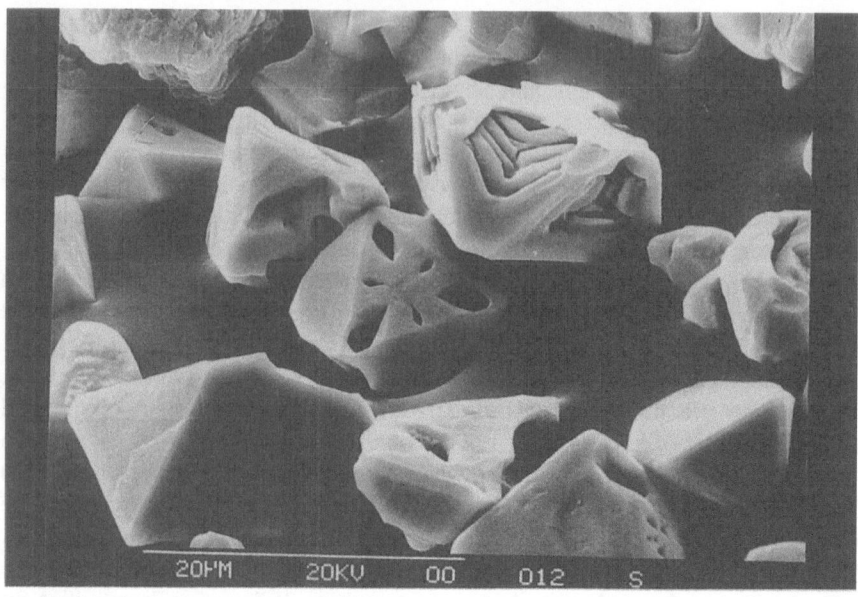

Fig. 2.2.1. Secondary electron microscope photo of spinel (magnesioferrite) from the K/T boundary in Caravaca, Spain, which is a reliable impact marker (photo courtesy B.F. Bohor).

Not all markers that are characteristic of extraterrestrial material are impact-derived. Extraterrestrial material accretes in the form of meteorites and cosmic dust onto the Earth (e.g., Kyte and Wasson 1986; Kyte et al. 1993), and it is possible that there are times at which this accretion rate increases significantly (e.g., Schmitz et al. 1997). The recent discovery of cosmic spherules in 1.4 Ga sandstone from Finland (Deutsch et al. 1997) leaves open the question of whether or not these spherules are just well preserved by chance or if they are the result of an increase in the accretion rate. Of particular interest are high ^{3}He concentrations that are known to be associated with interplanetary dust, which have been identified, for example, in upper Eocene sediments from the Pacific and Italy (Farley 1995; Farley et al. 1998; see Sect. 5.8.4). These data indicate a general and significant increase of the flux of extraterrestrial matter onto the Earth during an extended period of a few million years. This period coincides with at least two large impact events (Popigai and Chesapeake Bay), but the ^{3}He does not result directly from the impacts. The elevated ^{3}He flux was ascribed by Farley and co-workers to increased amounts of interplanetary dust in the inner solar system as a result of a cometary shower, which also included the impact of the bodies that formed the Chesapeake Bay and Popigai structures.

Thus, the study of impact markers in the sedimentary record, and, more specifically, at various paleontological boundaries, is an important component of impact-related research and may bring about the discovery of several previously unknown impact events and structures. Analysis of impact markers yields important information regarding the physical and chemical conditions of their formation, such as temperature, pressure, oxygen fugacity, composition of the atmosphere. These data are necessary to understand the mechanisms of interaction of impact events with the environment (see also Sect. 5.1).

2.3
Importance of the K/T boundary

We have already mentioned in several other places in this book that the study of rocks from the K/T boundary layer has been one of the most important driving forces for impact research in the past decades (cf. Grieve 1982). The K/T boundary can also serve as an example of how the various types of distal ejecta can be correlated with proximal ejecta and, eventually, an impact structure. It was by measuring the concentration of the element iridium, a specific marker of extraterrestrial material, in the sediments from the K/T boundary that provided the first direct evidence for a large impact event at the end of the Cretaceous. This led, in turn, to the theory that the mass extinction of the end of the Cretaceous was caused by a huge asteroid or comet impact. Furthermore, by studying the various impact markers at K/T boundary locations around the world, the Chicxulub impact structure on the Yucatan Peninsula, Mexico, was discovered. The study of shocked zircons and of impact glasses from the K/T boundary helped to establish the link to the Chicxulub crater (see Sect. 3.2). Studies of impact markers, together with high-resolution paleontological investigations, indicate that it is quite possible that there is a causal link between the impact event and the biologic crisis of the end of the Cretaceous.

Studies of the K/T boundary event provided data that helped with the understanding of the origin and significance of other, possibly impact-related, boundaries and exotic layers in the stratigraphic record. The impetus and publicity derived from K/T boundary research helped to introduce impact research to the wider geological community, and planetary geologists, who had been working on similar topics for decades (but outside the scope of the general geological community), were able to provide their results (e.g., from tektite studies or shock experiments) for the interpretation of the events that marked the end of the Cretaceous. As such, K/T boundary research played several important roles: increase the visibility and acceptance of impact research, and provided applications for decades of impact research. Data gained from, for example, research on tektites were useful in interpreting some aspects of the K/T boundary, and, in reverse, methods and concepts resulting from K/T studies aided with the interpretation of tektite data. In addition, the evidence that an impact event is associated with at least one major geological boundary has led to speculation – and more research – about the causes of other geological boundaries (see Sect. 3).

2.4
Tektites and Microtektites

Tektites are a group of natural glasses that have puzzled mankind for many centuries. After centuries of collecting, and decades of study, we are now closer to an understanding of their origin. First, though, we need to describe what tektites are. They are chemically homogeneous, often spherically symmetric objects that are in general several centimeters in size, and occur in four known strewn fields on the surface of the Earth. Strewn fields can be defined as geographically extended areas (in the case of tektites larger than just a few square kilometers) over which tektite material can be found.

Tektites found on land have traditionally been subdivided into three groups: (a) normal or splash-form tektites, (b) aerodynamically shaped tektites, and (c) Muong Nong-type tektites (sometimes also called layered tektites). The first two groups differ only in their appearance and some of their physical characteristics (see, e.g., O'Keefe 1963; Chao 1963; O'Keefe 1976). The aerodynamic ablation results from partial re-melting of glass during atmospheric re-entry after it was ejected outside the terrestrial atmosphere and solidified through quenching. Aerodynamically shaped tektites are known mainly from the Australasian strewn field, where they occur primarily as flanged-button australites (Fig. 2.4.1a, b). The shapes of splash-form tektites (spheres, droplets, teardrops, dumbbells, etc., or fragments thereof; Fig. 2.3.2a–c) have sometimes erroneously been described as aerodynamical forms; they, however, result from the solidification of rotating liquids, and not atmospheric ablation. Muong Nong type tektites were named after the type-locality in Laos (Lacroix 1935). They are usually considerably larger than normal tektites (samples of up to 24 kg have been described; Koeberl 1992a) and are of chunky, blocky appearance. Muong Nong-type tektites show a layered structure with abundant vesicles (Fig. 2.4.3). Microscopic examination of the layers (Fig. 2.4.4) shows bands of lighter and darker color (which should not be confused with schlieren) and mineral inclusions, such as zircon, baddeleyite, chromite, rutile, corundum, and cristobalite (Glass 1970, 1972; Glass and Barlow 1979; Glass et al. 1990, 1995) and coesite (Walter 1965) have been described from Muong Nong-type tektites.

Mainly due to chemical studies, it is now commonly accepted that tektites are the product of melting and quenching of terrestrial rocks during hypervelocity impact on the Earth. The chemistry of tektites is in many respects identical to the composition of upper crustal material (Taylor 1973; Koeberl 1986). Trace elements are very useful for source rock comparisons: many trace element ratios in tektites are indistinguishable from those in upper crustal rocks. The chondrite-normalized REE patterns of tektites are very similar to those of shales or loess, and have the characteristic shape and total abundances of the post-Archean upper crust. The determination of the exact source rocks of tektites is complicated because a variety of target rocks was apparently sampled by each impact event.

The discussion in the following chapters follows mainly Koeberl (1994b), with recent updates.

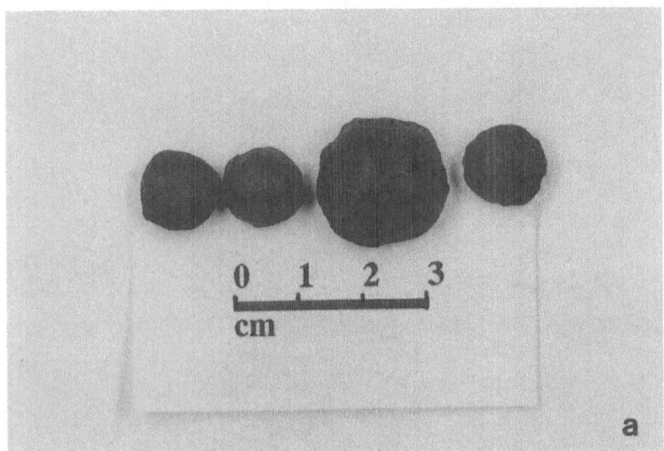

Fig. 2.4.1. a) Four small, aerodynamically shaped tektites from Australia (australites). The samples on the right and the second from the left show the typical button-shape with an ablation flange (samples courtesy H. Stehlik). **b)** Broken flanged australite, seen from the underside (sample USNM 3468).

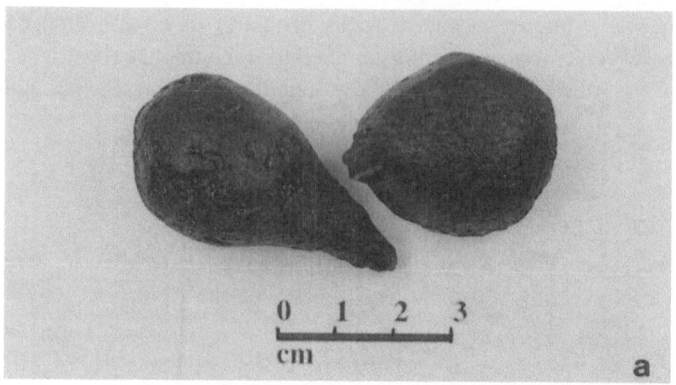

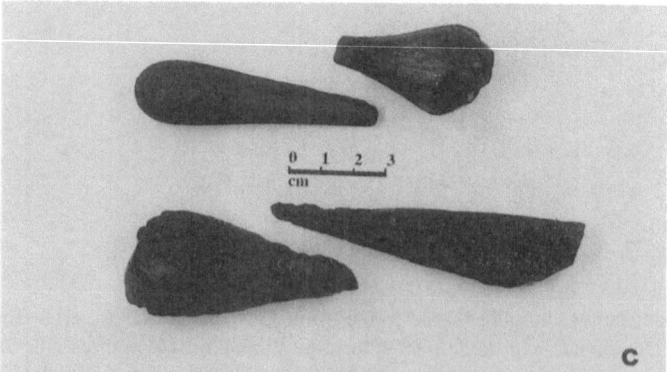

Fig. 2.4.2. Variety of splash-form tektites from the Australasian strewn field: **a)** teardrops from Guangxi (China); **b)** spherical and dumbbell tektites from Thailand; **c)** (next page) four drop-shaped thailandites (samples courtesy H. Stehlik).

Fig. 2.4.3. Muong Nong-type (or layered) tektite from Indochina. The scale bar (1 cm) is barely visible below the center of the sample. The layered structure with different colors and large vesicles is obvious (sample courtesy H. Stehlik).

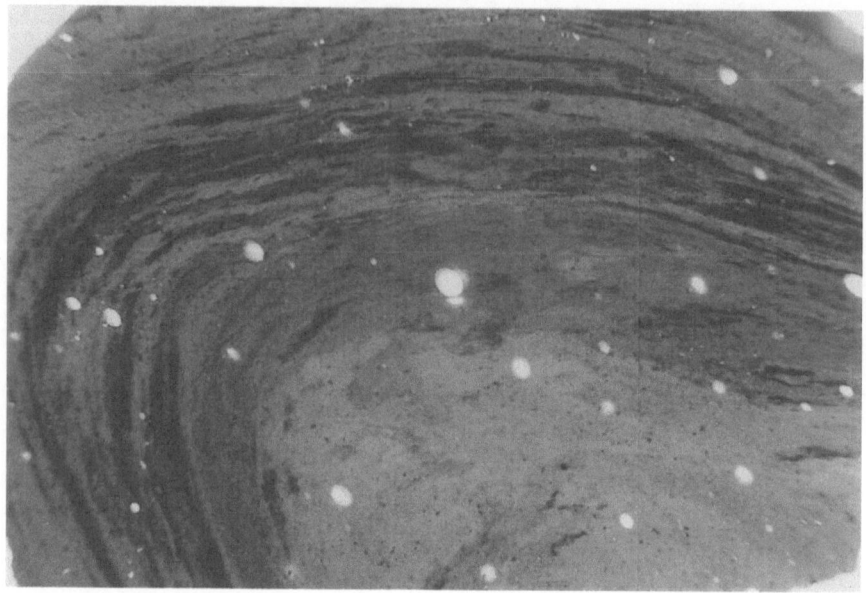

Fig. 2.4.4. Muong Nong-type indochinite (from Thailand). Thin slice (ca. 1.5 mm thickness) seen in transmitted light, showing the inhomogeneity and the different colored layers that are somewhat deformed. Bubbles are cut and, therefore, are slightly overexposed. Width of image about 10 cm.

2.4.1
Geographical Distribution: Strewn Fields and Age

So far, tektite occurrences are known from four geographically extended strewn fields (Fig. 2.4.1.1): the North American, Central European (moldavite), Ivory Coast, and Australasian strewn fields (Barnes 1963a; O'Keefe 1963). Tektites found within each strewn field are related to each other with respect to their petrological, physical, and chemical properties as well as their age. Ages of tektites have usually been determined by the K–Ar method (e.g., Zähringer 1963; Gentner 1969a), fission-track dating (e.g., Gentner et al. 1969b, 1970; Storzer and Wagner 1969, 1971, Storzer et al. 1973), or the Ar–Ar technique (e.g., Glass et al. 1986b; Izett and Obradovich 1992; Bollinger 1993). Koeberl (1994b) provided a summary of the properties of tektites found in the four known strewn fields. Any discussion of the origin of the tektites needs to explain the similarity of tektites in respect to age and certain aspects of isotopic and chemical composition within one strewn field as well as the existence of tektite material with different compositions present in each strewn field (Taylor 1973; King 1977; Koeberl 1986, 1988, 1989a, 1990).

The occurrence of tektite glasses is not restricted to the continents. Since the mid-1960s, microtektites from three of the four strewn fields have been found in deep-sea cores (see, e.g., Glass 1967, 1968, 1969, 1972; Cassidy et al. 1969). They are generally less than 1 mm in diameter and show a somewhat wider variation in chemical composition than tektites on land but with an average composition that is very close to that of "normal" tektites. Microtektites have been very important for defining the extent of the strewn fields (Fig. 2.4.1.1), as well as for constraining the stratigraphic age of tektites, and to provide evidence regarding the location of possible source craters. Some new finds of tektite material on Barbados and at DSDP Site 612 samples (see, e.g., Thein 1987; Koeberl and Glass 1988; Glass 1989) have yielded larger "micro"tektites as well as what seem to be fragments of tektites, and blur the traditional distinction between microtektites and "macro"tektites.

2.4.2
Definition and Classification of Tektites

The recent discoveries of impact-derived glasses in various geological contexts, and from various sources, led to the unfortunate habit of some researchers to term all impact-derived glasses (and even non-glassy, recrystallized objects) "tektites" (e.g., Claeys et al. 1992; Izett 1991; Smit et al. 1992a, b). However, there are some differences between tektites and impact glasses from known impact craters. Table 2.4.2.1 provides a comparison of the characteristics of tektites and impact glasses. Although tektites have of course formed during an impact event, they are a sub-group of impact glasses in general; maybe a possible distinction would be to call tektites "distal impact glasses", in contrast to "proximal impact glasses" that are found directly at a source crater.

The "definition" of a tektite has a somewhat historic aspect, but a short summary of the characteristics of tektites would include the following points: 1) they are glassy (amorphous); 2) they are fairly homogeneous rock (not mineral) melts; 3) they contain abundant lechatelierite; 4) they occur in geographically extended strewn fields (not just at one or two closely related locations); 5) they are distal ejecta and do not occur directly in or around a source crater, or within typical impact lithologies (e.g., suevitic breccias, impact melt breccias); 6) they are very poor in water (except some microtektites) and have a very small extraterrestrial component; and 7) they seem to have formed from the uppermost layer of the target surface (see Sects. 2.4.4 and 2.4.6). Thus, it is recommended to use the term "tektite" only for glasses that fulfill (most) of the above points, and, if in doubt, use the (probably much better) general term "impact glass".

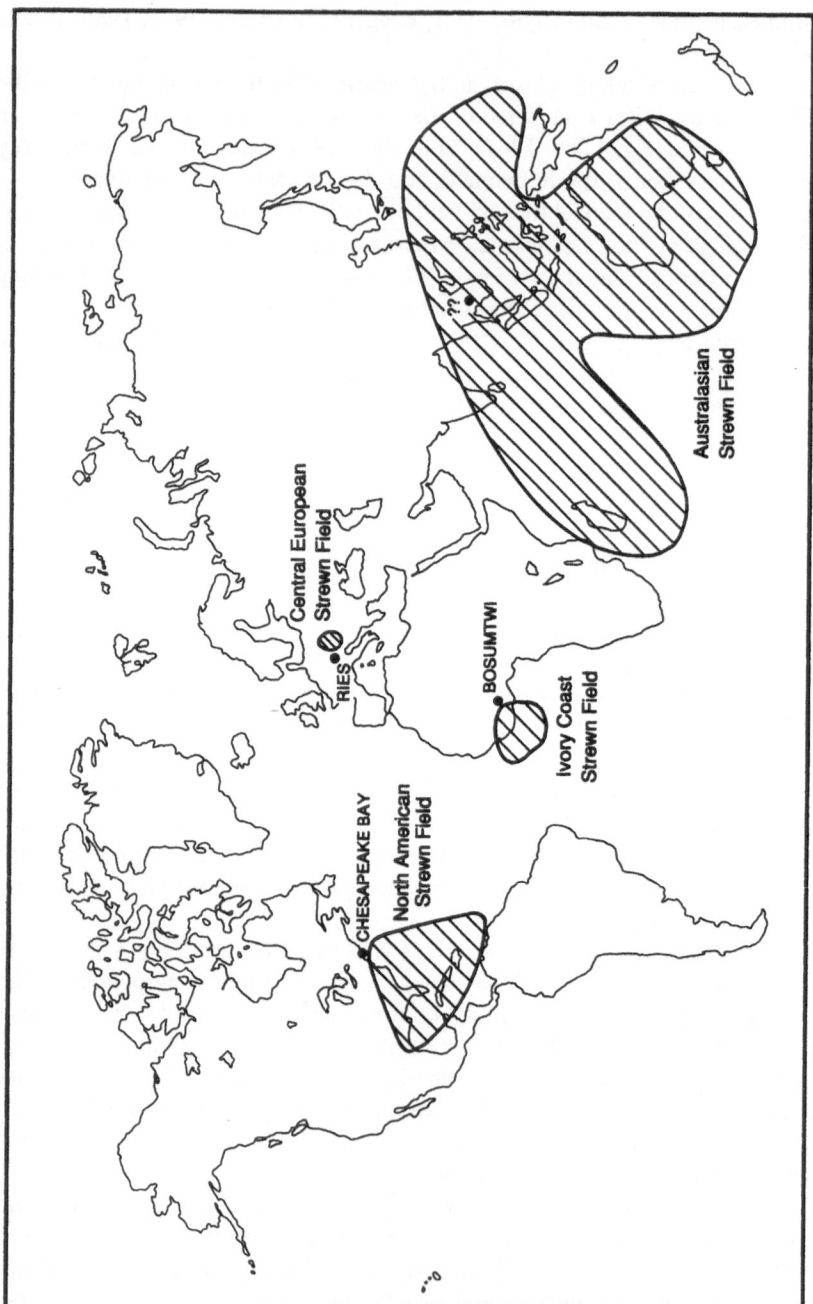

Fig. 2.4.1.1. Location and extension of the four tektite strewn fields on Earth. The solid circles mark the location of the known source craters (Chesapeake Bay, Ries, and Bosumtwi craters) and the suspected crater location for the Australasian strewn field.

Table 2.4.2.1. Comparison of tektites and "normal" impact glasses.

	Tektites	"Normal" Impact Glasses
Occurrence in strewn field	Yes	no
Source crater known	yes/no*	yes
Occurrence at source crater	No	yes
Chemical homogeneity (100 μm-mm)	large-scale homogeneity	usually inhomogeneous
Water content (wt.%)	0.002–0.02	0.02–0.07
Mineral inclusions	very rare	generally abundant (include partly digested quartz)
Shape	mostly regular spherically symmetric	mostly irregular
Ablation shapes	some do	no
Meteoritic component	very small	generally higher; variable
Heavy noble gas content (Ar, Kr, Xe)	generally lower	generally higher

Note: *knowledge of source crater reflects present knowledge; all tektite strewn fields have a source crater. Table after Koeberl (1994b).

2.4.3
Petrology and Geochemistry of Tektites

The geochemical and petrological characteristics of tektites provide an important background for the debate about source and formation mechanism of these glasses. We follow the assumption that the geochemistry of tektites provides information on the geochemistry of the source rocks, from which the tektites were produced. Reviews of this subject, as well as detailed tables of the major and trace element composition of all tektite groups, have been published by Chao (1963) and Koeberl (1986, 1990, 1992a, 1994b, and references therein); thus, we will only provide a short review here.

In general, the major and trace element composition of tektites is almost identical to the composition of the terrestrial upper crust. This fundamental observation was made mainly by Taylor and co-workers, who, during the 1960s and 1970s, have performed detailed geochemical studies of many tektite types (see, e.g., Taylor 1962a, b, 1966, 1973; Taylor and Sachs 1964, Taylor and Solomon 1964, Taylor and Kaye 1969, and Taylor and McLennan 1979). Of particular use in establishing such a relationship are trace elements: the ratios of, e.g., Ba/Rb, K/U, Th/Sm, Sm/Sc, Th/Sc, K vs. K/U, in tektites are practically the same as those in upper crustal rocks. As a result of mainly these chemical studies, it is now commonly accepted that tektites are the product of melting and subsequent quenching of terrestrial rocks during hypervelocity impact on the Earth.

One group of elements that have great genetic significance are the rare earth elements (REE), because geochemically they behave very similar to each other. Their absolute abundances, as well as their chondrite-normalized abundance patterns, are characteristic for different rock types of different provenance (e.g., Taylor and McLennan 1985). Thus, they can be used to infer the type and composition of the tektite parent rocks. The chondrite-normalized REE patterns of tektites are very similar to those of shales or other upper crustal sedimentary rocks, and have the characteristic shape and total abundances of REE distributions in the post-Archean upper crust. Fig. 2.4.3.1 shows an example of such a chondrite-normalized REE pattern.

It has been discussed before (e.g., Koeberl 1986, 1990, 1992a, 1994b) that the average REE patterns of post-Archean upper crustal rocks are different from those of Archean upper crustal rocks, with post-Archean sediments showing a characteristic negative Eu anomaly that is absent from Archean sediments (e.g., Taylor and McLennan 1985). This is also visible in Fig. 2.4.3.1 and demonstrates that Archean sediments are implausible as source materials for the four known Cenozoic tektite strewn fields. More extensive discussions of the use of REEs in tektite studies were given by, e.g., Taylor and McLennan (1979) and Koeberl (1992a, 1994b). REE analyses also indicate that loess, which is not known from Indochina, but has been proposed by Wasson (1991) as a possible source rock for tektites, was not a prominent source of the Australasian tektites (as discussed in detail by Koeberl 1992a; this was also confirmed by the Rb–Sr and Sm–Nd isotope studies of Blum et al. 1992). Also, none of the trace element ratios or REE patterns are similar to lunar or other extraterrestrial values, eliminating the lunar origin theory of tektites (e.g., O'Keefe 1976) as a viable contender (e.g., Schnetzler 1970; Taylor 1973, 1982). O'Keefe and co-workers suggested that tektites are similar in composition to lunar granites. However, they are not even similar in composition to terrestrial granites, and, furthermore, granites require water to form, which does not exist on the Moon. Granite-like residual melts on the Moon ("felsites") are very rare and small (the largest one found so far is 1.8 g), and also of a composition different from that of tektites (see discussion in Taylor and Koeberl 1994).

Whereas most studies cited in the previous paragraphs deal with Australasian tektites, the same conclusions can be drawn from REE abundances and patterns of the other three tektite strewn fields. The REE patterns (as well as other trace element abundances and ratios) of North American tektites are in direct agreement with a derivation from upper crustal rocks (e.g., Koeberl and Glass 1988; Glass et al. 1995, 1998). In case of the strewn fields that have been linked to specific craters, trace element studies have strengthened the evidence for a link. For example, comparisons have been made (e.g., Engelhardt et al. 1987) between the trace element composition of upper freshwater molasse samples which were presumably present at the site of the Ries crater, and those of the moldavites (see Sect. 2.4.5).

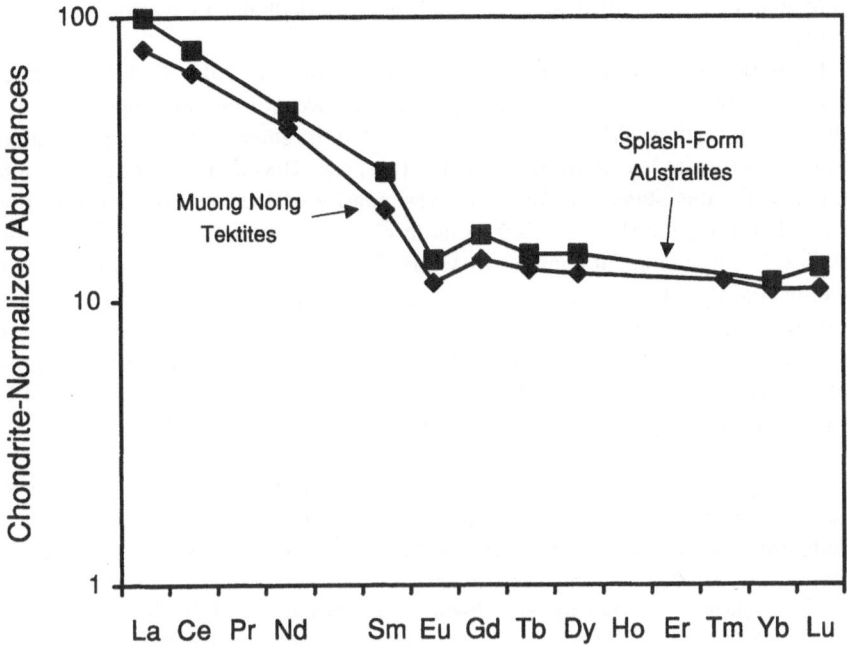

Fig. 2.4.3.1. Chondrite-normalized REE pattern of the average present (post-Archean) upper crust in comparison with the REE patterns of average Muong Nong-type tektites and average australites. The patterns are clearly very similar, with almost identical slopes, LREE/HREE or La/Yb ratios, and Eu anomalies (after Koeberl 1994b).

Muong Nong-type Tektites. Muong Nong-type tektites have first been described from the Australasian strewn field and differ from normal tektites in a few important parameters. They are larger, more heterogeneous in composition, of irregular shape, have a layered structure, and show a much more restricted geographical distribution (see, e.g., Chapman and Scheiber 1969; Futrell and Fredriksson 1983; Koeberl 1986, 1992a; Schnetzler 1992; Fiske et al. 1999; Schnetzler et al. 1999). Their average major element chemistry and their age are (within normal variations) identical to those of normal tektites. Muong Nong-types are abundant in the Australasian strewn field, but recently a few samples that are most probably of Muong Nong-type have also been found from the Central European (e.g., Glass et al. 1990) and North American (Wittke and Barnes 1988; Koeberl and Glass 1988; Glass et al. 1995) strewn fields. Muong Nong-type tektites differ in some characteristics from "normal" tektites: they are chemically less homogeneous (including differences in composition between layers). Furthermore, water (Koeberl and Beran 1988; Beran and Koeberl 1997) and other volatile elements, such as the halogens or Zn (see, e.g., Müller and Gentner 1973; Matthies and Koeberl 1991; Koeberl 1992a) are enriched in Muong Nong-type tektites compared to splash form tektites – or just less depleted in comparison to the source rocks.

Koeberl (1992a) and others have speculated that the volatile element enrichment in Muong Nong-type tektites results from lower formation temperatures than splash-form tektites. They probably formed from the same or a very similar source material, but were less thoroughly heated (see, e.g., Barnes 1990, for petrographic evidence for this statement, based on vesicularity of lechatelierite particles) and have, therefore, lost less of their volatile element content than the splash form tektites. Volatile element contents in Muong Nong-type tektites are comparable to, or only slightly lower than, the ones in upper crustal sedimentary rocks (e.g., Koeberl 1992a), while splash-form tektites show considerable depletions. Wasson et al. (1990) suggested that uranium may behave as a volatile element during tektite formation, but the observed U depletion can also result from oxidation of U^{4+} to U^{6+} and removal as soluble $[UO_2]^{2-}$ (McLennan and Taylor 1980).

Table 2.4.3.1. Major element composition of Muong Nong-type indochinites compared to tektites from all four known strewn fields, and average continental crust.

	Muong Nong Indochinites	Australasian Tektites	North American Tektites	Central European Tektites	Ivory Coast Tektites	Average Continental Crust
SiO_2	77.1–81.7	64.8–79.7	71.9–83.6	74.9–85.1	66.2–69.3	66.0
TiO_2	0.53–0.72	0.49–1.00	0.42–1.05	0.24–1.40	0.52–0.61	0.50
Al_2O_3	8.58–11.4	8.90–17.7	9.50–17.6	7.32–13.8	15.8–17.7	15.2
FeO	3.18–4.15	3.57–8.63	1.83–5.75	1.08–3.50	5.84–6.80	4.50
MnO	0.07–0.10	0.07–0.21	0.02–0.08	0.03–0.09	0.04–0.07	0.08
MgO	1.19–1.65	1.31–7.95	0.37–0.95	1.13–2.74	2.64–4.39	2.2
CaO	1.03–1.63	1.37–9.77	0.40–0.96	0.95–3.17	0.71–1.61	4.2
Na_2O	0.77–1.07	0.62–1.56	0.70–1.84	0.20–1.08	1.53–2.44	3.9
K_2O	2.24–2.55	1.34–2.81	1.60–2.51	2.23–3.81	1.70–2.08	3.4

For data sources, see Koeberl (1986, 1992a) and Koeberl et al. (1997b). All data in wt.% and all Fe as FeO. Australasian tektites include splash-form and ablation-shaped samples from Indochina, Australia, and the Philippines (except HNa/K australites, see Sect. 2.5); North American tektites include only bediasites and georgiaites.

Muong Nong-type tektites occur mainly as silica-poor and silica-rich varieties, with some samples having intermediate compositions. Table 2.4.3.1 gives the major element compositions of tektites from the four known strewn fields, as well as for different layers in Muong Nong-type tektites, and for the average upper continental crust. Koeberl (1992a) concluded that it is unlikely that Si-poor

Muong Nong-type tektites originated by vapor fractionation from Si-rich Muong Nong-type tektites silica, but rather that they are the result of mixing of different source rocks (which is also indicated by different relict mineral contents in Si-rich an Si-poor Muong Nong-type tektites).

Muong Nong-type tektites may be the key to some important questions. Their inhomogeneity may indicate that they have preserved the original target rock compositions much better and therefore they may be the missing link between target rocks and tektites (e.g., Koeberl and Fredriksson 1986). They are also important because they contain relict mineral grains that indicate the nature of the parent material and contain shock-produced phases that indicate the conditions of formation.

Microtektites. Microtektites from the North American, Ivory Coast, and Australasian strewn fields have been recovered from deep-sea sediments (see, e.g., Glass 1967, 1968, 1969, 1972; Cassidy et al. 1969). Microtektites are, by definition, less than 1 mm in size, and show a variety of shapes (Fig. 2.4.3.2.), ranging from spherical to dumbbell, disc, oval, and teardrop. In color they range from colorless and transparent to yellowish and pale brown. They often contain bubbles and lechatelierite inclusions. Microtektites occur in the stratigraphic layers of the deep-sea sediments that correspond in age to he radiometrically determined ages of the tektites found on land. Thus, they are distal ejecta and represent an impact marker.

The geographical distribution of microtektite-bearing cores defines the extent of the respective strewn fields, as tektite occurrences on land are much more restricted. Furthermore, microtektites have been found together with melt fragments, high-pressure phases, and shocked minerals (e.g., Glass 1989; Glass and Wu 1993) and, therefore, provide confirming evidence for the association of tektites with an impact event. In addition, the geographic variation of the microtektite concentrations in deep-sea sediments is not irregular, but increases towards the assumed or deduced impact location. For example, Glass and Pizzuto (1994) determined the geographic variation in abundance of Australasian microtektites (ranging from <1 to 3255 microtektites per cm^2 in the >125 µm size range), and found that the abundances increased towards Indochina; and deduced a possible source region in Cambodia. Also, the distribution of microtektites in the strewn field is not uniform, but may follow some kind of ray-like pattern. Recently, microtektites have been discovered on land in loess at Luochuan in China, slightly north of the commonly drawn limits of the strewn field (Li et al. 1993, 1995). Their stratigraphic occurrence near the Brunhes-Matuyama boundary and their composition indicate that they are indeed part of the Australasian strewn field.

The composition of most microtektites shows a similar (but somewhat more extended) range than tektites on land of the same strewn field. Besides the wider compositional range there is also a slight difference in the mean composition between the two groups (e.g., Bohor and Koeberl 1996), but it is not clear if this results from a difference in the origin (or location in the target stratigraphy), or could be due to different analytical methods used for their study. The analyses of normal tektites are usually performed as bulk analyses, whereas microtektites are analyzed by electron microprobe. On a micrometer-scale, larger tektites are often more inhomogeneous than individual microtektites (Delano 1992; B.P. Glass, pers. comm. 1993).

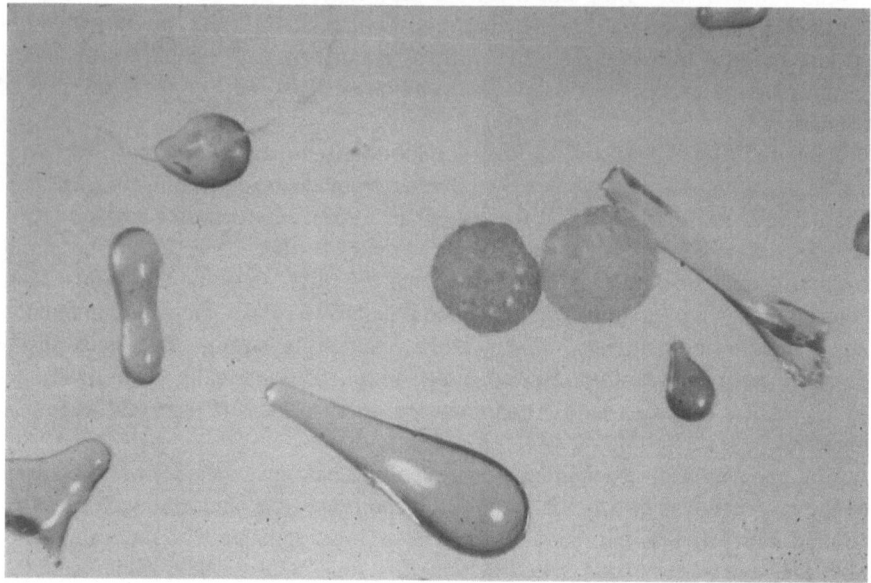

Fig. 2.4.3.2. Microtektites from the Australasian tektite strewn field, showing a variety of shapes. A so-called bottle-green microtektite is near the center of the image. Width of image 2.7 mm (photo courtesy B.P. Glass).

There is a group of highly unusual microtektites, the so-called bottle-green microtektites or high-Mg (HMg) microtektites (Glass 1972), which have a composition that is different from that of "normal" microtektites. They are green, corroded in appearance (Fig. 2.4.3.3), do not contain bubbles or lechatelierite inclusions, and are poor in Si and high in Mg. These glassy spherules are generally smaller than normal microtektites, but do not seem to show a different spatial distribution. The average composition of BGMTs is different from that of normal microtektites, tektites on land, or the average continental crust. Koeberl et al. (1999a) found not only significant enrichments in Mg, but also enrichments in Ti, Ca, Sc, Sr, Zr, Ba, Hf, Ta, W, Th, and the rare earth elements, and depletions in Si and the alkali elements, in comparison with average tektites. Importantly, the BGMTs show very high abundances of the siderophile elements (Cr, Co, Ni, and Ir). Up to about 10 ppb Ir were found (Koeberl et al. 1999a). These values are taken as a clear indication of an extraterrestrial component. Boron isotope measurements show vapor fractionation effects and a starting composition that seems to be compatible only with the B isotopic composition of soil. From these data, Koeberl et al. (1999a) proposed that the BGMTs formed in the earliest phases of the tektite-forming impact, possibly by some form of jetting, from the uppermost part of the target stratigraphy.

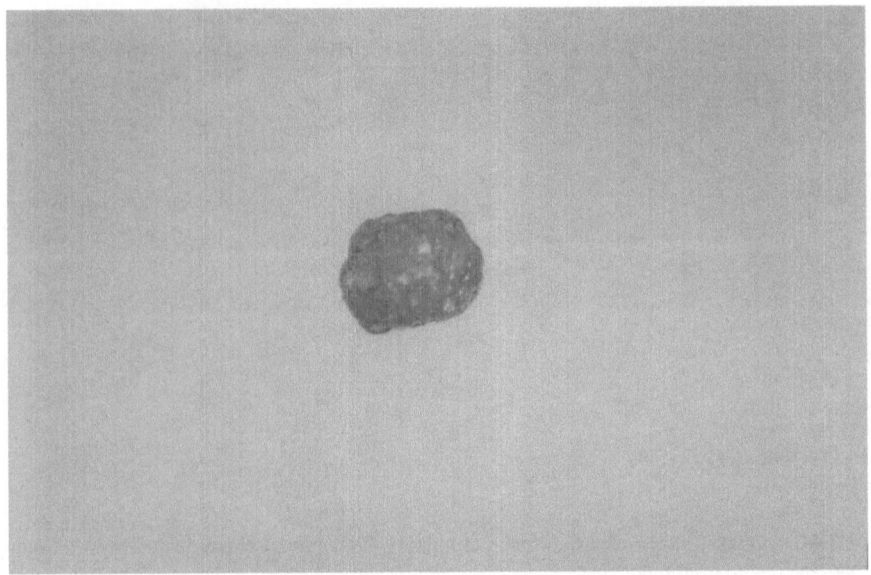

Fig. 2.4.3.3. Microphotograph of a bottle-green microtektite from the Australasian strewn field. Note the corroded appearance of the sample, in contrast to the smooth surfaces of normal microtektites (sample size about 250 μm; photo courtesy B.P. Glass).

Inclusions of Relict Minerals and Shock Phases. The occurrence of relict minerals in some tektites points to sedimentary source rocks. Lechatelierite – the amorphous remainder of partly digested quartz grains – is well-documented in tektites (e.g., Chao 1963) and indicates quartz-rich precursor rocks. Lechatelierite is also common in microtektites. Muong Nong-type tektites contain unmelted relict inclusions (Barnes 1963b; Glass 1970, 1972; Glass and Barlow 1979). The minerals identified so far include zircon, chromite, quartz (Glass 1970), rutile, and monazite (Glass 1972), all showing evidence of various degrees of shock metamorphism. They also contain high-temperature breakdown products, such as corundum (plus SiO_2), which formed from an Al_2SiO_5 phase, and baddeleyite (plus SiO_2), which formed from zircon (e.g., Glass and Barlow 1979; Glass et al. 1990, 1995). Muong Nong-type indochinites can be divided into two groups: those with low refractive indices (<1.503) contain mineral inclusions while those with high refractive indices (>1.513) do not (Glass and Barlow 1979; Glass and Koeberl 1989). The type of mineral inclusions present, as well as their size and shape, suggests that a fine-grained, well-sorted sediment was the tektite parent material.

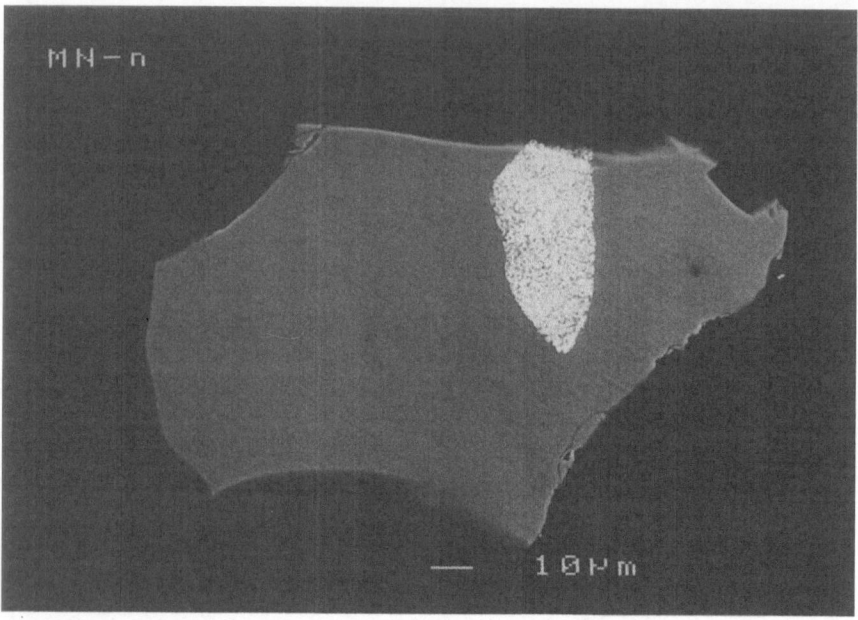

Fig. 2.4.3.4. Backscattered electron image of a partly decomposed zircon inclusion in a Muong Nong-type tektite from Indochina.

The high-pressure quartz polymorph coesite has been reported from Muong Nong tektites by Walter (1965) and was confirmed by Glass et al. (1986a). It also had been found in Darwin glass (Reid and Cohen 1962). Glass (1987) and Bohor et al. (1988) found coesite and shocked minerals associated with tektites in a layer containing an impact debris, tektite fragments, and microtektites at DSDP Site 612 in the North American tektite strewn field, supporting the association of tektites with shocked minerals. Stishovite has also been reported from the North American and Australasian microtektite layers (Glass 1989; Glass and Wu 1993).

2.4.4
Source Rocks of Tektites

The discussion presented above on the composition of (and inclusions in) tektites and microtektites makes it clear that only terrestrial upper crustal rocks are plausible as source rocks of tektites. Trace element data have narrowed the choice to post-Archean sediments. Further refinement is possible from the determination of abundances and ratios of various isotopes in tektites. One of the first isotopic studies was performed by Tilton (1958), who analyzed the isotopic composition of lead in tektites and found it to be in agreement with terrestrial values. Similar conclusions were obtained by Schnetzler and Pinson (1964) for Rb/Sr and by Taylor and Epstein (1969) for oxygen isotopes. There are indications that the isotopic composition of oxygen in some tektites (e.g., moldavites; see Engelhardt et al. 1987) is skewed towards the lighter isotopes in tektites compared to source rocks. This effect may be due to incorporation of isotopically light oxygen from meteoric water which resides in pores of sediments (Engelhardt et al. 1987). In contrast, Koeberl et al. (1998b) found no evidence for any modification of oxygen isotope values for Ivory Coast tektites. These tektites have $\delta^{18}O$ values that are virtually identical to those of target rocks from the Bosumtwi crater (see below).

Shaw and Wasserburg (1982) demonstrated that all tektites have distinct negative ε_{Nd} and large positive ε_{Sr} values, which are uniquely characteristic of old terrestrial continental crust. The Nd model ages of tektites reflect the age of the source terrain of the target rocks, i.e., the age during partial melting of the mantle to form new crust. Surficial processes, such as weathering, do not disturb the Sm–Nd isotopic system. However, the Rb–Sr isotopic system is severely disturbed by increasing the Rb/Sr ratio during weathering and sedimentation. The Rb–Sr sedimentation age of tektites shows a wide variation, ranging from young sediments for the Central European tektites (recent at the time of the impact; i.e., at 15 Ma) to rocks of about 0.95 Ga for the Ivory Coast tektites. For the North American tektites, the Rb–Sr/Sm–Nd systematics exclude most of the North American Precambrian shield areas as well as areas that are covered by sediments derived from them. Shaw and Wasserburg (1982), Ngo et al. (1985), and Stecher et al. (1989) concluded that sediments derived from, or incorporated in, the Appalachian orogeny provide the best match for North American tektites. This fits very well with the recently discovered Chesapeake Bay crater, which was recently concluded to be the source crater of the North American tektites (see Sect. 3.3.1).

From a study of the Rb–Sr and Sm–Nd isotopic systematic of Muong Nong-type and splash-form tektites from the Australasian strewn field, Blum et al. (1992) found that the source material was derived from Precambrian crustal terrane with Nd model ages ranging from 1040 to 1190 Ma. The Rb–Sr system indicates that the sediments that were later melted to form tektites were weathered and deposited about 167 Ma ago and probably comprised Jurassic sediments, which are not uncommon throughout Indochina. The Nd and Sr isotopic data suggest that all Australasian tektites were derived from a single sedimentary formation with a narrow range of stratigraphic ages close to 170 Ma, more or less ruling out multiple impact hypotheses (e.g., Wasson 1991). Blum et al. (1992) also concluded from their data that the Rb–Sr isotope systematics are inconsistent with soil or loess as source materials.

Other isotopic systems, such as boron, carbon, nitrogen, magnesium, or sulfur, have not yet been explored in detail or at all, mainly due to analytical problems caused by the low concentrations of these elements. A study of $\delta^{11}B$ in tektites by (Chaussidon and Koeberl 1995) showed most tektites have a limited range (-9.3 ± 1.5‰ to +2.7 ± 1.5‰), which is small compared to the range observed for common terrestrial rocks (-30 to +40‰). The B abundance and isotopic data can be used to place constraints on the tektite source rocks. Australasian tektites have high B and Li abundances; only clay-rich sediments, such as pelagic and neritic sediments, as well as river and deltaic sediments have B contents (up to 100 ppm) and $\delta^{11}B$ values that are in agreement with the range shown by Australasian tektites (-4.9 to +1.4‰). ^{10}Be and Rb–Sr data indicate continental crustal source rocks and exclude pelagic and neritic sediments. However, deltaic sediments, e.g., from the Mekong river, which are of continental crustal origin, agree with ^{10}Be, Rb–Sr, and B data, and support a possible source locality close to the coast of SE Indochina in the South China Sea.

The examples given above demonstrate that there is great potential in the use of various isotope system for tektite studies.

2.4.5
Source Craters of Tektites

Initially, when tektites were first discovered in the various strewn fields, their origin was a mystery. The studies discussed in the previous sections lead to the conclusion that tektites were formed during hypervelocity impact into post-Archean upper crustal rocks. This conclusion raised the question regarding source craters. During the past half century, numerous suggestions and educated guesses have been made regarding the location of the possible source craters for the tektite strewn fields. Relatively reliable links between craters and tektite strewn fields have been established between the Bosumtwi (Ghana) and the Ries (Germany) craters and the Ivory Coast and the Central European fields, respectively. Only recently, a plausible link was established between the newly discovered Chesapeake Bay crater and the North American strewn field. No large crater with an appropriate age has yet been identified for the Australasian strewn field.

2.4.5.1
North American Strewn Field

This is the oldest strewn field, at about 35 Ma (Gentner et al. 1969a, b; Storzer and Wagner 1969, 1971; Storzer et al. 1973; Glass et al. 1986b). The first North American tektites were known from an area near Bedias, Texas (since 1936), and some rare specimens from Georgia (e.g., Barnes 1963a). Two individual finds were reported to be from Martha's Vineyard and from Cuba (Garlick et al. 1971);

the first location is doubtful, but the latter probably represents a genuine tektite location (see Koeberl 1988, 1989a, for discussion). The extent of the North American strewn field was defined by findings of microtektites in a number of deep sea sediments (see, e.g., Glass et al. 1973, 1979, 1985; Glass and Zwart 1977). On the basis of stratigraphic, compositional, isotopic, and age data they were identified to be part of the North American tektite field. Other spherules of clinopyroxene composition, which were found in several cores throughout the Caribbean Sea and even the Pacific, were earlier thought to be associated with the North American tektite strewn field, but later shown to belong to a different event (e.g., Glass et al. 1985; D'Hondt et al. 1987; see Sect. 3.3 for more details).

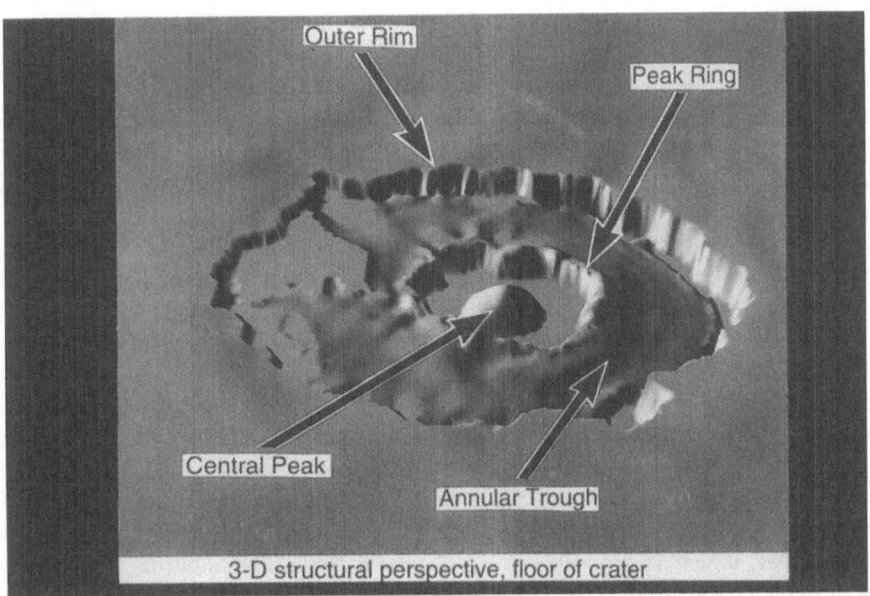

Fig. 2.4.5.1.1. Three-dimensional view of the Chesapeake Bay impact structure crater floor, with crater fill and post-impact sediments removed (computer image courtesy C.W. Poag).

The discovery of tektite fragments associated with microtektites in deep-sea deposits (that are now on land) at Barbados extended the occurrences of "normal" tektites within the strewn field (Glass et al. 1984). Chemical analyses (Koeberl and Glass 1988), isotopic studies (Ngo et al. 1985) and age data (Glass et al. 1986b) showed that the Barbados tektites belong to the North American strewn field. From their Rb–Sr/Sm–Nd isotopic studies, Shaw and Wasserburg (1982) concluded – as mentioned before – that the source rocks from which the North American tektites were derived were crustal material that formed very late in the Precambrian (Sm–Nd model age 0.62–0.67 Ga), which excludes most of the Precambrian shield areas of North America, as well as sediments derived from these areas.

Microtektites, tektite fragments, and impact debris occur together in relatively thick layers at DSDP Site 612 and ODP Site 904A (e.g., Thein 1987; Glass et al. 1998). The chemical composition of the DSDP 612 and ODP 904 tektites and microtektites is similar to that of other North American tektites, with some differences, e.g., lower Na and higher K contents in the 612 tektites (Koeberl and Glass 1988). Stecher et al. (1989) found that the Rb–Sr/Sm–Nd isotopic compositions of the 612 tektites show a wider scatter than that of the North American tektites. It was also debated if the 612 tektites are in a layer identical in age to that of the other North American microtektites (as proposed mainly by Glass 1989), or if they are in a different layer, which seems highly unlikely, given the chemical similarities. Keller et al. (1983, 1987) suggested that there are least three different Late Eocene impact layers (including the North American tektites), and Miller et al. (1991) additionally maintain that the Site 612 tektites are 0.5–1 Ma older than the other North American tektites and propose 3–4 Late Eocene impacts. However, this has been disputed by Glass (1990) and Wei (1995). Recently Glass and Koeberl (1999a) and Vonhof and Smit (1999) have reported microtektites, probably of North American strewn field association, from ODP Site 689B at Maud Rise, Southern Ocean (near Antarctica) and possibly even in the Indian Ocean (Glass and Koeberl 1999b). These findings would mean that the North American tektite strewn field is much larger than thought before.

Over the years, many locations have been suggested as the North American tektite source crater, mainly based on similarity in age. Among the suggestions were, for example, Popigai, Siberia (Dietz 1977); Wanapitei, Canada; Mistastin, Canada; and Bee Bluff, Texas (Wilson and Wilson 1979; King 1979), all of which were later discounted on basis of distance, exact age, isotopic constrains, size, and other criteria (e.g., Storzer and Wagner 1979; Shaw and Wasserburg 1982; Ngo et al. 1985; Koeberl 1989a; Glass 1989; Stecher et al. 1989). More recently, the Montagnais crater, located 200 km southeast of Nova Scotia and the first underwater crater ever discovered (Jansa and Pe-Piper 1987), was considered as a candidate, but neither its age (51.5 Ma; Bottomley and York 1988) nor isotopic data (Stecher et al. 1989) agree with those of the North American tektites. From the distribution of the tektites in the strewn field, Koeberl (1989a) and Glass (1989) suggested that the North American tektite source crater may be located at or near the eastern coast of the North American continent, maybe underwater.

This proposal may have been confirmed by the recent discovery of the Chesapeake Bay structure (centered at 37°16.5' N and 76°0.7' W). It is a complex peak-ring feature with a diameter of about 85 to 90 km (Fig. 2.4.5.1.1). The structure is currently buried 300–500 m beneath lower Chesapeake Bay, its surrounding peninsulas, and the adjacent inner continental shelf (Poag et al. 1994; Koeberl et al. 1996c; Poag 1997b; see also Sect. 3.3.1). The structure is partly filled with a unit termed the Exmore breccia (Poag et al. 1992), which is mainly composed of autochthonous sedimentary clasts in a sandy matrix, but also containing millimeter- to centimeter-sized basement clasts (Fig. 2.4.5.1.2). An impact origin was confirmed for the Chesapeake Bay structure by Koeberl et al. (1996c), based on the discovery of evidence for shock metamorphism (in granitic clasts within Exmore breccia samples) in the form of shock mosaicism in quartz, and abundant occurrences of shocked quartz, K-feldspar, alkali feldspar, and plagioclase grains.

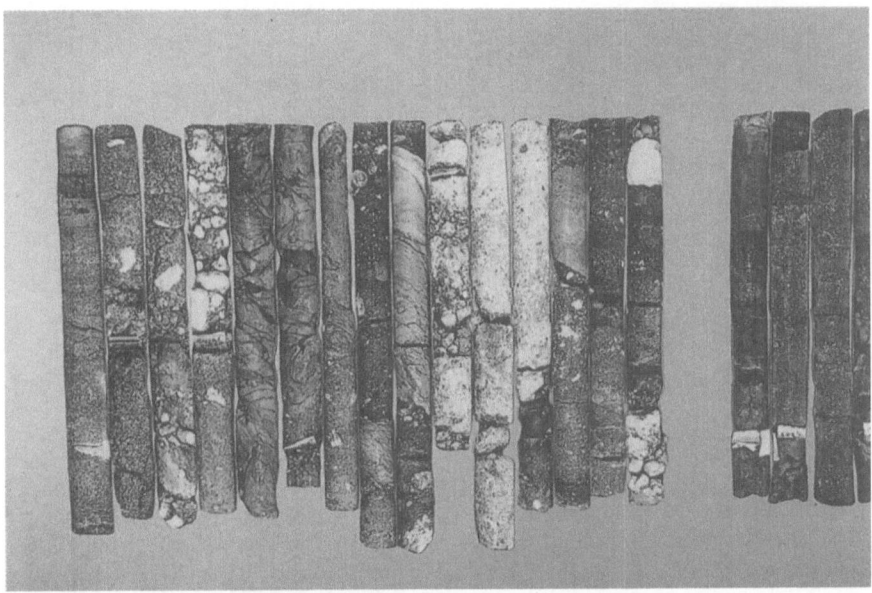

Fig. 2.4.5.1.2. Drill cores (from the Windmill Point core; see Poag et al. 1994; Koeberl et al. 1996c) showing (left part) the Exmore Breccia unit beneath the Chesapeake Bay impact structure; on the right side are four core sections of the Paleocene Aquia Formation (photo courtesy C.W. Poag).

The age of the structure has been estimated at 35.5 Ma, based on micropaleontological studies of the breccia and correlation with near-by impact deposits (Poag and Aubry 1995). The age, location, and geochemical composition of some basement clasts found within the breccia, which is similar to that of North American-tektites, led Koeberl et al. (1996c) to suggest that it might represent the source crater of the North American tektites. This conclusion also agrees with the Rb–Sr and Sm–Nd data (Show and Wasserburg 1982; Stecher et al. 1989), which indicated that their source rocks are likely to be derived from the Appalachian orogeny, in agreement with lithologies present beneath Chesapeake Bay.

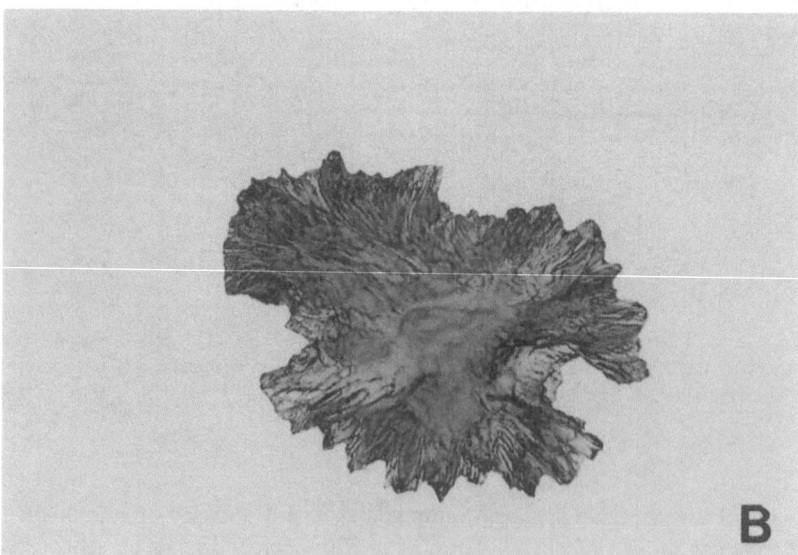

Fig. 2.4.5.2.1. Moldavite tektites (from Besednice, Czech Republic) from the Central European strewn field. **a)** Assortment of four samples, showing the variety in shapes and the deeply corroded nature of the tektites (in contrast to the much less corroded tektites from the Australasian strewn field; compare Figs. 2.4.1 and 2.4.2). **b)** Closeup of one moldavite sample (about 3 cm in long dimension) in transmitted light (samples courtesy H. Stehlik).

2.4.5.2
Central European Strewn Field

The first tektites from this strewn field have been described from Bohemia and Moravia (now in the Czech Republic) as early as the 18th Century. They were noted for their green color ("Bouteillensteine" – "bottle stones"). These were the first tektites to be analyzed for chemical composition (by the German chemist M. Klaproth in 1816). "Moldavites" were named after the first place (near the river Moldau – which is German for Vltava) they were found: this name has become the most widely used name for this group of tektites (2.4.5.2.1a, b). Within the last two decades, though, moldavites have also been found near Dresden (Germany), and in northern Austria. No microtektites are known from this strewn field. A more detailed account of historical aspects and distribution of the tektites is given by, e.g., Barnes (1963a), O'Keefe (1976), and Koeberl et al. (1988).

Fig. 2.4.5.2.2. Composite aerial view (from the southwest) of the Ries impact crater in southern Germany. The rim of the 24-km-diameter crater, which is the source crater of the Central European tektites, is marked by clouds (photo courtesy Rieskratermuseum Nördlingen).

The Ries crater in southern Germany was found, within errors of the measurement, to be of the same age (15 Ma) as the moldavites (Staudacher et al. 1982). Studies of the stratigraphy of the crater (see review by Hörz 1982) have shown that different ejecta from the Ries crater are related to a stratigraphic succession of the target rocks. From chemical and isotopic studies, the moldavites were interpreted to be high-speed ejecta derived from the uppermost layers of rock, the so-called Upper Freshwater-Molasse unit, but not from any of the deeper strati-graphic units at the Ries crater (see, e.g., Delano and Lindsley 1982; Shaw and Wasserburg 1982; Horn et al. 1985; Delano et al. 1987; Engelhardt et al. 1987).

2.4.5.3
Ivory Coast Strewn Field

Ivory Coast tektites were first reported in 1934 from a geographically rather re-stricted area in the Ivory Coast (Cote d'Ivoire), West Africa. Although some addi-tional specimens have been found later, the total number remains small (a few hundred). Microtektites (Fig. 2.4.5.3.1) were reported from deep-sea sediments of corresponding age from the eastern equatorial Atlantic Ocean west of Africa (e.g., Glass 1969; Gentner et al. 1970; Glass and Zwart 1979; Glass et al. 1991).

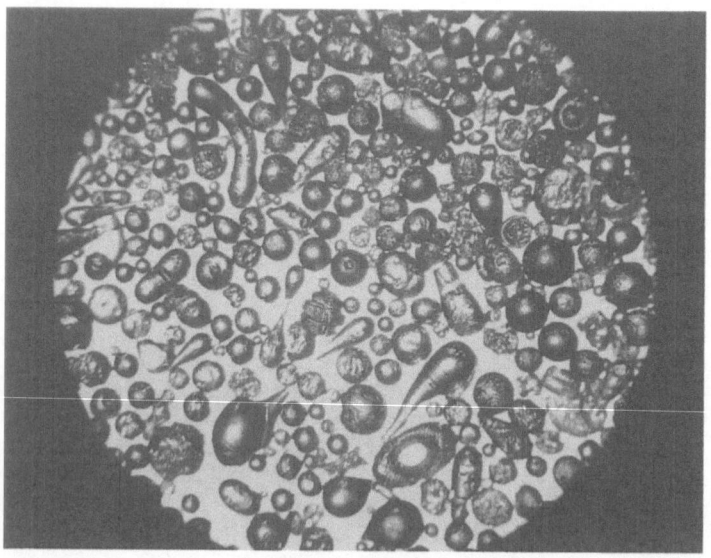

Fig. 2.4.5.3.1. Microphotograph of Ivory Coast microtektites from core K9–56 (see Glass et al. 1991), showing the variation in size and shape that is typical for microtektites. Spherical shapes are most common, but teardrops, dumbbells, and fragments are also found. Diameter of image 6.7 mm (photo courtesy B.P. Glass).

The first suggestions that the 10.5 to 11 km diameter Bosumtwi (or Ashanti) crater in Ghana (Fig. 2.4.5.3.2) is the source crater for the Ivory Coast tektites were made in the early 1960s (e.g., Barnes 1961). The Ivory Coast tektites and the Bosumtwi crater have the same age (1.07 Ma; Koeberl et al. 1997b), and there are close similarities between the isotopic and chemical compositions of the tektites and crater rocks (e.g., Fleischer and Price 1964; Kolbe et al. 1967; Schnetzler et al. 1966, 1967; Taylor and Epstein 1966; Gentner et al. 1969a, b; Storzer and Wagner 1977; Jones 1985; Koeberl et al. 1997b, 1998b). These observations strongly support a connection between the crater and the tektites. The crater is excavated in 2.1–2.2 Ga old metasediments and metavolcanics of the Birimian Supergroup (e.g., Jones et al. 1981). Koeberl et al. (1998b) found that the target rocks do not show any unambiguous evidence of shock metamorphism, but that distinct impact-characteristic shock effects (PDFs) were identified only in clasts within suevite-derived melt fragments. The compositional range of the target rocks is significantly wider than that of the Ivory Coast tektites, but overlap the tektite compositions. These authors found initial $^{87}Sr/^{86}Sr$ ratios of 0.701 and ε_{Nd} values of -17.2 to -25.9‰ in the Bosumtwi rocks.

Fig. 2.4.5.3.2. Aerial photograph of the 11-km-diameter Bosumtwi impact structure in Ghana, from the north rim. The central part of the crater is filled by the about 8-km-diameter Lake Bosumtwi, which has a maximum depth of about 80 m (photo courtesy L. Pesonen).

Using mixing calculations, Koeberl et al. (1998b) were able to reproduce the composition of Ivory Coast tektites from a mixture of Bosumtwi country rocks that includes about 70% phyllite-graywacke, 16% granitic dike, and 14% Pepiakese granite. The oxygen isotopic composition of the metasedimentary rocks and granitic dikes ($\delta^{18}O$ = 11.3–13.6‰) and the tektites ($\delta^{18}O$ = 11.7–12.9‰) agree fairly well. The large variation in Sr and Nd isotopic compositions of the target rocks do not allow the unambiguous identification of distinct endmember compositions, but in both a $^{87}Sr/^{86}Sr$ vs. 1/Sr plot and an ε_{Sr} vs. ε_{Nd} diagram, the tektites plot within the area occupied by the metasedimentary and granitic Bosumtwi crater rocks. These data support the interpretation that the composition of the Ivory Coast tektites is similar to that of rocks exposed at the Bosumtwi impact structure, indicating formation during the same impact event.

2.4.5.4
Australasian Strewn Field

For the Australasian field, no crater of the appropriate age is known, although many proposals for possible source craters from all around the world were made and later discounted (e.g., Weihaupt 1976; Bentley 1979; Dietz 1977; Glass 1979; Koeberl and Fredriksson 1986; Shaw and Wasserburg 1982; Storzer and Wagner 1979; Hartung and Rivolo 1979; Ford 1988). In contrast, Barnes (e.g., Barnes and Pitakpaivan 1962; Barnes 1989, 1990) suggested that the Australasian tektites have originated from an impact of a "diffuse object such as a comet", obviating the need for a distinct large source crater. In a similar vein, Wasson (1991) suggested that the tektites originated from a multitude of small craters scattered over all of Indochina. There are numerous problems with this model (see Blum et al. 1992; Koeberl 1992a; and Koeberl 1994b for detailed discussions). Bunopas et al. (1999) suggested that there is evidence for large-scale forest fires in Thailand that are coeval with the tektite formation event. Most workers agree that a single large impact crater is a much more plausible source crater of the Australasian tektites. Recently, Howard et al. (2000) reported on the possible presence of shocked quartz from a layer immediately above a tektite-bearing horizon in Thailand. If confirmed, this observation would support the single impact hypothesis.

Fission-track data suggested that australites may have a slightly different age than other Australasian tektites (Storzer and Wagner 1980a, b), but Glass (1986) found no evidence for a second, older (0.8–0.9 Ma), microtektite-bearing layer in deep-sea sediments. Bollinger (1993) and Izett and Obradovich (1992) found no age difference between australites and other Australasian tektites in their detailed Ar–Ar studies. Their best-fit age is 0.77 ± 0.02 Ma, which agrees well with the revised age for the Brunhes-Matuyama geomagnetic polarity reversal with which the Australasian microtektite layer is closely associated; however, the microtektite layer does not directly coincide with the Brunhes-Matuyama reversal, but the Australasian microtektites were deposited about 15,000 years prior to the geomagnetic reversal (Schneider et al. 1992).

The geographic distribution of Australasian tektites and microtektites is not homogeneous: there are radial and concentric patterns, and zones that do not contain microtektite-bearing deep sea cores (e.g., Stauffer 1978; Schnetzler 1992; Glass and Wu 1993). From an analysis of the radial distribution patterns, Stauffer (1978) suggested a crater that may be concealed beneath alluvial deposits of the lower Mekong Valley area. Schnetzler et al. (1988, 1999) suggested a possible off-shore impact location (about 175 km to the east of the Vietnam seashore) from satellite gravity data. Hartung and Koeberl (1994) proposed that the Lake Tonle Sap (about 100 km long and up to 35 km wide) in Cambodia was created by the Australasian tektite event. The dimensions are probably minimum values as the structure is almost completely filled with alluvium. The location of Tonle Sap is in agreement with chemical and isotopic (Rb–Sr/Sm–Nd) data for tektites, but more detailed studies are necessary. A first expedition to the Tonle Sap area has identified some possible source rocks, but no proof for impact has yet been found (partly due to the inaccessibility of the area). The quantity of both impact debris and microtektites in cores all over the Australasian strewn field increases towards Indochina (Glass and Wu 1993; Glass and Pizzuto 1994), which supports the locations proposed by Stauffer (1978) and Hartung and Koeberl (1994). The same conclusion (i.e., a possible single source crater in Cambodia, about 100 km in diameter) was also obtained in a recent study of microtektite distributions by Lee and Wei (2000). However, the Australasian tektite source crater remains, for the time being, elusive.

2.4.6
Tektite Formation: Facts and Open Questions

Our current understanding of the tektite-producing impact processes is still riddled with open questions. An attempt to present physical (or mathematical) models of a tektite-forming impact may not be successful at this time, because only educated guesses, but no detailed models, exist (Melosh 1989, 1998). First, it is useful to review processes that we know to have operated during tektite formation.

Volatilization. There has been a debate about the extent of volatilization and vapor fractionation that have affected tektites, ranging from very important (e.g., Walter and Carron 1964; Walter 1967; Walter and Clayton 1967; Ridenour 1986) to minimal (e.g., Molini-Vesko et al. 1982; Koeberl 1992a; Chaussidon and Koeberl 1995). Most chemical variations can be explained by a variation in source rock composition. There is little doubt, however, that the most volatile elements (e.g., the halogens, Cu, Zn, Ga, As, Se, Pb) were volatilized from the source rocks upon melting because the Muong Nong-type tektites, suspected to have experienced the lowest temperatures of all tektites, contain higher abundances of these elements than the splash-form tektites (Koeberl 1992a).

Water in Tektites. Tektites are, in general, very poor in water, with contents ranging from about 0.002 to 0.02 wt.% (20–200 ppm); impact glasses contain up to about 0.06 wt.% (600 ppm) water (see, e.g., Gilchrist et al. 1967; Engelhardt et al. 1987; Koeberl and Beran 1988; Beran and Koeberl 1997). The low water content is typical for impact glasses and can be used as convincing evidence for an origin by impact, as has recently been demonstrated by Koeberl (1992b) for the glasses from the Haitian K/T-boundary. It also easily distinguishes tektites from lunar materials, which have at least six orders of magnitude less water than tektites (e.g., Taylor 1982). Contrary to earlier belief, experimental evidence shows that it is possible to drive water out of parent sediments (containing up to several percent water) during the short time available for the tektite production (e.g., Glass et al.

1986a, 1988). In preliminary calculations on diffusion coefficients governing the water depletions in tektites, Vickery and Browning (1991) conclude that the calculated depletions are more than sufficient to account for the observed water contents in tektites. More recently, Melosh (1998) found that when silicates are shocked to 100 GPa, they almost instantaneously reach temperatures in excess of 50,000 K. Pressure release first leads to a supercritical silicate fluid and, once pressures drop below about 0.1 GPa, appearance of both a liquid and a vapor phase. Temperatures at this time are still around 5000 to 10,000 K and may remain in this range for several tens of seconds. Silicate viscosities at these temperatures are very low (about five orders of magnitude less than that of liquid water), and it is very likely that at these conditions water and other volatiles, including the noble gases, are rapidly lost.

Gases in Tektites. Müller and Gentner (1968) found that bubbles in tektites contain residues of the terrestrial atmosphere at low pressures. Jessberger and Gentner (1972) measured the gas content and composition in bubbles in Muong Nong-type tektites and found that the N_2/Ar ratio, as well as the isotopic ratios of $^{40}Ar/^{36}Ar$, $^{36}Ar/^{38}Ar$, $^{82}Kr/^{84}Kr$, $^{129}Xe/^{132}Xe$, $^{84}Kr/^{132}Xe$, and others, agree well with the respective atmospheric ratios, providing further evidence for an origin of tektites within the terrestrial atmosphere. Matsuda et al. (1989) and Matsubara et al. (1991) showed that the ratios of rare gases in impact glasses and tektites are consistent with those of the terrestrial atmosphere, and that, because of its higher mobility, during the long terrestrial residence times, Ne has predominantly diffused into the glass (Matsubara and Matsuda 1991). Tektites have low contents of dissolved heavy noble gases (Matsuda et al. 1993; Palma et al. 1997), and calculations show that, in order to explain the noble gas contents, tektite glass must have solidified at low ambient pressure, equivalent to a height of about 40 km in the atmosphere (Matsuda et al. 1993). Matsuda et al. (1996) studied a philippinite with a large (5 cm^3) bubble. The total gas pressure in this bubble was very low – about 10^{-4} atm. The gas in the bubble contained higher He and Ne abundances, compared to the other noble gases, and unusually high $^{20}Ne/^{22}Ne$ and $^{21}Ne/^{22}Ne$ ratios, which were interpreted as the result of very fast non-steady state diffusion in the early stages of tektite formation. Furthermore, Matsuda et al. (1993) showed that Muong Nong-type tektites contain higher abundances of the heavy noble gases than splash-form tektites (closer to values observed in "normal" impact glasses), and that the isotopic composition of Ne in vesicles in Muong Nong-type tektites is practically identical to that of the terrestrial atmosphere.

The "Age Paradox". The so-called age paradox (i.e., the observation that the stratigraphic age of many Australasian tektites found on land was reported to be much younger than their radiogenic age), has been discussed in detail in previous publications (e.g., Koeberl 1994b; Taylor and Koeberl 1994; Taylor 1999). In short, some authors have used this apparent discrepancy to argue against a relationship between microtektites and tektites, and that tektites only fell around 7000–24,000 years ago (e.g., Lovering et al. 1972; Chalmers et al. 1976). On the other hand, Gentner et al. (1970), and others, showed that microtektites found in the Australasian tektite strewn field have the same fission-track age of 0.7 Ma, which is in perfect agreement with the stratigraphic age of Australasian microtektites (e.g., Glass 1978; Glass et al. 1979). Glass and Wu (1992) have not found any indication for a younger event in deep-sea sediments less than 20,000 years old from 46 deep-sea drill cores. Chemical, isotopic, and age data confirm that microtektites and tektites have been produced in the same event. The discovery of shocked minerals and impact glasses together with microtektites in several deep sea sediments of the Australasian and North American strewn fields (e.g., Glass and Wu 1993) confirm the link of tektite production to impact events. Fudali (1993) showed that the proposed stratigraphic age of 5000–15,000 years for some australites is demonstrably incorrect. Shoemaker and Uhlherr (1999) found tektites weathering out of strata at Port Campbell (Victoria, Australia) that agree in age with the radiometric age determinations for tektites, demonstrating that the young stratigraphic age for tektites is only apparent at the result of reworking from older strata.

Meteoritic Component in Tektites. The geochemistry of tektites from all strewn fields is indistinguishable from that of the recent terrestrial upper crust. Meteoritic components (see Sect. 1.8) in tektites are very low or below detection limit (e.g., Morgan et al. 1975; Morgan 1978; Palme et al. 1978, 1979). A few philippinites contained some Ni-rich iron spherules, but this could have been due to in-situ reduction of target rocks (Ganapathy and Larimer 1983). The Ivory Coast tektites are the only ones that have a detectable extraterrestrial component (Koeberl and Shirey 1993). Morgan (1978) found only one HMg australite (out of six analyzed) that was slightly enriched in platinum-group elements. Koeberl et al. (1999a) found a significant meteoritic signature in bottle-green microtektites. Impactoclastic layers associated with tektite and microtektites have been only slightly more rewarding. Asaro et al. (1982) and Ganapathy (1982) reported enrichments of Ir in drill-cores from the North American strewn field at the same position where clinopyroxene spherules (but not microtektites; Keller et al. 1987) had been found, but no meteoritic component was found in the tektites; however, we now know that the Ir is associated with the Late Eocene clinopyroxene layer (cf. Glass and Burns 1987; Montanari 1990; Montanari et al. 1993). Schmidt et al. (1993) and Koeberl (1993a) found evidence for a minor Ir anomaly in deep-sea sediments associated with the Australasian microtektite-bearing layer in ODP core 758B (east of Malaysia, in the southern part of the Bay of Bengal). The low meteoritic contents indicate that no intimate mixing between projectile and target seems to have taken place during tektite formation.

Cosmogenic Radionuclides and Target Stratigraphy. The study of cosmogenic radionuclides provides evidence on the nature of the sedimentary precursor and the target stratigraphy. Pal et al. (1982) demonstrated that the ^{10}Be content of Australasian tektites cannot have originated from direct irradiation with cosmic rays in space or on Earth, but can only have been introduced from sediments that have absorbed ^{10}Be that was produced in the terrestrial atmosphere. This conclusion was supported by several additional studies (e.g., Yiou et al. 1984; Englert et al. 1984). The concentration of ^{26}Al is of crucial importance: if tektites were ever exposed to cosmic radiation in space to produce the observed ^{10}Be concentration, then the $^{26}Al/^{10}Be$ ratio must be between 2.7 and 5.4, depending on the details of the irradiation. Middleton and Klein (1987) found that the ^{26}Al concentration in tektites is so low that they were able to measure its concentration in only one sample, giving a $^{26}Al/^{10}Be$ ratio of about 0.07. Ivory Coast tektites and impact glasses (e.g., Aouelloul) were also found to contain ^{10}Be (Tera et al. 1983a, b; Raisbeck et al. 1988).

Cosmogenic radionuclides can also help in understanding of the target stratigraphy. The distribution of ^{10}Be in the environment is a strict function of the depth from the surface. Pavic et al. (1983) and Valette-Silver et al. (1983) demonstrated that most ^{10}Be is concentrated in the upper 20 m of soils and sediments. This limit may be variable as a function of different rock types, rainfall rate, and porosity, but probably by less than a factor of 10. Blum et al. (1992) have calculated that mixing of a 200 m column of bedrock into the surficial cover that contains the ^{10}Be explains the ^{10}Be concentrations observed in Australasian tektites. The absolute concentrations of ^{10}Be in the various types of Australasian tektites allow another important conclusion. The ^{10}Be concentrations in australites are higher than those in indochinites and philippinites, which in turn are higher than those in Muong Nong-type tektites, with average contents in australites, splash-form indochinites, and Muong Nong-type indochinites being $150 \cdot 10^6$, $100 \cdot 10^6$, and $57 \cdot 10^6$ atoms $^{10}Be/g$, respectively (Tera et al. 1983a, b; Pal et al. 1982; Raisbeck et al. 1988; Aggrey et al. 1998). For these data it can be concluded that australites (which experienced the highest temperatures and are the most distant ejecta) originated from close to or at the surface, whereas the Muong Nong-type tektites originated from a somewhat deeper unit, in agreement with impact mechanics and compositional data.

Open Questions. Despite the identification of three tektite source craters (out of four), the exact target rocks from which tektites have been produced are not known. There are numerous reasons for this: e.g., during the impact, melting of a variety of target rocks occurred; we still do not understand the exact physical and chemical processes which may alter the chemical composition during impact (i.e., unspecific fractionation); impact mixing is a non-equilibrium and heterogeneous process; also, tektites are most probably produced in the earliest stages of impact, which are poorly understood. Jakes et al. (1991, 1992) and Melosh (1998) have discussed the importance of superheating, and that superheated melts are reduced. This model may very well be valid, although we do not know in which state the tektite material is after impact melting, or, what the size distribution of the melt droplets is. Thermal data for tektites are not abundant (Wilding et al. 1996). Engelhardt et al. (1987) suggested that the tektite material is completely vaporized to a plasma state, and then recondensed in the form of coalescing droplets. It seems, however, that in such a model the relatively close compositional similarity between the tektites and their source rocks would be lost; also, the presence of lechatelierite particles in tektites cannot be explained by this model. It may also be important that tektites are more reduced in composition compared to impact glasses (e.g., Fudali et al. 1987; Yakovlev et al. 1993).

The formation of tektites, in which up to 10^9 t of glassy material is distributed over distances of up to 12,000 km (or about 800 km from the proposed source region, in the case of the Australasian tektites), must require distinct conditions, because otherwise more than just four tektite strewn fields would be known on Earth, considering the number of craters known. Maybe low angle impact is important because of the asymmetric distribution of tektites within a strewn field. This is true for all four cases: where craters are known to be associated with tektite fields, the craters are never in the center of the strewn field. Tektites and microtektites are not uniformly distributed within the strewn fields, but may follow some kind of ray-like pattern (similar to rays around lunar craters, but asymmetric); this may be related to the formation process. Tektite production has to occur before the main excavation phase of the crater formation has started. Numerous arguments have been presented here that show that tektites had to originate from target rock layers close to the surface. For example, it is otherwise not possible to explain their ^{10}Be content. The finding that bottle-green microtektites contain a significant meteoritic component indicates that these microtektites may have formed in the earliest phase of tektite formation. The quest to deduce the exact process of tektite formation, if by bow-shock melting (cf. Bohor and Koeberl 1996; Scott 1999), melting induced by an incandescent atmosphere (Wasson 1991; Wasson and Moore 1998), or a direct impact process, and the ejection mechanism (e.g., jetting – see Vickery 1993, for a discussion; or by ballistic shadowing – see Sugita et al. 1998), will provide material for much more research in the future.

2.5
Other Impact Glasses: Relation to Source Craters

Glasses have been found at several impact structures around the world, such as at the Aouelloul (Mauritania), Darwin (Tasmania, Australia), Henbury (Australia), Lonar (India), Wabar (Saudi Arabia), or Zhamanshin (Kazakhstan) craters (e.g., Taylor 1966; Kieffer et al. 1976b; Morgan 1978; Taylor and McLennan 1979; Koeberl and Fredriksson 1986; Koeberl and Storzer 1988; Meisel et al. 1990; Attrep et al. 1991a; Garvin and Schnetzler 1994; Koeberl et al. 1998a). All of these have in common that the craters are relatively young (ranging from about 6000 years or less for Wabar to 3.1 million years for Aouelloul), that the glass is found directly at the crater (less than one crater radius from the rim), and that the glasses provide the only confirming evidence that the host craters are of impact origin. For example, Aouelloul impact glasses (Fig. 2.5.1) are water-poor, have abundant schlieren of different chemical composition, partly digested quartz and feldspar grains, and contain lechatelierite and baddeleyite. The composition of the glass is similar to that of the sandstone in which the crater is exposed, but some siderophile elements are enriched in the glass. Koeberl et al. (1998a) performed Re–Os isotope studies (cf. Sect. 1.8.3) of the target sandstone and the impact glass and demonstrated the presence of a distinct extraterrestrial component in the glass. Whereas studies of such impact glasses are interesting and provide important information about cratering processes, they are not distal ejecta and, therefore, only of peripheral interest here. It should be emphasized that the same applies also to impact glasses from the Zhamanshin crater (zhamanshinites and irghizites), which are often erroneously called "tektites".

Fig. 2.5.1. Fragment of dark impact glass at the Aouelloul crater in Mauritania, on the south flank of the crater rim, about 100 m outside the rim. The glass provides the only physical evidence exposed on the surface that this structure is an impact crater.

Chapman and Scheiber (1969), in their detailed study of the composition of Australasian tektites, identified nine australite samples (from several locations in Australia) that have much higher Na and lower K contents than any other australites and were, thus, termed HNa/K australites (because they have a high Na/K ratio). Table 2.5.1 gives data for the major elements and selected trace elements for the HNa/K tektites in comparison with average indochinites. Fleischer et al. (1969) found that the age of samples from this australite subgroup is distinct from that of the other australites (they obtained ages around 4 Ma). Whereas the oxygen isotopic composition of these tektites is not unusual (Taylor and Epstein 1969), they do have distinct Sr (Compston and Chapman 1969), Nd and B isotopic and trace element compositions (Table 2.5.1; Koeberl et al. 2000b) that set them apart from other Australasian tektites. New age determinations by fission track, K/Ar, and Ar–Ar dating indicate ages around 10 Ma (Storzer 1985; Storzer and Müller-Sohnius 1986; Koeberl et al. 2000b), confirming that these nine tektite samples do represent a different impact event, but the source crater is, so far, unknown.

Other glasses of obvious impact origin, but unknown source include the so-called urengoites, a group of three samples, and the single (unrelated) south Ural Glass, which were analyzed by Deutsch et al. (1997). The urengoites, tektite-like objects from western Siberia, have high SiO_2 contents of 89 to 96 wt.%, crustal REE patterns as well as Sr and Nd isotope values, and ages of about 23 Ma. No source crater has yet been identified. A few samples of tektite-like glass of unusual composition were found among the ruins of the Mayan town of Tikal, Guatemala (Hildebrand et al. 1994). This glass is unrelated in age and composition to North American tektites. At present, the source of the glass is unknown.

Table 2.5.1. Composition of HNa/K australites in comparison to "normal" indochinites.

	HNa/K australites (range)	Indochinite (average)
SiO_2	61.7–64.0	72.7
TiO_2	0.49–0.61	0.78
Al_2O_3	15.7–16.4	13.4
FeO	5.59–6.88	4.85
MnO	0.09–0.12	0.08
MgO	3.44–4.43	2.14
CaO	4.49–6.27	1.98
Na_2O	2.71–3.98	1.05
K_2O	0.44–1.12	2.62
Sc	10.5–14.5	10.5
Cr	205–385	63
Co	45.3–68	11
Ni	650–875	19
Rb	29.6–39.3	130
Sr	435–550	90
Zr	140–168	252
Th	3.62–5.31	14.0
U	0.52–0.88	2.07
ε_{Sr} (‰)	-12.2	+200
ε_{Nd} (‰)	+4.7	-11.5
Age (Ma)	10	0.78

Major element data in wt.%, trace elements in ppm. All Fe as FeO. Indochinite data from Taylor and McLennan (1979) and Blum et al. (1992); HNa/K data from Chapman and Scheiber (1969) and Koeberl et al. (2000b).

2.5.1
Libyan Desert Glass

Libyan Desert Glass (LDG) is an enigmatic natural glass found in an area of about 6500 km^2 between sand dunes of the southwestern corner of the Great Sand Sea in western Egypt, near the Libyan border. The glass occurs as centimeter- to decimeter-sized irregular and strongly wind-eroded pieces. The total quantity of the glass present has been estimated at $1.4 \cdot 10^9$ grams, with a much larger original mass assumed (Barnes and Underwood 1976; Diemer 1997). In terms of chemical composition, LDG is very silica-rich and has low abundances of most major oxides (Table 2.5.1.1). The age of the LDG was determined by K-Ar and fission-track methods. While it is possible to calculate a K-Ar ages for LDG (58.3 ± 16.4 Ma; Matsubara et al. 1991), these values suffer from the very low K content of the glass, and other methodological problems (Horn et al. 1997). Better suited for an age determination of the LDG is the fission-track method, which gave ages ranging from 28.5 ± 2.3 Ma (Storzer and Wagner 1971) to 29.4 ± 0.5 Ma (plateau age; Storzer and Wagner 1977), which was recently confirmed (28.5 ± 0.8 Ma) by Bigazzi and de Michele (1996); see also Horn et al. (1997).

The origin of LDG has been the subject of much debate since it was discovered early in the 20th century (cf. Clayton and Spencer 1934; Weeks et al. 1984; Diemer 1997). Many workers were of the opinion that LDG is an impact glass (e.g., Kleinmann 1969; Jessberger and Gentner 1972; Barnes and Underwood 1976; Fudali 1981), but were deterred by the lack of a suitable impact crater. To avoid this problem, Seebaugh and Strauss (1984) suggested that the glass formed in a cometary collision, where shock-melting occurred without crater formation. Various other suggestions for the origin of LDG were made (e.g., a sol-gel process, or a sedimentary origin), but none of these suggestions were confirmed by any evidence.

Most researchers have now accepted the geochemical and geological evidence for an impact origin of LDG; see, e.g., Diemer (1997); Storzer and Koeberl (1991); Koeberl (1997b); Murali et al. (1989, 1997); Rocchia et al. (1996a, 1997); Abate et al. (1999). Evidence for an impact origin includes the presence of schlieren and partly digested mineral phases, such as lechatelierite (a high temperature melt of quartz); baddeleyite, a high temperature break-down product of zircon (Kleinmann 1969; Storzer and Koeberl 1991; Horn et al. 1997); and the likely existence of a meteoritic component (see below). Storzer and Koeberl (1991) suggested from their Zr/U and REE data that none of the sands or sandstones from various sources are good candidates to be the sole precursors of LDG. Compositional data for surface sands (Koeberl 1997b) show significant differences to the average LDG composition. Some admixture of monazite and/or apatite (e.g., from some type of sediment) seems likely to explain both the excess U and the REE patterns, although a better correlation is found with rocks from the BP and Oasis impact structures (see below).

More importantly, though, is the existence of a meteoritic component in LDG (Murali et al. 1989; Rocchia et al. 1996a; Koeberl 1997b). These authors have found that the contents of siderophile elements, such as Co, Ni, and Ir, are significantly enriched in some rare dark bands that occur in some LDG samples (Figs. 2.5.1.1 and 2.5.1.2). Koeberl (1997b) studied such dark bands and found that the contents of Fe, Mg, and Ni are high in the dark zones and low in the "normal" LDG. This co-variation can only be explained by a common source for those elements. Together with the observations of Murali et al. (1989), Rocchia et al. (1996a), and Koeberl (1997b) of high Ir contents in the dark zones, these data are only consistent with the presence of a meteoritic component. This is also in agreement with Os isotopic data of dark bands in LDG (C. Koeberl; unpublished data). The presence of a meteoritic component in LDG samples makes the Tunguska-like airburst hypothesis advanced by Wasson and Moore (1998) rather unlikely, as an extraterrestrial component in the glass would have to be the result of actual physical contact between the impactor and the target rocks.

Table 2.5.1.1. Major and selected trace composition of Libyan Desert Glass.

	Average	Range
SiO_2	98.4 ± 0.89	96.6–99.3
TiO_2	0.12 ± 0.06	0.05–0.24
Al_2O_3	1.19 ± 0.69	0.54–2.58
FeO	0.12 ± 0.09	0.05–0.39
MgO	0.011 ± 0.002	0.008–0.015
CaO	0.01 ± 0.005	0.01–0.02
Na_2O	0.005 ± 0.001	0.003–0.007
K_2O	0.009 ± 0.003	0.006–0.014
P_2O_5	0.01 ± 0.005	0.005–0.012
Sc	0.79 ± 0.42	0.26–1.66
Cr	5.74 ± 2.64	1.8–10.4
Mn	13.2 ± 3.62	7.6–19.1
Co	0.26 ± 0.14	0.11–0.58
Ni	2.5 ± 2.0	1–8
Rb	0.46 ± 0.12	0.28–0.68
Sr	24 ± 7	17–37
Zr	196 ± 47	105–270
La	6.91 ± 1.99	3.52–11.1
Th	2.59 ± 0.83	1.0–3.84
U	0.96 ± 0.28	0.41–1.45

Major element data in wt.%, trace element data in ppm. Data from Koeberl (1997b).

The geographic proximity of the LDG site to the BP and Oasis impact structures (French et al. 1974; see Koeberl 1994a for a review) has previously led to speculation that one of these two structures might be the LDG source. Diemer (1997) suggested that the BP and Oasis structures, which occur in the Nubian sandstone, of Early Cretaceous age, are older than the LDG. However, recent geochemical investigations related to the origin of Libyan Desert Glass (Barrat et al. 1997; Horn et al. 1997; Koeberl 1997b; Abate et al. 1999) indicated that the target rocks of the BP and Oasis structures are good candidates for the source material of LDG. Abate et al. (1999) conducted petrographic studies and, for the first time, major element, trace element, and isotopic analyses of rocks from the

BP and Oasis impact structures, for comparison of their geochemical characteristics with those of LDG.

Fig. 2.5.1.1. Libyan Desert Glass sample (about 2 mm thick slice) in transmitted light, showing dark streaks, which contain evidence for an extraterrestrial component. Such dark bands are very rare.

Abate et al. (1999) found that most of the target rocks at both structures have somewhat lower SiO_2 contents than LDG, but this distinction becomes less important when the composition of the sandstones is recalculated to a water-free basis. As for the major elements, there is a good correlation between the trace element compositions of samples from the target rocks of BP and Oasis structures and samples of LDG, but the refractory trace element content of the LDG is generally somewhat higher than that of the target rocks of the two Libyan craters. Rare earth elements (REE) from the BP and Oasis impact structures and from LDG have similar abundances and display a similar chondrite-normalized pattern. The isotopic ratios of Nd and Sr for samples of the two structures and for the LDG are in a similar range. They are characterized by negative ε_{Nd} values and positive ε_{Sr} values, which are characteristic of upper continental crustal rocks. In an ε_{Nd} vs.

ϵ_{Sr} diagram, LDG values plot within the field defined by the BP and Oasis rocks, and in a 1/Sr vs. ϵ_{Sr} plot LDG is also within the range defined by BP and Oasis target rocks.

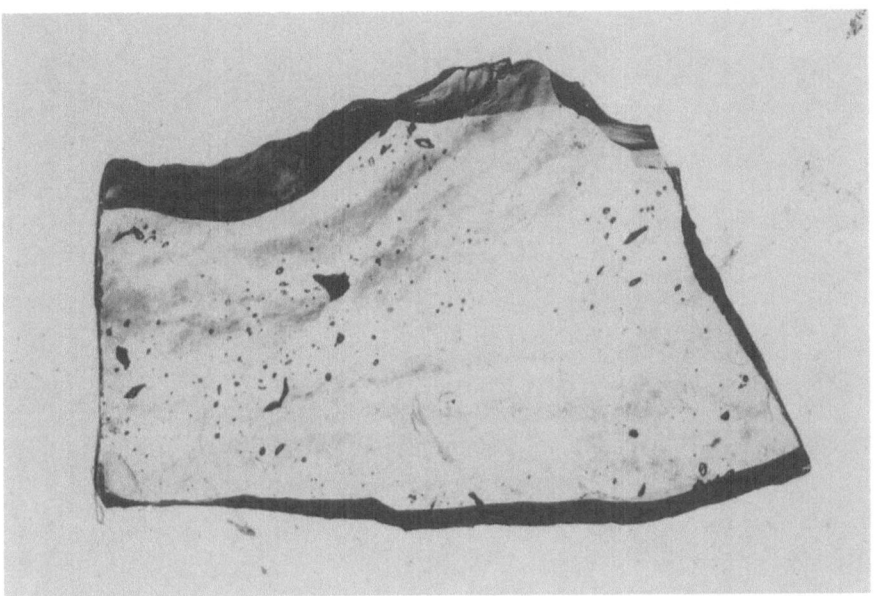

Fig. 2.5.1.2. Photograph of a polished "thick section" (about 0.5 mm thick) of a Libyan Desert Glass sample with dark streaks. Elongated bubbles are evidence for glass flow. Sample wide dimension about 10 mm.

All these petrographic, geochemical and isotopic data suggest that rocks found at the BP and/or Oasis structures could represent the parent material for LDG, but without further age information on the two impact structures it is not possible to unambiguously conclude that the BP or Oasis structures are the source craters for the LDG.

All this puts Libyan Desert Glass in a unique place: they are unlike tektites in many ways (e.g., they appear to occur only in a geographically restricted area, not an extended strewn field; they have higher water contents than tektites; some have significant meteoritic components), and they are also unlike lithic distal ejecta in that they are not associated with any shocked minerals or rock fragments. If they really originated from the BP or Oasis structures (maybe in an oblique impact?), they would have to be classified as distal ejecta – but so far these glasses remain unique in the geological record.

3
Important Distal Ejecta Layers and Their Source Craters

3.1
General

In this chapter we describe some key characteristics of the most important distal ejecta layers (e.g., the K/T boundary layer and the Late Eocene layers) and discuss their source craters. We will also discuss some other, less well established, possible impactoclastic layers, and mention some cases where not enough information is yet available to decide if there is an impact connection or not. This information sets the stage for a detailed description of the evidence for impact in the Umbria–Marche Sequence, which is given in Chaps. 4 and 5. The definition of distal ejecta was given in Sect. 2.1 and their use as impact markers in general was discussed in Sect. 2.2.

3.2.
The K/T Boundary and Chicxulub

The K/T boundary may not be the first distal ejecta layer that was investigated in detail (microtektite-bearing layers were discovered in the late 1960s; see Sect. 2.4), but it is by far the most important one (see Sect. 2.3), and probably the best studied one (cf. Ryder 1996). The story of the discovery of the K/T boundary layer at Gubbio, Italy, and the subsequent recognition that it is impact-derived, is best told by Alvarez (1997) (see Sect. 5.7 for details and field descriptions). A typical K/T boundary layer has a striking and distinct appearance in the field, and can easily be recognized and identified: there is distinct break in lithology, with a thin (usually up to 1 or 2 cm thick) layer commonly composed of clay or claystone. Good descriptions of field relations are given, for example, by Izett (1990) and Smit (1999), and references therein. Smit (1999) provides a definition of what constitutes the K/T boundary: The Global Stratotype Section and Point (GSSP) of the K/T boundary is at El Kef in Tunisia, where a 2 mm thick ejecta layer overlies Maastrichtian calcareous marls. Directly above the thin ejecta layer follows a 25 cm thick, fossil-poor, clay layer. The GSSP is placed directly below the ejecta layer, marking the end of the Cretaceous. The ejecta layer contains shocked quartz, Ni-rich spinel, and high PGE abundances. A similar definition can be used for other K/T boundary locations.

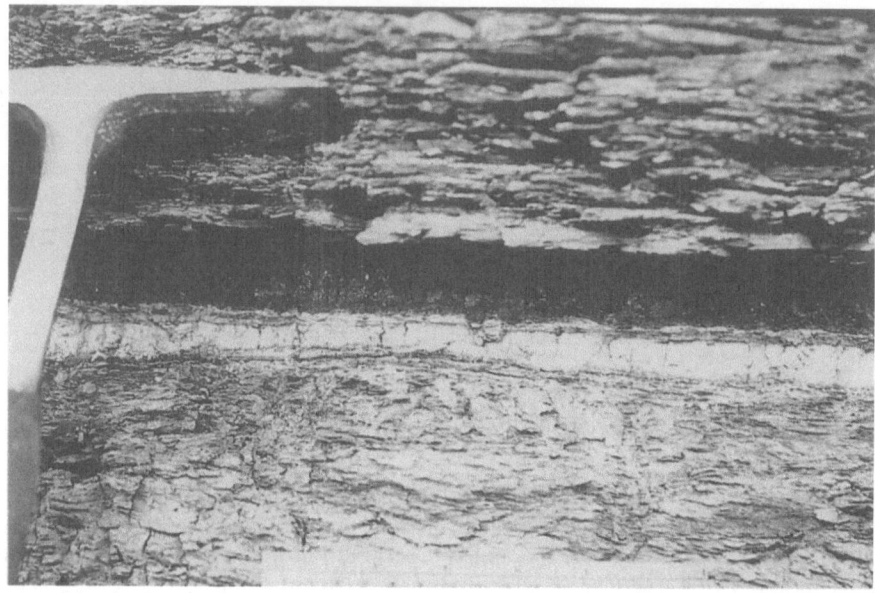

Fig. 3.2.1. K/T boundary layer at the Clear Creek North site, Raton Pass, Colorado, near the I-25 freeway. Scale bar on bottom is about 16 cm wide. This location represents a terrestrial site, in contrast to the marine sites discussed in Sect. 5.7. The boundary horizon is made up of two layers. The lower one is a light gray layer of kaolinite, about 2 cm thick, which is thought to have formed by alteration of glassy spherules. This is overlain by a thinner, laminated brownish "fireball" layer, which, in turn, is overlain by Tertiary coal, reflecting the swamp environment in which the material was deposited (photo courtesy B.F. Bohor).

The discussion of paleontological observations from different K/T boundary sites is mostly beyond the scope of this chapter. This is also the topic on which there is probably the largest variety of opinions, and it would not be possible to do justice to this subject in a short chapter. However, as far as Italian sites are concerned, the paleontological data are reviewed in Sect. 5.7. For other locations and a general discussion of paleontological aspects of the K/T boundary, the reader is referred to papers in Sharpton and Ward (1990), Ryder et al. (1996) and to the recent review of Smit (1999), and references therein. Paleobotanic studies provide also important data in terms of defining how severely the environmental changes induced by the K/T event affected numerous plant species (see, e.g., Tschudy et al. 1984; Spicer 1989). An interesting (but not undisputed) hypothesis was put forward by Wolfe (1991), who interpreted the paleobotanical data to indicate that the impact event occurred in June.

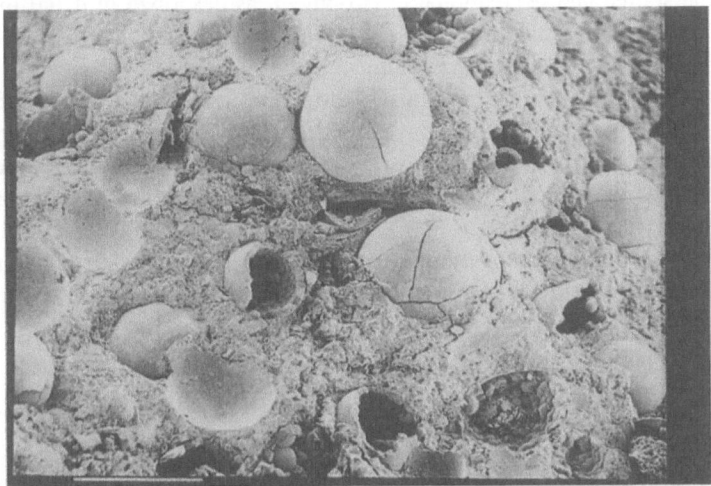

Fig. 3.2.2. Goyazite spherules in the lower layer of the K/T boundary at Teapot Dome, Wyoming, USA. The (mostly hollow) spherules are up to about 1 mm in diameter (white scale bar on bottom left is 1 mm wide) and have probably formed by alteration of originally glassy spherules (SEM photograph courtesy B.F. Bohor).

It should be noted that in undisturbed terrestrial K/T boundary locations (e.g., in the Western Interior of North America), the layer is up to 3 cm thick and has a dual nature (Fig. 3.2.1), whereas in most marine sections it does not display a dual-layered nature. At these sites, nannoplankton extinctions occur at the base of the clay layer (within the 29R magnetochron). At the Western Interior locations (Bohor et al. 1987b; Bohor 1990; Izett 1990), the lower layer is mainly of kaolinitic composition with some smectite; it contains hollow spheres up to 1 mm in diameter, which are most often also composed of (and filled by) kaolinite, but at some sites (e.g., in Wyoming) they are composed of goyazite (a hydrous aluminophosphate; an alteration product) and are easily discernable (Fig. 3.2.2). Bohor and Glass (1995) provide a detailed description of the formation and alteration of these spherules. The upper (thinner) layer is commonly laminated and consists of smectite and kaolinite clay with variable amounts of organic matter, giving it a dark gray to black appearance (Bohor 1990). Most of the shocked minerals and the siderophile element anomaly (see below) is restricted to the upper layer; it has been called the "fireball" layer (e.g., Bohor 1990). Shoemaker and Izett (1992) thought that the dual nature of the boundary layer may be the result of multiple impacts at the end of the Cretaceous, but that was at a time when the Manson impact structure in Iowa was still considered to be of K/T boundary age (see

Sect. 3.4). Despite related suggestions (Koutsoukos 1998), no subsequent evidence to support this suggestion has been found.

The composition and internal structure of the K/T boundary layer varies with distance from the source and depositional environment, as described in detail by Smit (1999) in a recent review. However, Smit (1999) unfortunately confuses the issue somewhat by using inappropriate nomenclature: he calls K/T ejecta at <2500 km from the source crater "proximal", those that are 2500 to 4000 km "intermediate", and those at >7000 km distance "distal". As described in Chap. 2.1, the definition of proximal and distal ejecta (which existed long before K/T boundary studies were made) is based on a changeover distance of 5 crater radii. Assuming a radius of about 100 km for the Chicxulub crater (see Sect. 3.2.2), all sites >500 km from the crater rim should be called "distal"; for further distinction between sites it would be better to use terms in accordance with their proper definition.

3.2.1
Evidence for Impact at the K/T boundary

Following the brief review of the appearance of the K/T boundary layer and its geographical variation, it is pertinent to summarize the most important arguments in favor of an impact origin; however, space limitations prevent a detailed discussion of these arguments (see, e.g., the reviews of Kring 1993 and Koeberl 1996, and papers in Silver and Schultz 1982; Sharpton and Ward 1990; Dressler et al. 1994; Ryder et al. 1996; Dressler and Sharpton 1999). For an explanation of the significance of the evidence, see the various sections in Chap. 1. In this section, we first discuss the origin of the rocks that mark the K/T boundary, whereas the link with the Chicxulub structure in Mexico is addressed in Sect. 3.2.2. Despite some claims to the contrary, the K/T boundary provides several independent lines of evidence in favor of the conclusion that an impact event was responsible for the end of the Cretaceous.

In the following sections, we describe the observations that are indicative of an impact event 65 Ma ago.

3.2.1.1
PGE Enrichments

The first physical evidence pointing to a contribution of extraterrestrial material that was discovered was the presence of unusually high PGE abundances in K/T boundary clay in Italy (Alvarez et al. 1980) and other locations around the world (e.g., Smit and Hertogen 1980; Ganapathy 1980; Kyte et al. 1980); this point has been made repeatedly in Chaps. 1 and 2 and is also discussed in more detail in Sect. 5.7. Iridium and other PGEs were found to be enriched in these K/T boundary clay layers by up to four orders of magnitude compared to average terrestrial crustal abundances. Early in the discussion about the origin of the Ir enrichments, terrestrial sources were suggested as possible causes for these enrichments. For example, some volcanoes were found to emit aerosols that are enriched in Ir and other siderophile elements (e.g., Zoller et al. 1983; Koeberl 1989b; Toutain and Meyer 1989), but in none of these cases were the interelement ratios (see next paragraph) similar to those observed at unaltered K/T boundary locations. Such models also fail to explain the overall amount of PGEs found worldwide at the K/T boundary (see, e.g., Kring 1993, for a more detailed discussion). See Sect. 5.7 for a discussion of Ir abundance profiles across typical K/T boundary sections, showing the significant Ir enrichment at the boundary.

There is a variation in the absolute amount of Ir found at K/T boundary locations around the world. Values quoted by various authors range from a few hundred ppt (ppt = 10^{-12} g/g) to a few hundred ppb. However, direct comparison of the data is very difficult, as the results critically depend on which type of sample was analyzed, and how the sample was treated. For example, Eugster et al. (1985) give Ir data for a Stevns Klint K/T boundary sample without treatment (4.6 ppb) and for one that was treated with acids (33 ppb). The absolute value depends in either case on how effectively only the fireball layer was collected, and on how much other material was incorporated in a bulk sample. At some locations (especially those that are close to the impact site, e.g., Gulf of Mexico), the K/T boundary layers are fairly thick and, therefore, the Ir abundance in a bulk sample can be rather low. For example, Rocchia et al. (1996b) found 0.7 ppb Ir in K/T boundary samples from ODP Hole 536 in the Gulf of Mexico. A better measure for the Ir distribution would be the fluence of Ir, which at ODP 536 is 55 ng/cm^2 (Rocchia et al. 1996b); this is on the order as the global average for K/T boundary sites. Kyte et al. (1996) found an Ir fluence of about 90 ng/cm^2 from K/T boundary samples over most of the Pacific plate.

3.2.1.2
PGE Abundance Patterns; K/T boundary Meteorite?

It was found that the interelement ratios of the PGEs in the K/T boundary clays are very similar to the values observed in chondritic meteorites (e.g., Alvarez et al. 1980; Ganapathy 1980; Kyte et al. 1980; Smit and Hertogen 1980; Palme 1982; Tredoux et al. 1989a). Terrestrial sources do not easily explain cosmic interelement ratios. Typical siderophile element data (including PGE data) for two K/T boundary locations are given in Table 3.2.1.2.1. These data, as well as those in the literature cited above, show that redistribution of the PGEs during alteration played some role in determining the present-day abundances, as, for example, Re and Au are fairly mobile under conditions of aqueous alteration (e.g., Colodner et al. 1992; Wallace et al. 1990b).

Table 3.2.1.2.1. Abundance of PGEs and other siderophile elements in samples from K/T boundaries.

Location	Co ppm	Ni ppm	Pd ppb	Re ppb	Os ppb	Ir ppb	Pt ppb	Au ppb
Caravaca	626	2580	110	2.7	46	56.9	131	43
Stevns Klint	146	1370	46	9.4	35	47.4	78	12.5

Data from Kyte et al. (1985).

The host phase of the PGEs is still not well known. Several proposals have been made over the years. For example, Doehne and Margolis (1990) reported on micrometer-sized Pt grains, whereas Martinez-Ruiz et al. (1997) described carbon-rich cores in some K-feldspars from the Agost and Caravaca K/T boundary sections that contain Ir, Pt, Pd, and Ni. Robin et al. (1993) found that some spinel-bearing spherules from DSDP Site 577 have Ir contents that range up to meteoritic values, but otherwise very few trace element analyses of spinel crystals are available (which is not surprising, given their small sizes).

Recently, Kyte (1998) described a 2.5 mm diameter inclusion recovered from K/T boundary sediments at DSDP drill core 576 (western North Pacific), and interpreted this fragment to be a fossil meteorite. The implication is that this would be a fragment of the K/T boundary impactor. Unfortunately the piece seems to be completely altered, but may have preserved some primary textures. The main minerals present are hematite and clays, with some of the hematite containing submicrometer-sized Ni–Fe metal and Ni–Fe sulfide grains. In a bulk sample, this altered fragment has Ir, Fe, and Cr abundances that are within a factor of about 2 of chondritic values. However, the explanation of the chemical composition of the fragment may be quite complicated, as other elements show non-chondritic abundances or ratios. The abundances of lithophile elements (e.g., Sc or the rare earth elements) are higher by factors of about 50 to 100 compared to chondritic values. Worse even, Au has an abundance of 1000 times the chondritic concentration (213 ppm), which is very difficult to explain by any straightforward alteration process (Kyte 1998). The severe alteration complicates the interpretation of this fragment as a fragment of the K/T boundary bolide and does not allow to use the chemical data obtained from this piece to constrain the composition of the impactor.

3.2.1.3
Meteoritic Os- and Cr-isotopic signature

Turekian (1982) proposed to measure the Os isotopic ratio in the K/T boundary clays, following the reasoning discussed in Sect. 1.8.3. Luck and Turekian (1983) subsequently analyzed marine manganese nodules and two K/T boundary clay samples. The manganese nodules had $^{187}Os/^{188}Os$ ratios of about 0.7 to 1, showing a clear continental crustal signature, while the K/T boundary samples from Stevns Klint (Denmark) and Starkville South (Colorado, USA) yielded values of 0.200 and 0.155, respectively (Luck and Turekian 1983; see also Koeberl and Shirey 1997). These results indicate clearly that the PGE signature at the K/T boundary is not the result of the concentration of PGEs from a crustal source by terrestrial processes. Subsequent analyses by Lichte et al. (1986) showed a $^{187}Os/^{188}Os$ ratio of 0.135 in clay for the Woodside Creek (New Zealand) K/T boundary. Krähenbühl et al. (1988), Esser and Turekian (1989), Schmitt (1990), and Meisel et al. (1995) determined comparable values, ranging from 0.137 to 0.212, for K/T boundary samples from Starkville South, Madrid, and Berwind Canyon (both Colorado, USA), Raton (New Mexico, USA), Shatsky Rise (DSDP 577), Stevns Klint, and Sumbar (Turkmenistan).

Peucker-Ehrenbrink et al. (1995) found that the $^{187}Os/^{188}Os$ ratio of seawater at the K/T boundary shows a sharp decrease, which they attributed to leaching of meteorite debris from the impact event. Meisel et al. (1995) measured the variation of the $^{187}Os/^{188}Os$ ratio across a K/T boundary section (at Sumbar, Turkmenistan) and found a significant and sudden decrease of the $^{187}Os/^{188}Os$ ratio from the end-Cretaceous rock layers to the actual K/T boundary clay that correlates with the maximum Ir (and Os) concentration; in the early Tertiary rocks, the $^{187}Os/^{188}Os$ ratio returns to higher values. Pearson et al. (1999) reported first Os isotope data on altered K/T boundary samples. All these studies provide clear evidence for an extraterrestrial component at the K/T boundary (and, together with other chemical and isotopic data, rule out a mantle source for the PGEs). Analyses of the PGE contents and the Os isotopic composition of a profile across the K/T boundary section at ODP 1049 (Blake Nose, western North Atlantic) are currently in progress (B. Peucker-Ehrenbrink and G. Ravizza, personal communication 1999). Unfortunately, despite the potential of the Os isotopic analyses, only very few K/T boundary sections, or any other boundary sections, have been studied in any detail.

Recently, Shukolyukov and Lugmair (1998) reported on Cr-isotopic studies on K/T boundary samples (see Chap. 1.8.4 for details on this isotopic system). These authors presented evidence for a meteoritic component that is in better agreement with a carbonaceous chondritic composition of the impactor, rather than an ordinary chondritic composition.

3.2.1.4
Soot and Organic Chemistry

Wolbach et al. (1985) found evidence for global wildfires in the form of a charcoal and soot layer at numerous K/T boundaries around the world, coinciding with the Ir-rich layer. The insoluble carbon fraction after acid dissolution is dominated by kerogen and elemental carbon. Kerogen is enriched about 15 times, and nitrogen is enriched 20 times, compared to the maximum abundances in the uppermost Cretaceous limestones (Gilmour et al. 1990). Both also show a marked change in their isotopic composition across the K/T boundary. No comparable soot enrichments of local or global distribution occur in the Late Cretaceous or in a wide range of other marine sediments (Wolbach et al. 1990). The presence of the hydrocarbon retene in the soot layer is diagnostic of wood fires in which resinous (coniferous) plants and trees were burning, indicating that most or all of the fuel was biomass (Wolbach et al. 1990). The isotopic composition of the carbon in the soot layer (average $\delta^{13}C = -25.8‰$) resembles that of natural charcoal and atmospheric carbon particles originating from biomass fires (Wolbach et al. 1990). From the isotopic composition of Chicxulub impact diamonds (see Sect. 3.2.1.7) we know that the isotopic composition of carbon in the target rocks is about -18 to -12‰; as the soot has a lighter isotopic composition, it should be derived from biomass burning rather than from the target rocks or the bolide (I. Gilmour, personal communication 2000). The total amount of soot in the atmosphere due to the global wildfires at the end of the Cretaceous has been estimated at $7\cdot10^{16}$ g, which must have had a large influence on the environment.

Fullerenes (large spherical molecules; the best known ones are made up of 60 or 70 carbon atoms) have also been detected in K/T boundary samples (Heymann et al. 1994, 1996). These authors found up to 5.4 ppb of C_{60} in K/T boundary clays from Woodside Creek and Flaxbourne River in New Zealand, which also have very high soot and elemental carbon contents (e.g., Wolbach et al. 1985; Gilmour 1998). Heymann et al. (1994, 1996) concluded that the fullerenes formed in global(?) wildfires that were ignited by radiation and hot melt from the K/T impact event (Melosh et al. 1990), rather than be the result of carbon chemistry in the hot vapor plume of the impact.

Further evidence for global wildfires comes from the detection of fossil charcoal at K/T boundary locations (Kruge et al. 1994) and from the discovery of polycyclic aromatic hydrocarbons (PAHs) at several boundary locations in Denmark, Italy, New Zealand, and Japan (e.g., Venkatesan and Dahl 1989; Gilmour et al. 1990; Mita and Shimoyama 1999). Gilmour et al. (1990) also point out that the significant input of the toxic and carcinogenic PAHs into the biosphere as a consequence of the impact event must have had negative effects on the environment (as they did in freshwater ecosystems following the Mount St. Helens volcanic eruption in 1980). Arinobu et al. (1999) concluded, from their study of carbon isotope stratigraphy and the detection of a spike of pyrosynthetic PAHs at the Caravaca K/T boundary, that about 20% of the terrestrial (aboveground) biomass available at that time must have instantaneously combusted. Few measurements exist so far on sulfur isotopes across the K/T boundary (e.g., Schmitz et al. 1988; Kajiwara and Kaiho 1992), but a shift in the isotope ratio at the boundary is consistent with the effects of bacteria feeding on dead biomass.

Another line of evidence in favor of an extraterrestrial event at the K/T boundary may the presence of large amounts of (supposedly) extraterrestrial amino acids that are associated with the boundary layer. They are absent in the boundary layer itself, but occur a few centimeters above and below (Zhao and Bada 1989). The two amino acids found – α-aminobutyric acid (AIB) and isovaline – are both common in meteorites, but uncommon in the terrestrial biosphere. However, isovaline is common in soil fungi and AIB is an essential part of ATP-synthase and, therefore, present in all cells, casting some doubt on the suggestion that these amino acids are of extraterrestrial origin (I. Gilmour, personal communication 2000). On the other hand, Zhao and Bada (1989) reported that the isovaline is racemic (i.e., a mixture of both left- [L] and right [R]-handed mirror forms of the molecule), whereas biogenic amino acids only occur in one form (L). It is interesting that such fragile molecules would survive the firestorm during the impact event. This led Zahnle and Grinspoon (1990) to propose that the amino acids were delivered from dust of a huge disintegrating comet, of which the K/T bolide was just the largest fragment. The dust would have slowly accreted onto the Earth. However, there is no evidence from ^{3}He studies for an enhanced influx of extraterrestrial dust around the time of the K/T impact event (Farley et al. 1999), in contrast to a broad peak of influx in the Late Eocene (Farley et al. 1998). On the other hand, Becker et al. (1999) report a high ^{3}He/^{4}He ratio in samples of the K/T boundary clay from an unspecified location at the Raton Basin (USA), and from Stevns Klint (Denmark), and speculate that some higher fullerenes might be present and host the ^{3}He. The study of fullerenes and ^{3}He is obviously a controversial topic. However, Pierazzo and Chyba (1999) found, in a series of calculations, that a certain fraction of extraterrestrial amino acids from the bolide would not be destroyed during a somewhat oblique impact. This would provide a possible mechanism for amino acid survival during the K/T impact event, and obviate the need for a ^{3}He-rich dust source.

3.2.1.5
Evidence of shock metamorphism

The first clear evidence of shock metamorphism at the K/T boundary was found by Bohor et al. (1984) in the form of shocked quartz grains in sediments from the Brownie Butte location. This landmark discovery confirmed the hypothesis of Alvarez et al. (1980) that a large-scale impact event occurred at the end of the Cretaceous. Shocked quartz and feldspar, and other shocked minerals (Figs. 1.6.4.1 and 1.6.4.2) were later found at practically all K/T boundary sites around the world (see, e.g., Bohor and Izett 1986; Bohor et al. 1987a; Bohor 1990). The shocked quartz grains show multiple intersecting sets of PDFs with shock-characteristic crystallographic orientations. As discussed above, such shocked minerals are associated with pressures far beyond those of any endogenic processes and are uniquely characteristic of hypervelocity impact. No quartz or other minerals showing true shock metamorphic characteristics have ever been found in a volcanic environment (e.g., Sharpton and Schuraytz 1989; de Silva et al. 1990). Furthermore, the modal abundance of shocked minerals (quartz-dominated), as well as the chemical composition of shocked feldspars at the K/T boundary, are indicative of a continental crustal source, and are not compatible with material derived from oceanic crust (Sharpton et al. 1990). Oceanic source rocks were initially proposed based on some mineralogical and geochemical data (e.g., Montanari et al. 1983; De Paolo et al. 1983). The discovery of clear evidence for impact in terrestrial K/T boundary locations also helped to negate arguments that marine sites represented condensed sections.

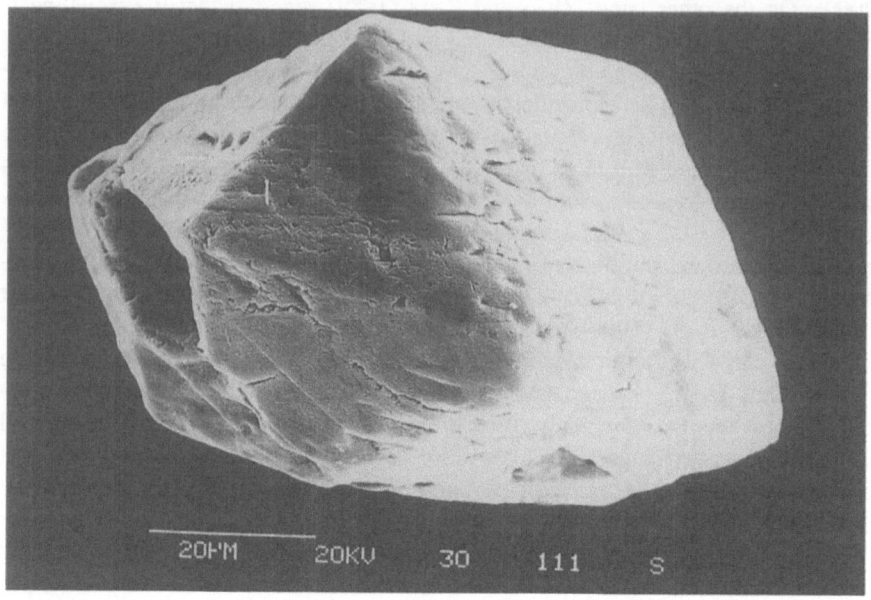

Fig. 3.2.1.5.1. Shocked zircon from the K/T boundary at Berwind Canyon, Colorado. The crystal was etched and shows planar features. Scale bar on lower left is 20 μm wide (SEM photograph courtesy B.F. Bohor).

Shocked quartz grains show a variation in their maximum grain sizes with geographic location. The largest grain dimensions (>0.6 mm) are found at K/T boundary sites in the Western Interior of North America, and decrease with increasing distance from North America (cf. Bohor 1990). For example, at Stevns Klint (Denmark) they are on the order of 0.15 mm in size, and in New Zealand they reach 0.11 mm. This size distribution provided an early pointer to the possible source crater location. Later, Alvarez et al. (1995) provided a quantitative model of the geographical distribution of shocked quartz.

Shocked zircons (Figs. 3.2.1.5.1 and 3.2.1.5.2), another signature of a continental crustal source, were discovered by Bohor et al. (1993). Shocked zircons display planar features (Fig. 3.2.1.5.1) with what appears to be multiple orientations, but it is unlikely that these features are PDFs *sensu strictu*, because PDFs are defined as planar features that contain amorphous material, which does not seem to be the case for shocked zircons (Leroux et al. 1999). Some shocked zircons, possibly those that experienced (incomplete?) melting, show a granular structure (Fig. 3.2.1.5.2). In addition to the finding of shocked minerals, the presence of the high-pressure quartz polymorph stishovite was also reported from K/T boundary sediments (McHone et al. 1989).

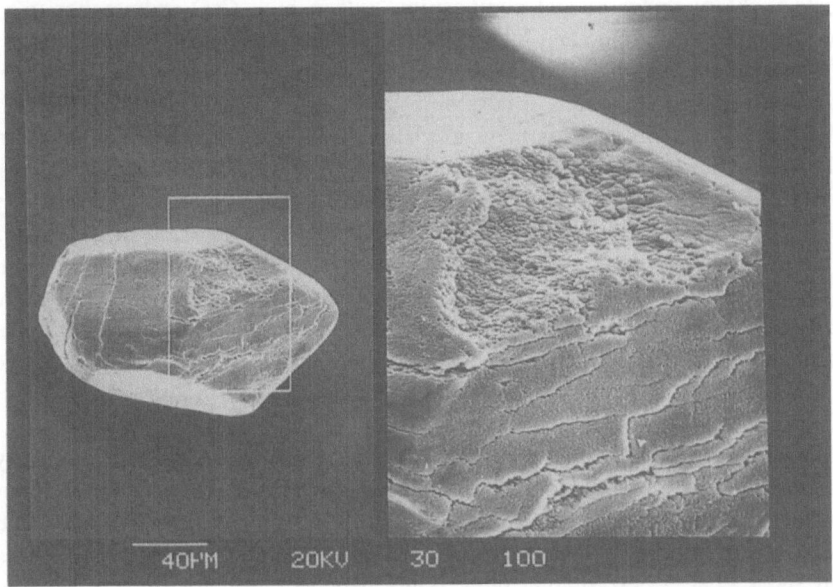

Fig. 3.2.1.5.2. Shocked zircon from the Berwind Canyon (Colorado) K/T boundary location. The sample was etched. The scale bar on the bottom of the left image is 40 μm wide. A close-up of part of the zircon crystal on the right side shows the internally granular texture, similar to what has been observed in Muong Nong-type tektites (see Fig. 2.4.3.4), and is thought to represent the result of higher shock pressures than those that form planar features (SEM photograph courtesy B.F. Bohor).

3.2.1.6
Impact glass

Sigurdsson et al. (1991a, b), Izett (1991), and Kring and Boynton (1991) described the presence of rare relict glass within alteration spherules (that are otherwise very common at numerous K/T boundary locations; e.g., Smit and Klaver 1981; Smit and Kyte 1984; Pitakpaivan et al. 1994) from the K/T boundary at Beloc, Haiti, and interpreted the material as impact glass. Sigurdsson et al. (1992) have shown, from comparison with experimental glasses, that the Haitian glasses have been quenched from temperatures much higher than common for volcanic processes. A detailed study by Koeberl and Sigurdsson (1992) provided not only geochemical arguments for the impact origin of these glasses, but also demonstrated the existence of rare inhomogeneous glasses with lechatelierite and other mineral inclusions, which are typical for an origin by impact. Blum and Chamberlain (1992) have obtained oxygen isotope data on Haitian glasses that specifically rule out a volcanic origin of these glasses. Blum et al. (1993) have confirmed this result with Rb–Sr and Sm–Nd isotopic data, showing that the Haitian glasses are mixtures of silicate rocks of upper crustal composition with a high CaO-endmember (e.g., limestone) (see also Premo and Izett 1992, and Bohor and Glass 1995).

Chaussidon et al. (1994) have shown that the sulfur in the yellow glasses occurs in the form of sulfate, which is not compatible with a volcanic source. Koeberl (1992b) measured the water content in glasses from Haiti and found a range of 0.013 to 0.021 wt.% H_2O (this was later confirmed by Oskarsson et al. 1996), which is further evidence for an origin by impact, as impact glasses are extremely dry. Koeberl et al. (1994c) have used Re–Os isotope systematics to find evidence for the presence of a small meteoritic component in the Haitian glasses. The Haiti glasses are different from tektites in some respects; for example, they are more oxidized than tektites (Oskarsson et al. 1996). Furthermore, high-precision age determinations on the Haitian glasses have shown that the materials have an age indistinguishable from that of the K/T boundary, at 65 Ma (e.g., Izett 1991; Swisher et al. 1992).

Glasses with similar properties were later recovered from some K/T boundary locations in Mexico (e.g., Mimbral) as well (e.g., Smit et al. 1992a, b). It is interesting to note that, despite all the data favoring an impact origin of the Haiti glasses, a few authors have suggested that the glasses are of volcanic origin (e.g., Stinnesbeck et al. 1993), or that there is no evidence at all for impact ejecta at Beloc (e.g., Jehanno et al. 1992). However, some of the co-authors of the latter paper seem to have subsequently accepted that impact did play a role, as Rocchia et al. (1996b) proposed a model that involves "fragmentation of a single-incident bolide, colliding with the Earth at grazing incidence, with ricochet and rebound of fragments". However, such a process seems unnecessarily complicated and difficult to reconcile with cratering mechanics. In summary, all available evidence supports an impact, and not a volcanic, origin for the glasses from Beloc and other K/T boundary locations.

3.2.1.7
Impact-derived diamonds

Small, nanometer-sized diamonds were first reported from K/T boundary sediments in Alberta, Canada, and were thought to be of extraterrestrial (meteoritic) origin (Carlisle and Braman 1991). More recently, larger diamonds have also been found at other K/T boundary locations (Gilmour et al. 1992), including some in Mexico (Hough at al. 1997). These diamonds, which have a unique C and N isotopic signature (cf. Gilmour 1998), are clearly connected to the impact process (and not of extraterrestrial origin as had been speculated by Carlisle and Braman 1991). Even though there is a similarity between the K/T diamonds and those found at known impact craters (e.g., Koeberl et al. 1997a; Langenhorst et al. 1999), it is, however, not yet clear if the K/T boundary diamonds also formed by shock processes (de Carli and Jamieson 1961), similar to those that are found in proximal ejecta (e.g., El Goresy et al. 1999). Chicxulub diamonds are, so far, the only ones found in distal ejecta deposits.

3.2.1.8
Occurrence of spinel

Magnetite contained in impact-derived spherules at the Petriccio section in Italy was first reported by Montanari et al. (1983), and later identified as magnesioferrite spinel by Smit and Kyte (1984). Spinels at the K/T boundary can be used as event markers, with abundance peaks similar to those observed for the PGEs. These spinels occur in a variety for morphologies (e.g., Fig. 3.2.1.8.1) and compositions, resulting from variable substitution of Mg for Fe^{2+}. The majority of the spinels describe a compositional range spanning the solid solution series between magnetite (Fe_3O_4) and magnesioferrite ($MgFe_2O_4$). Spinel crystals from Pacific sites display variable substitution of Al^{3+} for Fe^{3+}, ranging from magnesioferrite to pure spinel ($MgAl_2O_4$) composition (e.g., Kyte and Smit 1986). All K/T boundary spinels are highly oxidized (high Fe^{3+} content), and have high Ni (and Co) contents (a few measurements also indicate high Ir), but low Cr and Ti abundances. These compositional characteristics set them apart from most known igneous or metamorphic terrestrial spinels.

Numerous studies have been performed on the spinels (e.g., Montanari et al. 1983; Montanari 1991; Smit and Kyte 1984; Kyte and Smit 1986; Bohor et al. 1986; Robin et al. 1992, 1993; Kyte and Bostwick 1995; Rocchia et al. 1996; Kyte et al. 1996, and references therein). Three main theories to explain their origin were discussed. They could form as: ablation debris from the impactor, crystallites in melt droplets, or vapor condensates in the hot fireball. The first theory, that the spinels form by a process analogous to the formation of ablation spherules that originate from small meteors in the Earth's atmosphere, was mainly proposed by Robin et al. (1992, 1993) and Gayraud et al. (1996), but it seems difficult to reconcile the global distribution of spinel with ablation from a large impactor that traverses through the atmosphere in a few seconds. This may be one reason why Robin et al. (1993) and Rocchia et al. (1996b) resorted to a very complicated process, in which fragmentation and secondary impacts are necessary; however, all the evidence collected so far favors one single large impact event.

A more plausible origin for spinels is the formation in the hot vapor cloud of the impact. Kyte and Bohor (1995) provide a detailed assessment of this theory. They found that Ni-rich magnesiowüstite, a high-temperature mineral, is preserved in the cores of some larger spinel crystals, where it is intergrown with magnesioferrite. Kyte and Bohor (1995) suggested that this mineral, which does not form in terrestrial igneous systems, can only evolve in the hot fireball of the impact. In their model, magnesiowüstite-magnesioferrite intergrowths occur in droplets that equilibrate with the vapor at temperatures exceeding 2300 °C, where the refractory MgO is enriched in the liquid relative to the more volatile SiO_2. Kyte et al. (1996) cited further evidence in support of this model.

Fig. 3.2.1.8.1. Skeletal magnesioferrite (etched) from the K/T boundary at DSDP Site 596 in the southwestern Pacific (54–55 cm layer). Scale bar on bottom left is 20 μm (SEM photograph courtesy B.F. Bohor).

3.2.2
Chicxulub Impact Structure: Relation to the K/T Event

Despite substantial indirect and direct evidence for an impact event at the K/T boundary (as discussed in the previous sections), some individuals still refused to accept the idea of an impact and resorted to some equally non-uniformitarian and extremely rare endogenic processes to try explain the observations made. However, even though some of the more fanciful hypotheses could explain one or two features of the K/T boundary, they singularly failed to provide a complete explanation of all the evidence listed above. Therefore, all these points together provide rather straightforward evidence for an impact event at the end of the Cretaceous Period. Another point worth considering is that the deposition of the K/T boundary layer, with all its shocked minerals, spinels, spherules, etc., had to occur in a geologically very short timespan, probably on the order of a year or less (e.g., Preisinger et al. 1986).

In the late 1980s, despite all the evidence that had been accumulated in support of a K/T boundary impact event, a major point of contention was the lack of a suitable large impact structure with an age of 65 million years. Crater diameters, as derived from estimates of total meteoritic material present at the K/T boundary worldwide (e.g., Alvarez et al. 1980), were calculated to be on the order of at least 100 to as much as 200 km. Several craters were proposed, but a link with the K/T boundary event remained tenuous. Numerous proposals (some of them rather eccentric) were made before 1990 (see, e.g., Kring 1993, for a discussion). However, at that time at least two impact structures of medium to large size on Earth were thought to have an age identical to that of the K/T boundary. The first one, the Kara crater in Russia (estimated at about 65–80 km diameter) was thought to have an age of 65 Ma (Nazarov et al. 1989); however, more precise age determinations showed that it is older than 65 Ma (Koeberl et al. 1990; Trieloff and Jessberger 1992), with a most probable age of 70.3 Ma (Trieloff et al. 1998). Moreover, Nazarov et al. (1991) suggested that the Kara crater may be larger than 100 km in diameter, rather than the 65 to 80 km thought before, which would make it one of the largest impact structures known today. The other major contender for the title "K/T boundary crater" was the Manson structure in Iowa, which is discussed in Chap. 3.4; it turned out to be 74 million years old, and would have been too small anyway (37 km diameter) to explain the global effects.

Fig. 3.2.2.1. Geographical location of the Chicxulub impact structure on the NE part of the Yucatan (Mexico) peninsula.

As mentioned above, studies of K/T boundary distal ejecta around the world were used to constrain the location of the possible source crater. For example, from the analysis of the abundance and size variation of shocked quartz grains, it was concluded that the source crater had to be somewhere on or near the North American continent (see, e.g., Kring 1993, for a discussion). However, it was not until 1991 that Hildebrand et al. (1991) proposed that a large buried structure in NE Yucatan (Mexico; Fig. 3.2.2.1) might be the elusive K/T boundary crater. This structure had been recognized earlier by Penfield and Camargo (1981), who suggested it to be an impact structure, but unfortunately their abstract was presented at a meeting of economic geologists, who apparently were not interested in impacts and extinctions (even though impact structures have been shown to contain economically important deposits, ranging from ores to hydrocarbons; for a review, see Grieve and Masaitis 1994).

Following its confirmation as an impact structure, Chicxulub was studied extensively, mainly by geophysical methods. This was partly in response to a debate about the exact size of the structure (see below). In the following paragraphs, we will not discuss the whole literature that exists by now on Chicxulub, but just review the evidence that links Chicxulub to the K/T boundary event. This means that the geophysical data and the debate dealing with the diameter will only be mentioned briefly.

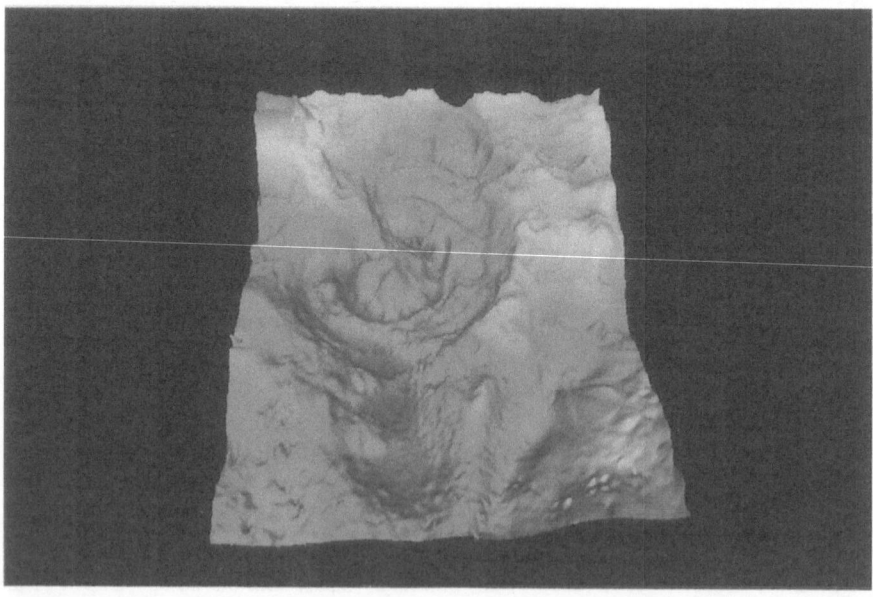

Fig. 3.2.2.2. Computer-generated image of the circular gravity anomalies at Chicxulub, showing the inner peak ring and several outer rings. The most conspicuous structure is about 180 km in diameter; the outer fourth ring is difficult to see (image courtesy V.L. Sharpton).

Penfield and Camargo (1981) first recognized a circular gravity anomaly, which is similar to that of known impact structures (cf. Pilkington and Grieve 1992). The structure is marked by an extensive circular -30 mgal negative Bouguer gravity anomaly of about 180 to 200 km diameter, with a central 20 mgal high (Hildebrand et al. 1991). Sharpton et al. (1993) compiled a new gravity map, resulting in a by now famous computer-generated image (Fig. 3.2.2.2), from which they deduced the existence of three major rings and evidence of a fourth (somewhat irregular) outer ring structure with about 140 km radius. These data, and a general spacing rule for rings of multi-ring impact structures, prompted Sharpton et al. (1993) to interpret the feature as marking the outer limit of the basin rim crest, yielding a diameter of about 280 km for the Chicxulub structure. In contrast, Pilkington et al. (1994) maintain that their modeling does not show any evidence of a fourth outer ring and, thus, that Chicxulub has a diameter of only 180 km. Some surface expressions that may be related to the Chicxulub structure (which does not have any true surface exposures), e.g., the distribution of sinkholes (Perry et al. 1995), seem to support a diameter larger than 180 km, probably on the order of 240 km (Pope et al. 1993, 1996). A part of the ring of cenotes (sinkholes) in Yucatan is shown in Fig. 3.2.2.3.

A recent study of the horizontal gradient of the Bouguer anomaly over the structure by Hildebrand et al. (1995) was interpreted to be consistent with a 180 km crater diameter. On the other hand, another interpretation of the gravity data, in combination with drilling information, was made by Sharpton et al. (1996). Total magnetic field data indicate an anomaly about 180 to 210 km in diameter (e.g., Pilkington et al. 1994). The geophysical data were used by Schultz and D'Hondt (1996) to infer that Chicxulub formed by a bolide that impacted from a southeast direction at an angle of 20° to 30° from the horizontal. According to Schultz and D'Hondt (1996), this interpretation is also supported by the extinction patterns in North America. However, Pierazzo and Melosh (1999), in model calculation of Chicxulub as an oblique impact, were not able to reproduce the results of Schultz and D'Hondt (1996). Kyte and Bostwick (1995), on the other hand, interpreted the distribution of K/T spinels in the Pacific basin as indicating an oblique impact from the east. It seems that it is difficult to arrive at a conclusion regarding impact angle and direction.

In view of the dispute over the diameter of Chicxulub, other data were necessary to decide between the opposing views. Early seismic data obtained during oil prospecting were only available in part (Camargo-Zanoguera and Suárez-Reynoso 1995). Thus, a new detailed offshore seismic study, performed in the mid to late 1990s by an international consortium, yielded important new data (Morgan et al. 1997; Morgan and Warner 1999). These data show that Chicxulub is a multi-ring impact basin – the only one on Earth that is well preserved. It is generally assumed that Vredefort in South Africa (e.g., Reimold and Gibson 1996) and Sudbury in Canada (e.g., Deutsch et al. 1995), which were both on the order of 250 km in original diameter before erosion and deformation, also represent multi-ring basins, but they are not well enough preserved for detailed studies. In such a case, the "diameter" of the structure is more or less a matter of definition (i.e., which is the ring is assumed to give the "diameter"?). Morgan and Warner (1999) placed the crater rim at a diameter of about 145 km, which is surrounded by an outer ring at 180 km diameter and an exterior ring at about 250 km diameter. The latter one agrees with the data of Pope et al. (1996). At the center of the structure is a peak ring with a diameter of about 80 km.

Whereas the size of the structure clearly has implications regarding the magnitude of the event and the amount of material ejected, petrographic and geochemical considerations are important to establish the link between the Chicxulub structure and the K/T boundary deposits. A major problem is that the structure is currently covered by several hundred meters of post-impact Tertiary sediments, which requires obtaining drill cores. Samples from early drill cores obtained from Petróleos Mexicanos in the 1950s and 1960s showed that two of the boreholes at the center of the structure penetrated a dense crystalline rock, which was initially thought to represent andesitic volcanic material, but is now recognized as impact melt rock (Hildebrand et al. 1991; Sharpton et al. 1992; Koeberl 1993b; Schuraytz et al. 1994). Samples from the Y6 borehole, located about 60 km from the center of the structure, and from the centrally located C1 borehole were studied in most detail and showed the presence of a well-sorted, graded polymict breccia sequence and coherent melt rocks with glass remnants (e.g., Sharpton et al. 1992). The most important findings in the breccia and melt rock samples are clear, abundant evidence for the presence of shock metamorphism, including the following: a) abundant planar deformation features (PDFs) in quartz and feldspar crystals from crystalline basement clasts within the breccias and melt rocks (Sharpton et al. 1992); up to five sets of PDFs were found and measured by universal stage methods, showing the presence of the impact-characteristic orientations, with $\{10\bar{1}3\}$, or ω, and $\{10\bar{1}2\}$, or π, orientations, being the most common; b) shock mosaicism in quartz and feldspar; c) diaplectic glass occurs as partly digested inclusions within glasses and melts; and d) impact melts, glassy and fine-grained recrystallization products. These findings provide unequivocal evidence for an impact origin of the Chicxulub structure.

The presence of a meteoritic component in some of the melt rocks from the C1 and Y6 drill cores was indicated by elevated Ir contents of up to 13.5 ppb (Sharpton et al. 1992; Koeberl et al. 1994c; Schuraytz et al. 1994). However, the meteoritic component seems to be heterogeneously distributed; for example, Schuraytz et al. (1994) found pyrite crystals with high and variable contents of Ni and Co, and some opaque mineral grains that contained high Ir, which they interpreted as evidence for extensive post-impact hydrothermal activity. Koeberl et al. (1994c) used the Os isotopic system in a study of melt rocks from the C1 core and found elevated Os contents and low $^{187}Os/^{188}Os$ ratios, indicating the presence of an extraterrestrial component. Schuraytz et al. (1996) found some micrometer-sized particles in Chicxulub impact melt rock samples that consist of almost pure Ir, which they interpreted as possible remnants of a carbonaceous chondrite-type impactor; this is at least in qualitative agreement with the Cr isotope data of Shukolyukov and Lugmair (1998).

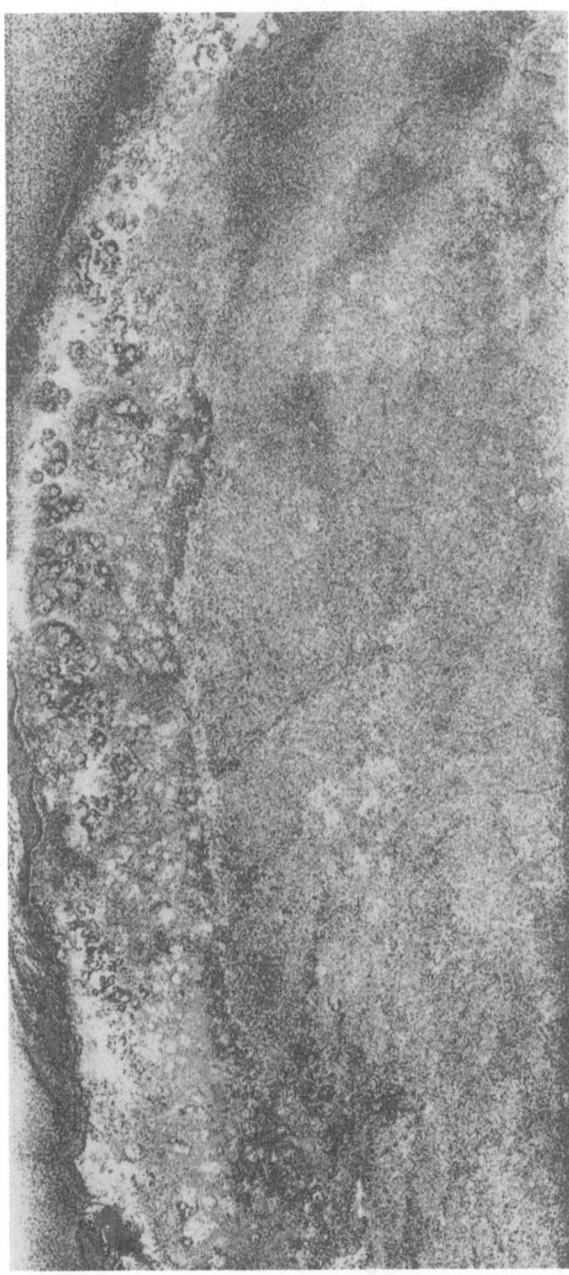

Fig. 3.2.2.3. Radar image of the southwestern part of the ring of cenotes in Yucatan that seem to mark the outer limit (exterior ring) of the Chicxulub impact structure. The image is centered at 20 °N and 90 °W; southwest is up. (SIR-C/X-Space Shuttle image, NASA).

Isotopic measurements were of crucial importance to establish a link between the Chicxulub crater and the impact deposits found at the K/T boundary. Blum et al. (1993) measured the Rb–Sr, Sm–Nd, and oxygen isotopic composition of melt rocks from Chicxulub and found that, in a diagram of the oxygen versus the strontium isotopic composition (Fig. 3.2.2.4), the data for the Chicxulub melt rocks fall on a mixing hyperbola between the various types of impact glasses from the Haitian K/T boundary and a carbonate endmember (representing the carbonate platform rocks at Yucatan). The data indicate a common source for the Haitian impact glasses (i.e., K/T boundary impact ejecta) and the Chicxulub crater rocks.

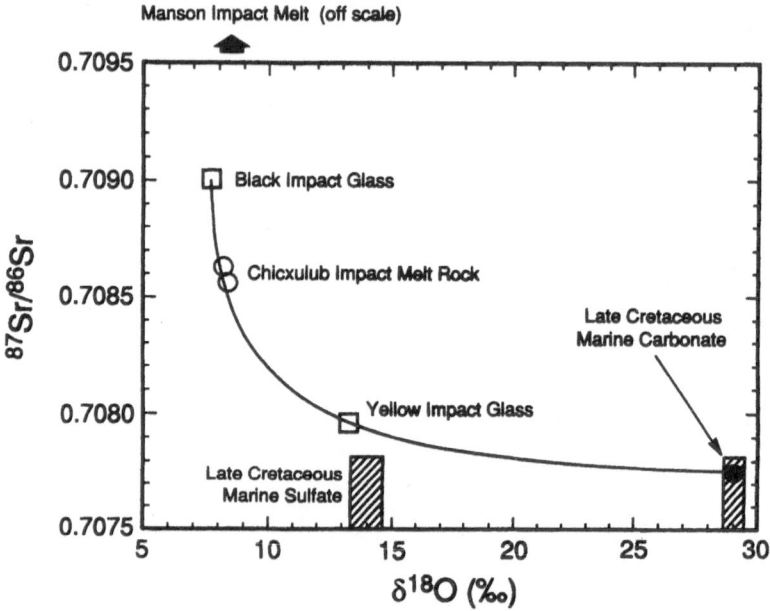

Fig. 3.2.2.4. Plot of $^{87}Sr/^{86}Sr$ (recalculated to 65 Ma) versus $\delta^{18}O$ for Chicxulub impact melt rock (open circles) and impact glass from Haiti (open squares). Also plotted are the fields for average carbonate (hatched area in lower right; average is shown by solid circle) and Late Cretaceous marine sulfate. The Chicxulub melt rock plots exactly on a mixing hyperbola defined by the two types of Haitian impact glass and the carbonate endmember, indicating a common source. Impact melt rocks from the Manson crater plot off scale and are unrelated (after Blum et al. 1993, and Koeberl 1996).

A similar result is obtained from plotting the Sr versus the Nd isotopic composition for Chicxulub melt rock samples and Haitian impact glasses (Fig. 3.2.2.5). The K/T boundary impact glasses and the Chicxulub melt rocks have similar values at about +58 and -3 for $\varepsilon_{Sr}^{65\,Ma}$ and $\varepsilon_{Nd}^{65\,Ma}$, respectively. The depleted mantle Nd model ages of the Chicxulub rocks fall in a tight range of about 1040–1080 Ma, suggesting that the silicate endmembers of these breccias and melt rocks had a source with a middle Proterozoic average crustal residence age, but younger sedimentation and crystallization ages (Blum et al. 1993). The data are also inconsistent with derivation from the mantle, because values for rocks derived from the upper mantle generally fall in a narrow range of ε_{Nd} of +4 to +10‰ and ε_{Sr} of -10 to -30‰.

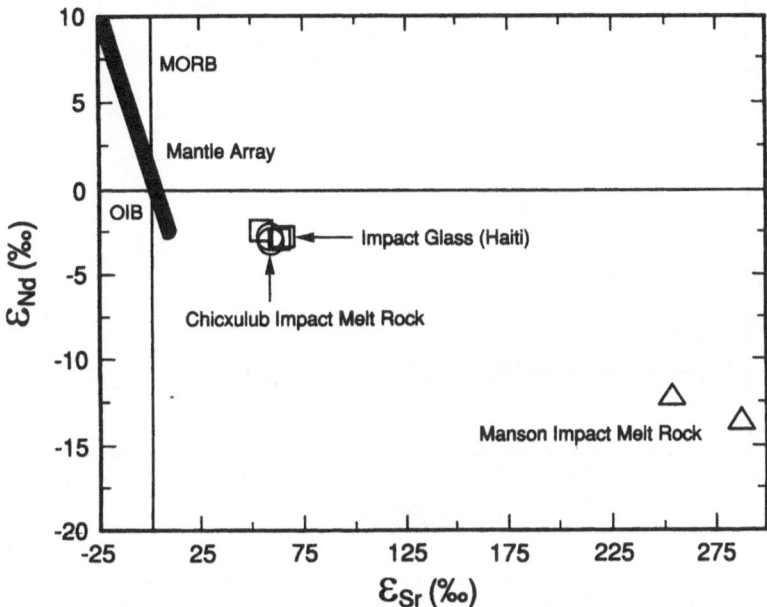

Fig. 3.2.2.5. Plot of $\varepsilon_{Sr}^{65\,Ma}$ versus $\varepsilon_{Nd}^{65\,Ma}$ for Chicxulub melt rock samples and Haitian impact glasses. The values for Chicxulub melt rock and impact glasses are similar at about +58 and -3 for $\varepsilon_{Sr}^{65\,Ma}$ and $\varepsilon_{Nd}^{65\,Ma}$, respectively, and significantly different from mantle compositions and Manson crater data (after Blum et al. 1993, and Koeberl 1996).

Shocked zircons that were found at distal K/T boundary sites (see Sect. 3.2.1.5) helped to establish another link between the K/T boundary and the Chicxulub structure. Krogh et al. (1993a, b) and Kamo and Krogh (1995) reported on single-zircon dating using the U–Pb isotopic composition of zircons from various K/T boundary sites and from the Chicxulub crater. Fig. 3.2.2.6 shows a concordia diagram for shocked zircons extracted from an impact melt breccia from Chicxulub and from K/T boundary deposits in Colorado and Haiti. The data show a major intercept at about 545 Ma, indicating a Pan-African basement age. Zircons from a Canadian K–T boundary site gave more or less identical results (Kamo and Krogh 1995). The more highly shocked zircons give a partially reset U–Pb age, with a lower intercept of 57 Ma. This is remarkably close to the impact age at 65 Ma (in fact, if only zircons from one location are used, the data show less scatter and the intercept is closer to 65 Ma). Thus, the U–Pb isotopic composition of zircons from the distal ejecta (from K/T boundary clay) agrees with that of zircons extracted from Chicxulub impact melt breccias.

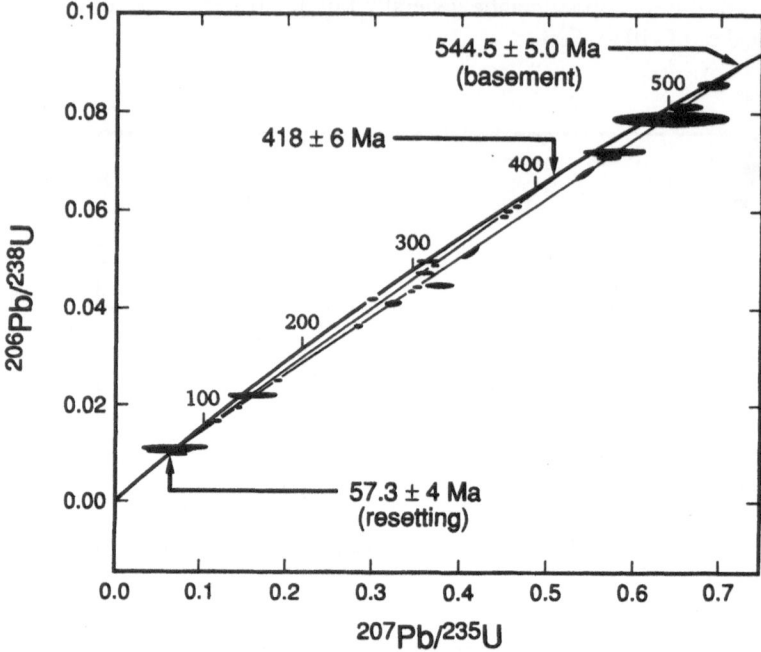

Fig. 3.2.2.6. Concordia diagram for U–Pb isotopic composition of shocked single zircons from K/T boundary sediments in Colorado and Haiti, and from Chicxulub impact breccia. The results for the three sites are indistinguishable. Most zircons define an upper intercept of about 545 Ma, with a lower intercept (defined by the most severely shocked zircons) of about 57 Ma, close to the impact age. A few zircons from the Chicxulub and Haiti locations have an upper intercept at about 418 Ma. The results demonstrate a common source for the K–T boundary ejecta and the Chicxulub impact melt breccias (diagram after Krogh et al. 1993, and Koeberl 1996).

The studies mentioned here have not only established Chicxulub as one of the largest impact structures known on Earth, but have also shown that there are several lines of evidence that link the distal K/T boundary ejecta with rocks from Chicxulub. Thus, an age determination for Chicxulub would confirm this link. Two detailed studies of the radiogenic age of the impact melt rock and impact glasses from the Chicxulub structure have been published early on. Sharpton et al. (1992) used hand-picked fragments of fine-grained impact melt rock (from the Y6 and C1 cores) for ^{40}Ar–^{39}Ar step-heating age determinations. They found evidence for low-temperature alteration, but high-temperature increments in the age spectra of a sample from C1 give an age of 65.2 ± 0.1 Ma, which they interpreted as the crystallization age. Swisher et al. (1992) were able to separate individual small glass fragments (0.4–0.5 mm in size) from the C1 impact melt rocks and were also studied with the ^{40}Ar–^{39}Ar step-heating technique. As Swisher et al. (1992) seem to have succeeded in obtaining unaltered samples, they obtained better plateau ages than Sharpton et al. (1992). Two typical examples of the Chicxulub glass age spectra are shown in Fig. 3.2.2.7, with plateau ages of 64.94 ± 0.11 and 65.00 ± 0.08 Ma. Swisher et al. (1992) found an average age of 64.98 ± 0.05 Ma for their samples, which is indistinguishable from the age of 65.07 ± 0.10 Ma that they obtained for impact glasses from the Beloc (Haiti) and Arroyo el Mimbral (Mexico) K/T boundary layers (cf. also Izett 1991).

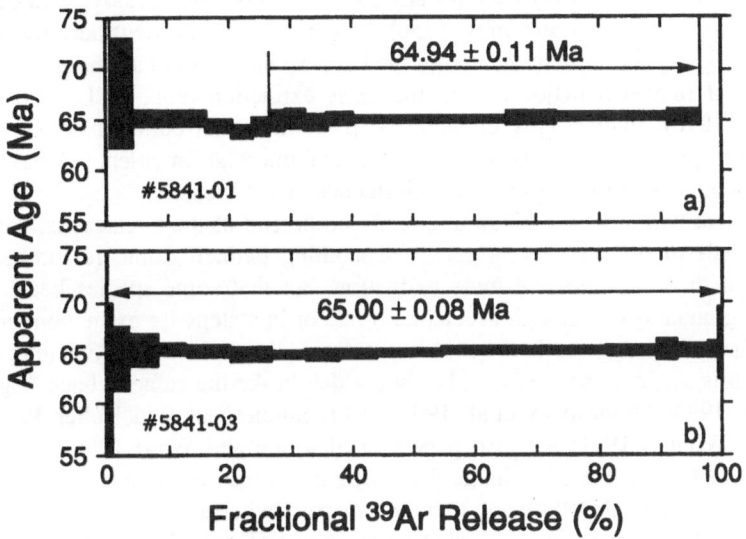

Fig. 3.2.2.7. ^{40}Ar–^{39}Ar age spectra, obtained by stepwise heating, of two impact glass samples that were extracted from the Chicxulub C1-N9 impact melt breccia. The patterns show good plateaus and the results are indistinguishable from those for the impact glasses from Haiti, and demonstrate a K/T boundary age for the Chicxulub impact crater (diagram after Swisher et al. 1992, and Koeberl 1996).

All the geochemical and geochronological data summarized in the previous paragraphs clearly confirm that Chicxulub is indeed the long-sought K/T impact structure. It is of the right size, location, and age, and rocks from the structure show geochemical similarities with distal ejecta from K/T boundary layers.

3.2.3
Chicxulub and the K/T Mass Extinction

We should at least briefly go back to the beginning of the whole story of the impact ejecta layer at the K/T boundary and the search for, and discovery of, the

Chicxulub impact structure, namely, that there was a major geological and bio-logical event 65 Ma that prompted geologists and paleontologists to mark the end of the Cretaceous Period in the first place. This major event is marked by one of the largest mass extinctions of species during the last several hundred million years (e.g., Raup and Sepkoski 1982; Sepkoski 1990, 1992, 1996; Hallam and Wignall 1997). Geologists have, for a long time, been searching for a cause (or causes) to explain the end-Cretaceous and other mass extinctions. Large-scale volcanic eruptions, sea level fluctuations, and climate changes have been popular explanations. Extraterrestrial explanations (ranging from supernova explosions to cometary or asteroidal impact) were usually assumed to be unnecessary, with just a few exceptions (e.g., de Laubenfels 1956). Now that we know that there was at least one major impact event at the end of the Cretaceous, it would seem logical to assume that it played a major role in the mass extinction (e.g., McLaren and Goodfellow 1990). Interestingly enough, the paleontological community is still split over this question, with many workers maintaining that, in essence, "maybe there was an impact, but it had little or no influence on the biosphere".

Clearly there was a mass extinction directly at the end of the Cretaceous, with more than half of all then living species becoming extinct. Some researchers maintain that there was no real mass extinction, but that some species became extinct in a gradual (even though accelerated) rate or in a stepwise extinction pat-tern, and that many species had become extinct well before the actual impact event, which is accepted as kind of a last straw that broke the camel's back (e.g., Keller et al. 1993; Stinnesbeck et al. 1993, 1997; Stinnesbeck and Keller 1996; Hallam and Wignall 1997; see also papers in Sharpton and Ward 1990, and in Ryder et al. 1996). On the other hand, the geochemical record indicates a sudden event (e.g., Alvarez et al. 1984), and there are workers who find no evidence for a stepwise extinction pattern in more or less the same K/T boundary sections, and who conclude that all extinctions were instantaneous (e.g., Pospichal 1994, 1996; Huber 1996; Smit 1982, 1999; Smit et al. 1988, 1996; see also other papers in Sharpton and Ward 1990, and Ryder et al. 1996).

To resolve the question regarding stepwise or sudden extinctions, a blind test was conducted using the El Kef K/T boundary layer (Ginsburg 1997a; Smit et al. 1997). However, the results of the study (Canudo 1997; Masters 1997; Olsson 1997; Orue-etxebarria 1997; Keller 1997; Smit and Nederbragt 1997) led to no agreement between the different investigators, as they all seem to have obtained slightly different results (regarding the presence of certain Foraminifera in each sample), and, therefore, more or less maintained their original positions. Thus, the El Kef study was not considered a full success (Ginsburg 1997b). There are sev-eral reasons for this; as Kouwenhoven (1997) pointed out, the test was not really blind, as, for example, the participants received unprocessed samples and, being familiar with K/T boundary samples, they would have known right away which samples are Maastrichtian and which are Danian. It seems that the resolution of the question regarding gradual versus sudden extinctions has to be addressed in a future "totally blind" test, such as a double blind test, similar to the ones that are commonly used in pharmaceutical studies.

The extinction of the dinosaurs at the end of the Cretaceous, which is probably the group of animals that caused the most publicity for K/T boundary studies, has been a similar point of contention. Many workers have maintained that they died out gradually over some time (which certainly must have happened), but also that there was an accelerated pace of extinction just before the impact event. However, the scarcity and large size of dinosaur bones in the geological record make it very difficult to obtain a high enough time resolution. Indeed, when very time-consuming and labor-intensive studies at dinosaur fossil-rich locations are done, any perceived gaps in the fossil record disappear and no decline in the number of dinosaur species can be found (e.g., Sheehan et al. 1999).

Early on, large-scale volcanism was cited as a probable main cause, not only for the end-Cretaceous extinctions in general, but also for the supposedly pro-longed period of extinctions. The eruption of the flood basalts at the Deccan Traps in India was found to be coeval with the K/T boundary (e.g., Courtillot et al. 1996; Courtillot 1999). However, the flood basalts were emplaced over a period of less or equal to one million years (Courtillot et al. 1996), which is several orders of magnitude longer than the time of deposition for the K/T boundary layer. Also, studies of the Indian K/T boundary sections showed that the Deccan volcanism was neither triggered by the K/T impact (Boslough et al. 1996 discussed a possible mechanism for an impact to trigger flood basalt eruptions), nor does there seem to be any other causal relation (Bhandari et al. 1995, 1996). Also, Galer et al. (1989), in a Pb isotope study of K/T boundary samples, found that the Pb isotopic compo-sition of the boundary layer does not isotopically resemble known Deccan Trap lavas. Thus, the Deccan volcanism may have contributed to the environmental conditions at the end of the Cretaceous, but it did not contribute to the sudden environmental changes that coincide with the K/T boundary.

However, here is not the place to provide an exhaustive discussion of the pros and cons of stepwise and sudden extinctions. Rather, we want to mention some of the environmental effects caused by the Chicxulub impact event. A major impact event at the specific geographic location of the Chicxulub crater, with abundant carbonate and anhydrite rocks of the Yucatan platform, must have had important consequences for the biosphere. During the impact event, thousands of cubic kilometers of rock and dust were ejected from the crater, and large amounts of water, CO_2, and SO_2 must have been released into the atmosphere. Preliminary estimates (assuming about 300–2000 km^3 of vaporized sediments) yield sulfur masses of $3.5 \cdot 10^{16}$ to $7.0 \cdot 10^{17}$ g, and about 10^{19} g of CO_2 that were released almost instantaneously into the atmosphere (Chen et al. 1994; Pope et al. 1994). Any detailed discussion or speculation regarding the long-term effects of such enor-mous amounts of gas in the atmosphere are beyond the scope of this volume (but see, e.g., Brett 1992; Chen et al. 1994; Pope et al. 1994; Lyons and Ahrens 1996; Baines et al. 1996).

Fig. 3.2.3.1. Quarry at Albion Island in northern Belize. The main layer of Chicxulub ejecta is visible as a distinct horizon in the middle of the quarry face. It is possible that the ejecta scoured the surface as they were moving away from the impact site in the form of a base surge (see Sect. 2.1) (photo courtesy J. Smit).

The exact amount of gases released varies between different model calculations (see, e.g., Ivanov et al. 1996), and some calculations indicate that the release of sulfur into the atmosphere had more severe effects than the increase in CO_2 (Kring et al. 1996; Pierazzo et al. 1998). There is abundant evidence for a short, but intense, period of acid rain directly following the K/T impact (e.g., Crutzen 1987; Prinn and Fegley 1987), which influenced the extinction rate (D'Hondt et al. 1994, 1996). In addition, the marine sediments that comprised the Chicxulub target rocks also contained halogens (chlorine, bromine), which, when released into the upper atmosphere, destroy the ozone layer, leading to a substantial increase in the flux of mutagenic ultraviolet radiation at the Earth's surface. The atmospheric effects of large impacts have been reviewed by Zahnle (1990), Covey et al. (1994), and Toon et al. (1997); see also Emanuel et al. (1995) for the discussion of the possible influence of hypercanes (mega-hurricanes) in the end-Cretaceous environmental catastrophe. It is obvious that there must have been short-term, as well as long-term climatic changes that most likely led to the mass extinction that marks the K/T boundary.

Fig. 3.2.3.2. K/T boundary section at El Mimbral, Mexico, about 700 km from the outer limit of the Chicxulub structure, showing a complex succession of layers. The person on the left side (J. Smit) is standing on Cretaceous limestones; his height marks the thickness of the spherule-rich layer. This is overlain by a solid layer of a possible tsunami deposit. On the bottom of this deposit is occasional plant debris, whereas ripple structures in sandstone cap this unit, which is overlain by less consolidated Tertiary rocks.

No matter what the extent of the global effects of the impact were, the immediate destruction within several thousand kilometers distance from Chicxulub must have been catastrophic. For example, an unusual deposit of breccia-like material with a thickness of several tens of meters was found at a quarry at Albion Island (not really an island; this is just the location name), in northern Belize, which was interpreted as the most proximal ejecta deposit of the Chicxulub crater found so far (e.g., Ocampo et al. 1996). Many K/T boundary sites in Mexico, Cuba, and even in the U.S. show thick and complexly layered deposits that have been interpreted as impact-induced tsunami deposits (e.g., Bourgeois et al. 1988; Iturralde-Vinent 1992; Hills et al. 1994; Smit et al. 1996). For example, the Peñalver Formation and its equivalents in northwestern Cuba, which contain Late Maastrichtian fossils, are characterized by a <180 m thick, normal-graded calcareous clastic deposit with a basal conglomerate, and have been suspected to represent a single event deposit, possibly as a result of a K/T impact-related tsunami. Takayama et al. (1999) reported on first evidence of shocked quartz in some of samples from these Cuban deposits. K/T boundary sections at most of the Mexican sites are characterized by a complex set of clastic sandstones, with coarse ejecta at the bottom and fine-grained ejecta, as well as a smeared-out Ir anomaly on top (Smit 1999).

Fig. 3.2.3.3. Plant debris at the base of the possible tsunami deposits at the El Mimbral K/T boundary location in Mexico.

One of the best known locations in Mexico is El Mimbral, which has been described in detail by Smit et al. (1992a, b, 1996). This section shows a succession from Cretaceous limestone into a spherule layer, which is overlain by a tsunami deposit and a zone with Ir enrichment on top, and capped by Tertiary units (Fig. 3.2.3.2). At the bottom of the tsunami deposit, plant debris, which was obviously ripped up and destroyed, has locally accumulated (Fig. 3.2.3.3). Deposits that are similar to the Mexican K/T sites were also found in Haiti (Maurrasse and Sen 1991), Guatemala (Stinnesbeck et al. 1997), and in Brazil (Albertao and Martins 1996) at the Poty K/T boundary section (cf. Albertao et al. 1994; Koutsoukos 1998). The top of the deposits that are interpreted to result from tsunamis is marked by current ripple structures that allow the measurement of paleocurrents (Smit et al. 1996). A particularly well developed exposure is shown in Fig. 3.2.3.4. In contrast, Bohor (1996) interpreted these deposits as an (impact-induced) Bouma sequence, whereas Stinnesbeck et al. (1993) and Stinnesbeck and Keller (1996) suggested that these horizons have been deposited over a longer time interval and represent sea-level changes. Smit (1999) summarized arguments why the latter two explanations are unlikely.

Fig. 3.2.3.4. Sedimentary structures in the K/T boundary deposit at La Lajilla, Mexico, showing abundant ripple marks that indicate about 200 individual current directions and are thought to have formed as a result of the interaction of large tsunami waves (e.g., Smit et al. 1996) (photo courtesy J. Smit).

Recently, K/T boundary sediments were studied at Blake Nose, off Florida in the Atlantic Ocean (Norris et al. 1999). Both seismic studies and analyses of cores from ODP Site 1049 indicate folding and slumping of sediments directly underneath the K/T boundary ejecta. Norris and Firth (1999) suggested that this was the result of a mass failure of the North Atlantic margin that was triggered by the K/T impact event, probably by seismic energy. The possibility of massive disruptions of the sea floor led Max et al. (1999) to speculate that this could result in the release of large amounts of methane gas from gas hydrates. The methane blowout may have added fuel for a global firestorm, which is documented in the soot layer at the K/T boundary. A similar methane release mechanism from ocean floor gas hydrates was recently proposed by Norris and Röhl (1999) and Katz et al. (1999) to account for climate changes at the end of the Paleocene, at about 55.5 Ma. At this time the Earth's climate and oceans warmed up significantly within a short time (10 to 20 k.y.), with deep-ocean and high- latitude surface water temperatures rising by 4 to 8 °C (e.g., Kennett and Stott 1991). In addition, the carbon isotopic record shows a significant excursion towards lighter isotopic compositions, similar to what has been observed at the K/T boundary. Confirmation of a massive methane contribution to the global fires at the end of the Cretaceous will require a detailed study of the carbon isotopic evidence.

3.3
Late Eocene Impact Layers

Late Eocene marine sediments around the world contain evidence for at least two closely spaced impactoclastic layers. Initially, it was thought that there is only one layer, namely the one known from the eastern U.S. coast, the Caribbean, and the Gulf of Mexico (e.g., Glass et al. 1973, 1979, 1984), which is correlated with the North American tektite strewn field (see above). This layer contains microtektites (i.e., glassy – not recrystallized – spherules), shocked minerals, and high-pressure phases (e.g., coesite) (e.g., Glass 1989), but no marked siderophile element anomaly. The presence of crystalline spherules composed mostly of clinopyroxene (cpx) was detected in the same deep sea sediments (e.g., Glass et al. 1985) and initially it was considered that these spherules also belong to the North American tektite strewn field (Glass and Burns 1987). The cpx spherules were found not only in the Caribbean and the Gulf of Mexico, but also in the Pacific Ocean (e.g., Glass et al. 1985; D'Hondt et al. 1987).

Glass and Burns (1988) proposed the name "microkrystites" for such impact-produced, crystallite-bearing spherules. More detailed work showed that the microtektite layer and the microkrystite (cpx) layer (both in middle to lower magnetic chron C15) are in fact separated from each other by about 25 cm, with the cpx layer being the lower (i.e., older) one (e.g., Keller et al. 1983, 1987; see also Obradovich et al. 1989). They also suggested that there is yet another, much older layer in middle to lower magnetic chron C16 at DSDP Sites 216 and 292 near Indochina. The separation between the two layers amounts to about 10 to 20 k.y. In contrast, Hazel (1989) argued for at least six different layers, i.e., that the impactoclastic layers at different sites represent different layers. This was disputed by Glass (1990), who concluded that there are only two layers. Miller et al. (1991) suggested that there are four layers and that the layer with microtektites and tektite fragments at DSDP Site 612 (e.g., Thein 1987; Koeberl and Glass 1988; Glass 1989) is much older than the North American tektite strewn field (and is correlated with the "old" layer at DSDP 216 and 292 of Keller et al. 1983, 1987). Such a proposal is not in agreement with the chemical and isotopic similarities between the DSDP 612 tektites and other North American tektites (e.g., Koeberl and Glass 1988; Glass 1989; Stecher et al. 1989). In a detailed assessment of the evidence, Wei (1995) came to the conclusion that there are only two layers (the older cpx layer and the about 10 to 20 k.y. younger North American microtektite layer), that those at DSDPs 216, 292, and 612 cannot be distinguished in age, and that all are in chron C16 rather than C15. The conclusion that there are only two distinct layers was recently supported by isotopic analyses reported by Whitehead et al. (2000) (see below). The assignment of absolute ages to the layers is difficult, as it depends on the time scale used (see Sect. 5.8 for more discussion).

The microkrystite layer has by now been found at numerous other locations, indicating that it has a more or less global distribution, and it seems to be associated at several locations with enhanced Ir abundances (see, e.g., Alvarez et al. 1982; Montanari et al. 1993). Impact ejecta that have been interpreted to be part of the cpx layer have also been found in rocks from the marine Umbria–Marche Sequence in Italy at the Global Stratotype Section and Point for the Eocene–Oligocene boundary at Massignano near Ancona (see Sect. 5.8.3). At this location the impactoclastic layer (dated at 35.5 ± 0.3 Ma) is present as a 5 cm thick marly horizon containing altered microkrystites and Ni-rich spinel (Pierrard et al. 1998), shocked quartz with multiple sets of PDFs (Clymer et al. 1996; Langenhorst 1996), and enhanced Ir levels (e.g., Montanari et al. 1993; Pierrard et al. 1998; Huber et al. 1999) (see Sect. 5.8.4 for details). The recent findings by Glass and Koeberl (1999a,b) and Vonhof and Smit (1999), which were mentioned in Sect. 2.4.5.1, also indicated that the North American microtektite layer seems to extend to the southern Ocean near Antarctica and into the Indian Ocean and may, therefore, also have a near-global extent. Liu and Glass (2000) were able to confirm that both the microtektite layer and the microkrystite layer are present at Site 689B near Antarctica and deduced a separation of the two events by about 8 to 12 k.y.

It is also interesting to note that Farley et al. (1998), in a study of the content of ^{3}He in the marine sediments at Massignano, found a broad peak of ^{3}He enrichment over a timespan of about 2 million years that coincides with the two Late Eocene impactoclastic layers. This isotope is a proxy for the influx of extraterrestrial dust, and is interpreted as indicating that during the Late Eocene there was a time of enhanced comet activity in the inner solar system, probably resulting in a higher impact rate than usual (see Chap. 5.8.4).

So far, two large impact structures of Late Eocene age have been confirmed, namely the Chesapeake Bay structure off the eastern coast of the United States, and Popigai in Siberia, Russia (see below). Wanapitei (7.5 km diameter) and Mistastin (28 km diameter), both in Canada, are slightly older than Chesapeake Bay and Popigai (Table 1.9.1). Another structure of similar age may be the underwater structure at Fohn-1, North Bonaparte Basin, Timor Sea, which has been described by Gorter and Glikson (2000) as a possible impact structure of Late Eocene to pre-Miocene age. However, both the identification of this structure as an impact crater, and its age, have yet to be confirmed. The evidence presented by Gorter and Glikson (2000) is insufficient, as no indication of shock metamorphism exists and the geochemical data that allegedly indicate the presence of a meteoritic component are of low analytical precision and too close to the detection limit to allow such a conclusion. This leaves Popigai and Chesapeake Bay as the only two significant impact structures that coincide in age with the Late Eocene microtektite and microkrystite layers.

3.3.1
Chesapeake Bay

As discussed in Sect. 2.4.5.1, the 35 Ma Chesapeake Bay impact structure has now been identified with a certain degree of confidence as the source crater of the North American tektite strewn field (Koeberl et al. 1996c; Poag 1997b). The structure, which is not exposed at the surface, has a diameter of about 85 to 90 km, based on geophysical studies (Poag 1997b). An impact event that created a crater of this size would be capable of globally distributing its distal ejecta. The story of the discovery of this crater, the largest one known in the United States, is told by Poag (1999). The outline of the structure was determined on the basis of multi-channel seismic-reflection profiles transecting the bay and nearly 60 bore holes drilled inside and outside the crater rim. The seismic profiles define the outer rim of the structure, which is 90 km in diameter (Poag 1997b). The gravity signature of the Chesapeake Bay structure (Koeberl et al. 1996c; Poag 1997b) is in agreement with observations from other complex impact structures. A flat-floored, 300 to 1200 m deep, annular trough separates the outer rim from an irregular, low-relief peak ring, with a 30 km wide inner basin in the center. Poag and Foster (2000) presented the first seismic evidence for 10 km diameter central peak. The pre-impact coastal plain rocks consist of a seaward-thickening wedge (300–1200 m thick) of mainly lower Cretaceous to upper Eocene age, poorly lithified, sedimentary rocks, which is underlain by a crystalline basement complex comprising granitic and metasedimentary rocks of Proterozoic to Paleozoic age.

Four drill cores have penetrated into the Exmore breccia: the Exmore, Windmill Point, Kiptopeke, and Newport News cores (Poag et al. 1992, 1994). The Windmill Point and Newport News cores were drilled just outside the crater rim, whereas the Exmore core was drilled near the outer rim of the crater, and the Kiptopeke core is located at the inner ring. No core is available from the center of the structure, however, or reaches basement (although a new drilling project is being planned to begin in 2000; C.W. Poag, personal communication, 1999). The Exmore samples contain a variety of lithologies. The section consists predominantly of poorly consolidated sedimentary units. In thin section, siltstones or sandstones (sometimes with glauconite) alternate with sedimentary and crystalline fragments derived from the basement. Most particulate samples consist of a mixture of sedimentary and crystalline rock types, as well as mineral fragments derived from granitoids. Intragranular deformation of felsic minerals (mainly quartz and feldspar) is often limited to non-diagnostic microdeformation features, such as poorly developed irregular fracturing, undulatory extinction, and occasional kink-banding of mica, and locally strained calcite (Koeberl et al. 1996c).

In a study intended to confirm the impact origin of the Chesapeake Bay structure, Koeberl et al. (1996c) reported the finding of evidence for shock metamorphism in 14 Exmore core breccia samples from 372.0 to 415.6 m depth. Some samples exhibit characteristic microfracture patterns, which indicate shock pressures of 5 to 10 GPa, and shock mosaicism in quartz. These authors also found abundant occurrences of shocked quartz grains with PDFs in individual quartz grains and crystals from granitic fragments. The PDFs in quartz from the in the Chesapeake Bay core samples showed up to 6 different sets per grain (Koeberl et al. 1996c). Shock deformation was also found in K-feldspar, alkali feldspar, and plagioclase, which show up to three sets of PDFs. Several shocked K-feldspar grains also display local, incipient melting. A number of particulates also contain individual granitoid-derived rock fragments, which are partially or almost totally melted. These fragments represent impact melt, because they commonly incorporate shocked quartz and feldspar clasts. The measurement of the crystallographic orientation of the PDFs in quartz showed the dominance of the shock-characteristic orientations and confirmed the impact origin of the Chesapeake Bay structure.

The chemical composition of breccia samples from the Exmore breccia (recalculated on a water-free basis) is in good agreement with the composition of North American tektites and was taken by Koeberl et al. (1996c) to suggest that the tektites were derived from the Chesapeake Bay structure. However, more detailed chemical and isotopic studies, including the determination of the Sr and Nd isotopic compositions of the target rocks at Chesapeake Bay, are still in progress. The results from these studies should be sufficient to confirm (or reject) the link between Chesapeake Bay and the North American tektites.

3.3.2
Popigai

There is a second large crater with an age that is indistinguishable from that of the Chesapeake Bay structure and the two ejecta layers, namely the 100 km diameter Popigai impact structure in Siberia, which has been dated by Bottomley et al. (1997) at 35.7 ± 0.8 Ma. The Popigai structure is exposed in Archean crystalline rocks of the Anabar Shield, with overlying Proterozoic to Mesozoic sedimentary sequences (e.g., Masaitis 1994; Vishnevsky and Montanari 1999), and is the largest Cenozoic crater on Earth. Archean basement rocks, which are exposed to the south of the structure, comprise the Verkhne-Anabar and the Khapchan series. These rocks are strongly deformed and belong mainly to amphibolite and granulite facies of regional metamorphism, with various gneisses being the dominant rock type. The average composition of impact melt rocks from Popigai is similar to that of the average composition of Anabar Shield gneisses (Vishnevsky and Montanari

1999). Minor occurrences of ultrabasic rocks (amphibolized and serpentinized peridotites and pyroxenites) occur within the gneisses, as do granitoids of various compositions that make up about 10% of the gneiss volume. The sedimentary cover has a total thickness of about 1.6 km and ranges from quartz arenites, dolomites, and sandstones of Proterozoic age to Cambrian dolomites, marls, and limestones, to Permian sandstones. A Mesozoic cover is exposed in a few locations to the northeast of the structure and includes Triassic volcano-sedimentary formation (which are not found in any impact breccias), some Jurassic marine and continental formations, and Cretaceous siltstones, sandstones, coal-bearing sands, and clays, all of which feature prominently in suevitic breccias at the crater (e.g., Vishnevsky and Montanari 1999).

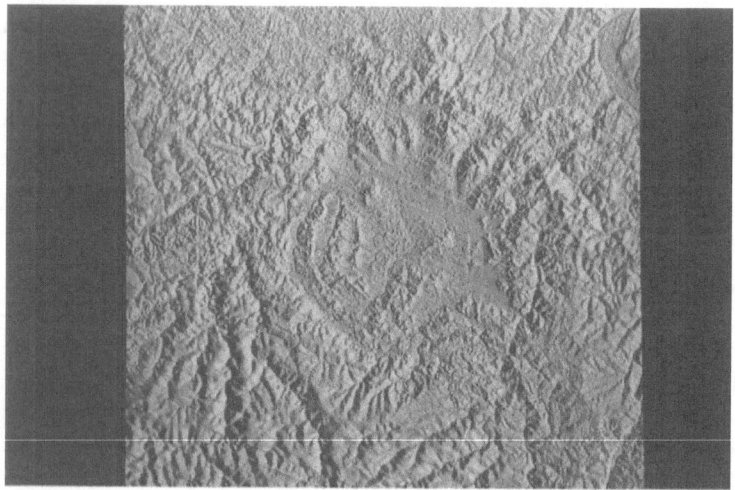

Fig. 3.3.2.1. Radar satellite image of the Popigai impact structure, showing the extent of the circular basin that makes up the structure with a diameter of about 100 km. (Source: gdcinfo.agg.cmr.ca/crater/world_craters.html)

Details about the geology of the Popigai structure are given in the reviews of Masaitis et al. (1980), Masaitis (1999), and Vishnevsky and Montanari (1999). The structure is a circular basin (Fig. 3.3.2.1) that is characterized by several ring features, but lacks a topographic outer rim. It is separated from the surrounding areas by a complex concentric zone of uplifted and overturned blocks and klippen of about 100 km in diameter. Slightly deformed rocks extend to a maximum distance of about 70 km from the crater center. The structure hosts a large variety of impactites, including suevites, impact melt rocks, and other impact breccias. These are described in detail by, e.g., Masaitis (1994) and Vishnevsky and Montanari (1999).

Fig. 3.3.2.2. Megabreccia at the Popigai impact structure, exposed on a cliff at a river. The polymict breccia contains clasts that range in size up to several tens of meters (photo courtesy P. Claeys).

Fig. 3.3.2.3. Thick layer of impact melt rock at the Popigai impact structure in Siberia. The melt rock overlies more fragile megabreccia deposits that is more susceptible to weathering (photo courtesy P. Claeys).

Popigai is characterized by thick breccia and impact melt deposits of impressive appearance. Fig. 3.3.2.2 shows a megabreccia deposit that is exposed at a river and comprises a polymict breccia with clasts that are up to several tens of meters in size. Fig. 3.3.2.3 shows a cliff made up of an almost 100 m thick layer of impact melt rock (locally named "tagamite"), having an appearance that is similar to that of a volcanic lava flow. These melt rocks, however, were found to contain evidence for an extraterrestrial component, and undigested or partly digested clasts within the melt rock contain shocked minerals. Impact-derived diamonds have been found in impact melt rocks within the crater (e.g., Koeberl et al. 1997a) and at distances of up to 500 km from the structure (e.g., Vishnevsky and Montanari 1999), but it is not clear if they have weathered out of ejecta or have been transported by glacial action.

It is now commonly assumed (e.g., Vonhof et al. 1995) that the global Late Eocene microkrystite layer originated from the Popigai impact event. Confirmation of such a link has to be done by using isotope geochemical methods, as radiometric age determinations do not allow to resolve an age difference of 10 or 20 k.y. First results of such a study were recently reported by Whitehead et al. (2000). These authors analyzed the Rb–Sr and Sm–Nd isotopic compositions of composite samples (1 to 9 mg) of microtektites and microkrystites from DSDP Sites 216, 462, and 612. Site 216 (eastern Indian Ocean) represents the slightly older layer, and Site 612 (eastern coast of the U.S.) represents the slightly younger layer that has been associated with the North American tektite strewn field and the Chesapeake Bay impact structure. Site 462 is located in the western central Pacific. The results of Whitehead et al. (2000) show that samples from Sites 216 and 462 have isotopic compositions that are similar to those of impact melt rocks from the Popigai impact structure, and are distinct from the Site 612 samples, which (in a ε_{Nd} vs ε_{Sr} diagram) plot within the field defined by North American tektites. Both are isotopically distinct from Mistastin and Wanapitei impact melt rocks. The Nd model ages of Site 216 and 462 samples also fit the Popigai impact melt rock model ages and distinct from those of the North American tektites. These results seem to confirm the association of the lower of the two Late Eocene impactoclastic layers with the Popigai impact structure and also that two large impact events occurred at that time within a few tens of thousands of years.

3.4
Manson Impact Structure and Ejecta Layer

The Manson impact structure is a nearly circular feature with a diameter of about 35 to 37.8 km, centered at 42° 35' 05" N and 94° 33' 33" W, slightly north of the town of Manson in north-central Iowa (Koeberl and Anderson 1996a). Although there is no surface expression of the structure, it was identified as a well-preserved complex impact structure that has a large central peak with a central depression. Although impact rocks are found at the bedrock surface, the region is covered by 30–100 m of Pleistocene glacial drift. The pre-impact surface of the region was covered by about 0.1 km of Cretaceous marine shale-dominated sediments, overlying about 0.8 km of carbonate-dominated Paleozoic strata, and a southeastward-thickening wedge of poorly-consolidated Middle Proterozoic sandstones, siltstones, and shales ranging in thickness for zero to about 3 km within the structure, all resting on an Early to Middle Proterozoic gneiss- and granite-dominated basement (see, e.g., Hartung et al. 1990; Anderson et al. 1996; as well as papers in Koeberl and Anderson 1996).

The area of the crater has been known as an area of anomalous geology since the beginning of the 20th century because of samples recovered during the drilling of a Manson town well and the unusually "soft" water produced by the well. Over the decades, several explanations were proposed for the origin of the Manson "anomaly", ranging from a cryptovolcanic feature, to a cryptoexplosive feature, and, finally, an impact structure (see Hartung and Anderson 1988, for a review). Early dating attempts of rocks from the Manson structure by Ar–Ar dating by Kunk et al. (1989) indicated that the structure may have formed at 65.7 Ma, an age indistinguishable from that of the K/T boundary. The possibility that a sizeable impact structure in the U.S. could be associated with the end-Cretaceous mass extinction led to an increase in research on the Manson structure (see Hartung and Anderson 1988, and Hartung et al. 1990, for reviews).

The Iowa Geological Survey Bureau and the U.S. Geological Survey began in 1991–1992 a joint research core drilling program at Manson; one of the driving forces was the possibility that there is a link between the structure and the K/T boundary. Over 1200 m of core was recovered from 12 locations that are representative for the different parts of the structure. One of the first products of the investigation of the core materials recovered was a new and more accurate age for the formation of the Manson structure. $^{40}Ar/^{39}Ar$ analysis of sanidine feldspar (recrystallized from impact melt) by Izett et al. (1993) showed that the age of the Manson structure was 73.8 ± 0.3 million years, which is 9 million years prior to the age of the K/T boundary. More recently, Izett et al. (1998) determined a more precise age of 74.1 ± 0.1 Ma. Papers by a number of authors, describing results dealing with 1200 m of core and numerous sets of samples from water well cuttings from the Manson impact structure, as well as geophysical studies of the crater structure, are collected in Koeberl and Anderson (1996a), and provide a fairly complete and detailed summary of our knowledge of the second-largest impact structure known in the United States.

Izett et al. (1993, 1998) and Witzke et al. (1996) also described the discovery of a distal ejecta layer related to the Manson impact structure in the Crow Creek Member of the Cretaceous Pierre Shale in South Dakota and Nebraska (Fig. 3.4.1). This unit, which consists of a silty and sandy laminated limestone and an overlying marlstone, and has a thickness of about 3 m. Izett et al. (1998) conclude that the stratigraphic age of this unit is identical of the (radiometric) age of the Manson structure, and supported this conclusion by Ar–Ar age determinations on bentonites from the Pierre Shale in central South Dakota. Izett et al. (1993, 1998) and Witzke et al. (1996) report the finding of shocked minerals (mainly quartz and feldspar, possibly a few zircons) from rocks of the Crow Creek Formation. Izett et al. (1998) noted that the shocked quartz grains with multiple sets of PDFs are most common in the lower part of the Crow Creek Member, but some are also found in the overlying marlstone. These authors also found that the largest shocked quartz grains, with diameters of 1.5 to 3.2 mm, are found closer to the Manson structure (215 to 315 km), than those with smaller sizes (0.1 to 1.2 mm), which are found 410 to 475 km from Manson. This represents a range of 12 to 26 crater radii from the source crater. The mode of origin of the Crow Creek Member is not yet clear. Witzke et al. (1996) favored normal marine transgression, whereas Izett et al. (1993, 1998) prefer an interpretation in which the Crow Creek Member was deposited during an impact-triggered tsunami. No matter which interpretation is correct, Manson is one in a short list of impact structures that has been linked to a distal ejecta layer.

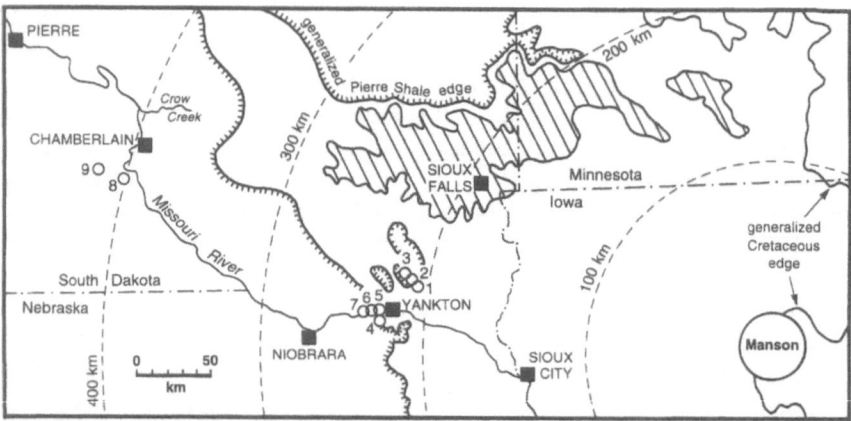

Fig. 3.4.1. Location of the ca. 37 km diameter Manson impact structure in Iowa, with the generalized paleogeographic outlines of the Cretaceous edge and Pierre Shale edge at about 74 Ma. Numbered open circles mark locations where distal ejecta of the Manson impact event have been found within the Crow Creek Member of the Pierre Shale (for details, see Witzke et al. 1996).

3.5
Acraman Impact Structure and Ejecta Layer

Gostin et al. (1986) reported on the discovery of an impactoclastic layer within Late Precambrian shales of the 590 Ma Bunyeroo Formation in the Adelaide geosyncline, South Australia. This layer contains abundant shocked quartz grains wit PDFs, shattered mineral grains, and small shatter cones (e.g., Wallace et al. 1990a). The ejecta were found in outcrops and drill cores over several hundred kilometers. At the same time, Williams (1986, 1994) identified the Acraman structure in South Australia as an impact structure, and confirmed it to be the source crater of the Bunyeroo impact ejecta layer. Gostin et al. (1989) and Wallace et al. (1990b) detected enrichments of the PGEs in the ejecta layer; however, post-formational redistribution had altered the PGE patterns. The diameter of the Acraman structure is at least 90 km, with some outer arcuate features at 150 km diameter (Williams 1994). Impact ejecta have been found at distances of up to 450 km from the Acraman structure (i.e., about 10 crater radii), making this a true distal ejecta layer.

The impact ejecta at the Bunyeroo Gorge have been subdivided by Wallace et al. (1996) into several sublayers. These are: a) the lowest one is discontinuous (occurs as lenses) and comprises a sandy clast-bearing horizon containing mostly centimeter-sized fragments (rarely up to 40 cm in diameter); it is coarse material that may have been transported entrained in an ejecta curtain and was deposited in the water column shortly after the impact; b) An overlying porous graded sandstone layer, which may finer material that settled from the air; c) on top is a greenish-gray mudstone host rock layer with dark red impact ejecta clasts that had sunk into the mudstone. The first two layers contain abundant evidence of shock metamorphism, including quartz grains with multiple sets of PDFs and recrystallized impact glass fragments. Gostin and Zbik (1999) also reported the presence of rare shocked zircons with distinct and closely spaced planar features from ejecta clasts. Compston et al. (1987) tried to use zircons from Acraman ejecta to date the age of the impact, but their zircons were much less shocked than the ones described by Gostin and Zbik (1999) and gave, therefore, only the age of the target rocks. The ejecta show abundant evidence of low temperature alteration, which helps to explain the redistribution of the meteoritic component in the ejecta (e.g., Wallace et al. 1990b).

3.6
Morokweng and the Jurassic–Cretaceous Boundary

A large, near-circular aeromagnetic anomaly in the region around the town of Morokweng, approximately centered at 23° 32' E and 26° 20' S, close to the border with Botswana in the Northwest Province of South Africa, had long been interpreted as the expression of an intrusive body. However, reinterpretation of regional gravity and aeromagnetic images, as well as the first reports of impact characteristic shock metamorphic effects in rocks from the Morokweng area, demonstrated the presence of a large meteorite impact structure (see, e.g., Corner et al. 1997). The size of the Morokweng impact structure is still the subject of debate, but it must have been larger than 70 km, which is the diameter of the present composite magnetic anomaly. Early estimates (e.g., Corner et al. 1997) proposed original diameter values in excess of 300 km, but more recent work on a deep drillcore near the structure (Reimold et al. 2000) failed to turn up evidence that this core penetrated through impact deposits. This may limit the diameter to about 80 km. On the other hand, circular paleogeographical features surrounding Morokweng have a diameter of about 200 km (Reimold et al. 1999).

Koeberl et al. (1997c) presented petrographic, chemical, and isotopic data for an impact melt body (which is also the cause of the magnetic anomaly) in the Morokweng structure. Disregarding a small number of obviously altered melt rock samples, the Morokweng melt body is extremely homogeneous in composition. Variations for major elements do not exceed 2–5 relative percent. Chemical analyses revealed that the impact melt rock contains significant abundances of siderophile elements with average values for Cr of 440 ppm, of 50 ppm for Co, 780 ppm for Ni, and up to 30 ppb for Ir. No variations with depth and no differences between drill cores were noted. In light of petrographic and isotopic evidence for the melt rock cited by Koeberl et al. (1997c) and Reimold et al. (1999), it can be excluded that mantle-derived sources are responsible for these high siderophile element concentrations. Abundances of the other platinum group elements confirmed a near-chondritic composition of this component. After correction for indigenous components, Koeberl et al. (1997c) concluded that a meteoritic (chondritic) component of 2–5% was present in the Morokweng melt rock. Chromium isotopic data of Shukolyukov et al. (1999) confirmed the presence of a meteoritic component and indicated a probable L-chondritic source.

Koeberl et al. (1997c) also reported single zircon U–Pb and Th–Pb ages for the Morokweng impact melt rock of 144.7 ± 0.7, and 146.2 ± 1.5 Ma, respectively. This age is, within error, indistinguishable from the currently accepted age for the Jurassic–Cretaceous boundary, which is placed at 145 Ma at the base of the Berriasian (Harland et al. 1990; Gradstein et al. 1994). In comparing these ages, however, one has to bear in mind that the age of the Morokweng structure is a radiometric age, whereas the Jurassic–Cretaceous (J/K) boundary age value is based on interpolated stratigraphic ages. This apparent coincidence suggests a causal relationship. However, the Morokweng situation is somewhat different from that of the Cretaceous–Tertiary (K/T) boundary, where a well-defined and firmly dated impact-derived layer marks the boundary. There is no equally well-defined stratotype for the J/K boundary and there are only a few well-documented boundary locations in the world. Whereas there is a report of some PGE enrichments in a J/K boundary site in Siberia, with associated microspherules (Zhakarov et al. 1993), these data have yet to be confirmed. It is also not sure that the Siberian layer is really marking the J/K boundary (Rampino and Haggerty 1996).

The exact stratigraphic placement of the J/K boundary is still a matter of debate, as the cross-calibration of the biostratigraphic time scale with the paleomagnetic time scale was marred by uncertainties in the numerical calibration of each time scale (Remane 1991). However, paleomagnetic analyses within the last 15 years (e.g., Ogg and Lowrie 1986; Bralower et al. 1989) have provided a more solid basis for global correlation of the J/K boundary event. Several authors have recently placed the boundary between the Tithonian (Uppermost Jurassic) and the Berriasian (Lowermost Cretaceous) stages at approximately 145 Ma (Harland et al. 1990: 145.5 ± 2.5 Ma; Gradstein et al. 1994: 144.2 ± 2.9 Ma).

Whereas earlier workers did not assign much importance to the J/K boundary (cf. Remane 1991), Raup and Sepkoski (1984), Sepkoski (1992), and Sepkoski (cited in Rampino and Haggerty 1996) now consider the J/K extinction as one of the large extinction events of that last few hundred million years. Rampino and Haggerty (1996) interpreted data from these papers to indicate that the J/K extinction is the fifth-largest extinction event during the last 300 million years, as mainly recorded by marine genera and reptiles. Some evidence has been cited to suggest that the extinction events occurred at slightly different times in the Boreal and Tethyan provinces (cf. Ogg and Lowrie 1986; and M. Rampino, personal communication, 1997). For example, the boreal J/K boundary (offshore Norway) has been placed between the Volgian and the Ryazanian stages at an interpolated age of 142 ± 2.6 Ma. It is interesting to note that the top of the Volgian offshore Norway is marked by impact ejecta from the nearby 40 km diameter Mjølnir impact structure (Dypvik et al. 1996). More recent data (H. Dypvik, personal communication, 2000) places the late Volgian at about 150 Ma. None of the data can be directly compared, as different timescales are used. It is also difficult to use biostratigraphical correlations to determine if the J/K boundary represents a global event, because of limited faunal exchange between different paleobiogeographic provinces. Gosses Bluff (Australia) has also a similar age (Table 1.9.1).

The formation of a large impact crater of about 80 km diameter (possibly even larger) may have had some environmental and biological consequences. It may be speculated if the lesser extinction rate at the J/K boundary compared to the K/T boundary could be the result of a difference in target rocks. Whereas the bolide that formed the Chicxulub crater in Mexico hit a sequence of crustal rocks overlain by carbonate and evaporite rocks, leading to the release of enormous quantities of CO_2 and SO_2 (with associated environmental consequences; Sect. 3.2.3), the Morokweng impact event occurred mainly in granitoids.

No clear evidence has yet been found for distal impact ejecta at the J/K boundary. Kudielka et al. (1999) presented some preliminary stable isotope data across the Bosso River section in the Umbria–Marche sequence, which seems to be one of the best preserved J/K exposures (see Sect. 5.4). However, in the section analyzed so far, neither the carbon nor the oxygen isotope ratios show any significant shifts, which implies that there was no apparent disruption of the carbon cycle at this locality. Thus, the search for possible distal ejecta deposits associated with the Morokweng impact structure has to go on. If none are found, several interpretations are possible. First, Morokweng might have been too small to cause distal ejecta, or the small amount initially deposited might be eroded. Second, and more likely, the bad stratigraphic control on the exact position of the J/K boundary could mean that it is very difficult to locate the boundary (and the probably very thin ejecta layer), or maybe the impact event was not quite coeval with the end of the Jurassic. In any case, the tantalizing prospect remains that the end of the Jurassic is marked by one or more large impact events.

3.7
Other Confirmed and Possible Ejecta Layers

There is a number of layers in the sedimentary record that have been interpreted, by some workers, as being the result of impact processes (e.g., Archean spherule layers in South Africa and Australia). In addition, evidence of impact events has either been confirmed (Late Devonian) or suggested (Permo–Triassic and Trias-sic–Jurassic boundaries). In none of these cases has it been possible (so far) to detect a source crater. The cases discussed in the next sections do not provide an exhaustive discussion of every impactoclastic layer ever proposed, but are in-tended to describe some of the more important layers or boundaries.

3.7.1
South African and Australian Archean Spherule Layers

Spherule layers in the approximately 3.4 Ga Barberton Greenstone Belt, South Africa, have been interpreted by Lowe and co-workers (e.g., Lowe and Byerly 1986; Lowe et al. 1989; Kyte et al. 1992; Byerly and Lowe 1994) as the result of large asteroid or comet impacts onto the early Earth. The enrichment of sidero-phile elements, especially the platinum group elements (PGEs), in these layers was taken by these authors as key evidence for the presence of an extraterrestrial component (Lowe et al. 1989), although some fractionations were noted (Kyte et al. 1992). The interpretation of these spherule layers as the result of impact processes has been challenged by Buick (1987), French (1987), Koeberl et al. (1993), and Koeberl and Reimold (1995). Fig. 3.7.1.1 shows a sample of the Barberton spherule layer.

Koeberl and Reimold (1995) reported on detailed petrographical and geo-chemical studies, from which they made a number of observations that are either not supporting the impact hypothesis, or even contradicting it. The textures of most spherules are not necessarily the result of impact, but could instead be the product of radial or intersertal growth of crystals during secondary mineral forma-tion (Fig. 3.7.1.2). There is no difference in the content of siderophile elements between spherule layers and layers devoid of spherules. The siderophile element abundances are high where sulfide minerals (e.g., pyrite, gersdorffite, and chal-copyrite) and/or chromite are present, independent of the presence or absence of spherules.

High contents of, e.g., Ir (up to 2700 ppb), Ni (0.96 wt.%), and Cr (1.6 wt.%) were found in various samples. Abundances of these elements in chondritic mete-orites are approximately 600 ppb, 1.4 wt.%, and 0.35 wt.%, respectively, resulting in respective meteoritic components in these samples of 450%, 70%, and 460% by weight. Such superchondritic concentrations cannot be primary meteoritic signa-tures (iron meteorites do not have high Cr or chondritic interelement ratios). Im-pact melt rocks have typically <<1% of a meteoritic component. Furthermore, spherules of any kind are very rare in known impact deposits (see, e.g., Graup

1981, for a discussion of spherules at the Ries impact structure). Even at the K/T boundary, spherules are rare and occur only in layers at distal sites with a thickness of less than a few centimeters. If present at all, they are not usually associated with any significant Ir or PGE anomaly.

Fig. 3.7.1.1. Polished slab of a sample (BA-1; from the Princeton section of the Agnes Mine) of one of the early Archean spherule layers in the Barberton Mountain, South Africa, area. The sample is described in detail by Koeberl and Reimold (1995) and shows several sublayers, in which spherules (some of them deformed) exhibit evidence of size sorting.

It is likely that the high abundances of the siderophile elements in some Barberton samples and their enrichment in secondary minerals resulted from remobilization and reconcentration of these elements. Due to different distribution coefficients of the PGEs in hydro- and mesothermal processes, the interelement ratios should have changed during remobilization as well. Thus, the PGE abundance patterns and ratios are not primary and cannot be used as an argument in favor of an impact origin either. Nickel-rich Cr-spinels have chemical compositions that are totally unlike those of (undisputed) impact-derived spinels, e.g., those from the K/T boundary or from Late Eocene impactoclastic layers. There is no indication of any evidence of shock metamorphism, the commonly accepted definitive criterion for recognition of impact, in any samples from the Barberton spherule layers, which is unusual, because such evidence is preserved even in heavily altered samples from deeply eroded Archean impact structures, such as the 2.02 Ga Vredefort structure in South Africa (e.g., Reimold and Gibson 1996).

Fig. 3.7.1.2. Thin section photograph of early Archean spherules from the BA-1 sample (see Fig. 3.7.1.1), showing the deformation of the spherules and the internal textures. The mineralogical composition of the spherules is not significantly different from that of the matrix and indicates severe alteration and recrystallization. Width of image 2.2 mm, crossed polarizers (photo courtesy W.U. Reimold).

These observations provide serious problems for the interpretation of the Barberton spherule layers as "impact spherules". On the other hand, recent Cr isotopic analyses of several spherule layer samples from the Barberton area (Kyte et al. 1999; Shukolyukov et al. 2000) are in agreement with part of the chromium in the samples being of extraterrestrial origin and suggested that (a) carbonaceous chondritic projectile(s) could have been involved in the formation of these layers. These data seem to provide good evidence for an impact origin, but the issue is probably more complicated. It also became clear that the initial petrographic and geochemical arguments of Lowe et al. (1989) were ambiguous and in conflict with what is known from practically any other confirmed impact structure or ejecta deposit on Earth (Koeberl and Reimold 1995).

No mechanism for the formation of such thick layers of spherules is known. It is equally peculiar that a meteoritic component would be present at levels that are four orders of magnitude above those normally observed in impactites and up to a factor of 5 above bulk meteorite compositions. Clearly this implies secondary alteration and redistribution, in which case there is no explanation why the PGE interelement rations would remain constant during the redistribution. If the observation by Shukolyukov et al. (2000) of extraterrestrial Cr in the spherule layers is confirmed and the link with the formation of the spherules themselves is substantiated, this would indicate that some very unusual and poorly understood processes occurred >3 billion years ago. Such a confirmation would have to include detailed analyses of a variety of local country rock samples and unusual PGE-rich occurrences, such as the Bon Accord body (Tredoux et al. 1989b). Many open questions remain and should be addressed in further studies.

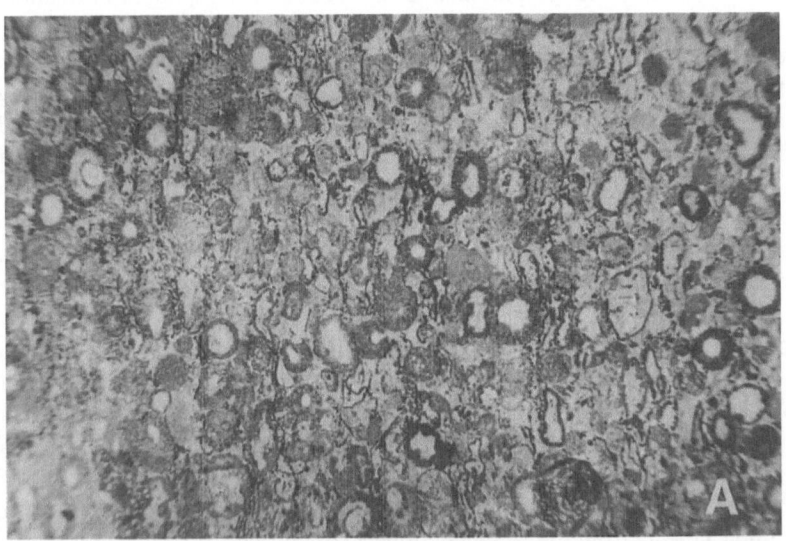

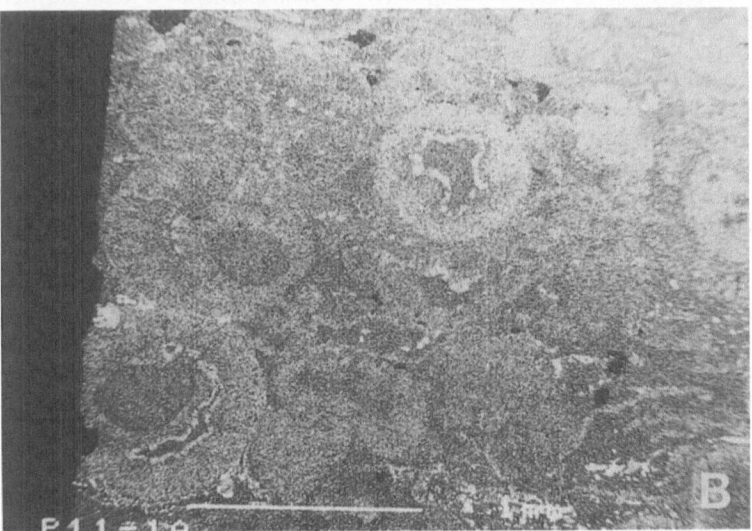

Fig. 3.7.1.3. Photographs of a thin section of the Monteville spherule layer (Transvaal Supergroup, South Africa). a) Overview of the spherule layer sample; width of image about 20 mm, crossed polarizers. b) Backscattered electron image of a group of spherules, showing the cores. The mineralogical composition of the spherules and the matrix is similar and indicates pervasive alteration. Scale bar on bottom is 1 mm wide.

Other occurrences of unusual spherule layers were reported by Simonson (1992) from the Hamersley Basin in Western Australia. This author described spherules that are, on average, about 0.5 to 1 mm in diameter, which consist mainly of K-feldspar, have mostly spherulitic, vesicular, and other crystalline textures, and occur as a stratigraphic marker in the 2.6 Ga Wittenoom Formation. On the basis of similarities to microtektites and microkrystites, Simonson (1992) interpreted the spherules as having formed in an impact even and having been redeposited in a sediment gravity flow. Later, three additional spherule-bearing layers were found in the Hamersley Basin sequence, which were also interpreted to be of impact origin (e.g., Simonson et al. 1997; 1998). Radiometric ages of 2541 +18/-15 Ma and 2548 +26/-29 Ma have been obtained directly on carbonates associated with two of these layers (cf. Simonson et al. 1998). None of these spherules are associated to with shocked minerals (Simonson and Davies 1996). The absence of shocked quartz led Simonson et al. (1998) to suggest that the spherule beds formed from an impact into an oceanic target, where quartz is not a major component. On the other hand, shock-characteristic planar deformation features develop in all rock-forming minerals, which would include those that make up oceanic crust. Simonson et al. (1998) found that the Hamersley Basin spherule layers show enrichments in the platinum-group elements (up to 1.7 ppb Ir), but that the interelement ratios are non-chondritic, probably as the result of low-temperature redistribution processes, similar to what has been observed for the distal impact ejecta from the Acraman impact structure in Australia.

Simonson et al. (1997) also reported on the discovery of similar spherule layer in the Monteville Formation of the Transvaal Supergroup in South Africa (Fig. 3.7.1.3a, b), which might be correlated with one of the Australian layers. In South Africa, Simonson et al. (1997) found spherules at a constant stratigraphic level in cores from two sites (the Pering mine and the Kathu core) and in surface exposures at two additional sites (Monteville farm and SW of Douglas) over an area of more than 10,000 km². The South African as well as the Australian layers contain abundant millimeter-size spherules that consist largely of K-feldspar, are locally replaced destructively by carbonate, and display external shapes and internal radiating fibrous textures similar to impact-derived spherules from, e.g., the Late Eocene microkrystite strewn field (Sect. 3.3). Thin tuff beds are present in the same stratigraphic succession as the spherule layer and contain millimeter-sized spheroidal clasts. However, these spherules display finely clastic textures and crude concentric layering diagnostic of accretionary lapilli. The Monteville spherule layer is similar in age to the Australian layers; tuffs in the general stratigraphic vicinity of the spherule horizon in the Griqualand West succession yielded dates of about 2.55 Ga (cf. Simonson et al. 1997). Recently, Koeberl et al. (1999b) found that the Monteville spherule layer is associated with elevated siderophile element abundances (Table 3.7.1.1). These values are, for Ni up to 167 ppm, Co up to 87 ppm, Cr up to 418 ppm, and, especially, Ir up to 6.4 ppb. No comparable values were measured in any of the other samples, including pyrite-rich country rock or tuffaceous samples. Subsequent analyses showed that the platinum-group element ratios in   the   spherule   layer   are   roughly   chondritic.   The

Table 3.7.1.1. Selected chemical data for some samples from the Monteville spherule layer (Pering Mine), South Africa (from Koeberl et al. 1999b).

	P11-1A SL	P11-1B SL	P11-1B1 SL	P9-1A SL	P11-27B Tuff	P11-31 S/T
SiO_2	52.5	n.d.	62.5	57.9	32.1	58.8
TiO_2	2.63	n.d.	1.63	2.36	1.09	1.35
Al_2O_3	24.9	n.d.	17.3	20.5	8.92	14.4
Fe_2O_3	0.55	2.54	6.68	0.55	8.16	6.76
MnO	0.01	n.d.	0.03	0.04	0.77	0.06
MgO	3.63	n.d.	1.88	2.42	7.37	2.74
CaO	0.35	n.d.	1.76	0.88	13.6	0.62
Na_2O	0.055	0.069	0.051	0.095	0.064	0.11
K_2O	11.1	8.56	6.29	11.5	6.55	10.3
P_2O_5	0.07	n.d.	0.04	0.05	0.20	0.14
LOI	4.38	3.59	3.88	21.35	4.30	n.d.
Total	100.2	99.7	100.1	100.1	99.7	
Sc	40.6	23.5	20.3	13.2	22.7	21.2
V	545	293	n.d.	378	160	177
Cr	385	264	180	418	104	152
Co	2.39	42.7	87.1	17.1	21.2	16.7
Ni	15	68	167	65	49	99
Zn	30	640	200	650	592	124
Ga	6.8	6.2	26	9	52	12
As	0.39	8.98	21.9	1.93	4.89	7.59
Se	0.75	0.72	0.91	0.29	0.47	0.42
Sb	0.049	0.95	1.45	0.38	0.33	0.61
Cs	7.38	4.92	3.51	19.1	2.14	14.6
Hf	5.17	8.83	12.6	5.38	2.45	5.43
Ir (ppb)	2.5	6.4	5.6	4.8	0.6	<1
Au (ppb)	4.1	3.9	8	1.2	6.1	5.3
Th	4.22	5.03	5.93	5.63	1.35	6.85

All Fe as Fe_2O_3; major elements in wt.%, trace elements in ppm, except as noted. SL = spherule layer, S/T = shale with tuff. Samples P11-1A, -1B, -1B1, and P9-1A are spherule layer samples, while P11-2B and -3B are carbonate beds immediately in contact with the spherule layer.

PGE abundance levels in these spherules are comparable to those found in "normal" impactites and ejecta and indicate the presence of about 1% by weight of a chondritic component. None of the tuff layers show enhanced PGE contents. The high siderophile element contents in the Monteville spherules do not correlate directly with abundances of some chalcophile elements (Table 3.7.1.1), and, unlike in the disputed Barberton Mountain Land spherule layers, are not hosted by secondary sulfides.

However, all in all the identification of Precambrian impact deposits (especially distal ejecta) remains a largely unresolved problem. Unfortunately, so far no definitive criteria for the identification of Archean impact deposits are known (cf. Weiblen and Schultz 1978). For none of the South African (Barberton and

Monteville) or Australian spherule layers has a source crater been found; given the scarcity of the geological record it is likely that it will never be found [if all these layers are indeed the result of impact events and sources craters exist(ed)]. It is not clear why impact events in the Archean would predominantly produce large volumes of spherules, which are mostly absent from post-Archean impact deposits (i.e., those for which source craters are known). On the other hand, none of these spherule layers is associated with any shocked minerals, which are the hallmark for all confirmed impact structures and ejecta. Even rocks from the 2 Ga Vredefort impact structure contain abundant shocked minerals, so it is unlikely that Archean impacts would, for some reason, not produce shocked minerals. Related to this topic is the suggestion by (for example) Rhodes (1975) that the Bushveld Complex in South Africa, a huge igneous province, was formed by multiple impacts. However, French (1990b), Buchanan and Reimold (1998), and Buchanan et al. (1999) did not find any evidence to support this contention. The question regarding how to identify Archean impact deposits remains open and will hopefully be addressed in future studies.

3.7.2
Late Devonian Impact Layer and Alamo Breccia

The Frasnian–Famennian (F/F) boundary in the Late Devonian is associated with one of the five largest mass extinctions in the geological record. Glassy spherules, probably similar to microtektites, have been discovered in Late Devonian sections in south China slightly above the F/F boundary (e.g., Wang 1992) and also in Belgium (e.g., Claeys et al. 1992; Claeys and Casier 1994). The spherules have all the characteristics of having formed by impact (Claeys et al. 1992; Wang 1992). In China and Australia the spherule layer seems to be correlated with a minor Ir anomaly (e.g., Playford et al. 1984), which could not be confirmed for the Belgian sections (Claeys et al. 1996). It is possible that the spherule layers in China and in Belgium do not belong to precisely the same layer, as conodont stratigraphy indicates a slight time difference, but this has not yet been confirmed. No source crater for the distal impact layer(s) has yet been identified, although the 54 km diameter Siljan impact structure in Sweden is of similar age. However, this structure may be too small to cause any mass extinctions; at any rate, the relation between the microtektite layer(s) and the F/F mass extinction has not been explored in any detail. Wang et al. (1991) presented evidence for a catastrophic biotic event at the F/F boundary, maybe associated with the impact spherules (Wang and Chatterton 1993), but Racki (1999) concludes that impacts had no distinct influence on the F/F mass extinctions.

Fig. 3.7.2.1. Alamo Breccia at the Hancock Summit location, showing the chaotic nature of the megabreccia and the large variation in clast size, with mainly small clasts being visible in the photograph. Long dimension of the card at center is about 10 cm.

There is evidence for another large impact event in the Late Devonian. A large-scale impact event, dated from conodont stratigraphy at about 367 Ma, occurred in a nearshore marine setting, and resulted in the deposition of the spectacular Alamo Breccia in Nevada (e.g., Warme and Sandberg 1996; Warme and Kuehner 1998). This megabeccia is composed predominantly of limestone with minor dolostone and rare quartzite (Fig. 3.7.2.1). It was found to contain rare shocked quartz with multiple sets of PDFs (Leroux et al. 1995), altered spherules, and possibly an Ir anomaly, and is spread discontinuously over a semi-circular zone of about 200 km diameter and has a total thickness of more than 100 m in some locations. The breccia show a variation in lithology and thickness as a function of increasing distance from the inferred center; no crater has yet been found. It may have been destroyed in (unrelated) post-impact tectonic activity and, thus, is no longer present.

The breccia seems to represent both, direct impact ejecta and redistributed tsunami deposits. As the diameter of the source crater is not known, it is not clear if the Alamo Breccia consists of proximal or distal ejecta, but the former is more likely. The age of the Alamo event does not seem to coincide with the ages inferred for the Belgian and Chinese microtektite horizons, being up to 3 Ma older, and the inferred diameter of the source crater (e.g., Warme and Kuehner 1998) may be too small to cause global effects. On the other hand, Morrow et al. (1998) suggested, based on observations of possibly also impact-related megabreccias in Western Europe, that the Alamo event may only have been one of several impact events in the Late Devonian, all of which led to environmental stress and a collapse of the ecosystem that culminated in the F/F mass extinction.

3.7.3
Permian–Triassic Boundary

The Permian–Triassic (P/Tr) boundary (Fig. 3.7.3.1) is associated with the largest mass extinction known in Earth history, during which about 85% of all living species went extinct (Erwin 1994). Following the association of the K/T boundary mass extinction with a large impact event, speculations bloomed that other major mass extinctions might also be related to impact events. However, so far the evidence in favor of such a proposal is controversial. Holser and Margaritz (1992) provided an instructive comparison between the K/T and P/Tr boundary events. The P/Tr boundary is marked by a distinct and sudden drop in the carbon isotopic ratio. Siderophile element anomalies (e.g., enhanced Ir contents) were found at some P/Tr boundary locations (e.g., Holser et al. 1989; Xu et al. 1993), but their source is not clear and confirmation of their extraterrestrial nature is still pending. Chai et al. (1992) found some minor Ir enrichments and concluded that an impact may have been involved, but in some cases it was not possible to confirm the PGE enrichments found by other authors (e.g., Zhou and Kyte 1988; see also Asaro et al. 1982; Yin and Tong 1998). Thus, it is presently not clear if the small Ir anomalies found at some locations (e.g., at the Gartnerkofel, Austria, section; Attrep et al. 1991b) have terrestrial or extraterrestrial sources. It might be possible to obtain evidence in one way or another from Os isotopic studies.

Recent research, however, succeeded in demonstrating the P/Tr boundary event was a much shorter event than thought before, and that the severe environmental changes that resulted in the mass extinction were brought on within less than a few hundred thousand years (Bowring et al. 1998). In addition to their geochronologic data, these authors also found that at Meishan, China, the negative excursion in the carbon isotopic composition had a duration of less than about 160,000 years, indicating a catastrophic addition of isotopically light carbon. Bowring et al. (1998) suggested that this addition could be the result of the impact of an icy, carbon-rich comet (but also scenarios involving volcanic processes are conceivable). Rampino and Adler (1998) used fossil data to show that the end Permian extinction pulse was very abrupt, with no indication of stepwise extinctions. Retallack et al. (1998) reported on the possible discovery of shocked quartz grains from P–Tr boundary locations in Australia and Antarctica, but the exact association of the quartz-bearing layers with layers that have enhanced Ir contents and with the P–Tr boundary is still unclear. Similarly, the importance of the possible discovery of fullerenes at the P/Tr boundary (Chijiwa et al. 1999) is not resolved. Other researchers concluded that the P/Tr extinctions have only endogenic causes (e.g., Schopf 1974; Hallam and Wignall 1997). It seems that the data available so far are not complete enough to allow any definitive conclusions regarding the cause of the P/Tr mass extinction, or the possibility of any impact involvement.

Fig. 3.7.3.1. Permo–Triassic boundary section in Sechuan, China, as pointed out by one of the authors (A.M.).

3.7.4
Triassic–Jurassic Boundary

The Triassic–Jurassic (TR/J) boundary is marked by yet another major mass extinction, which is listed as one the "big five" extinctions mentioned by Sepkoski (1992) and Rampino and Haggerty (1996). The first evidence for a possible impact component at the TR/J boundary came from Badjukov et al. (1987), who found possible shocked quartz at a TR/J boundary location at Kendelbachgraben in Austria. Later Bice et al. (1992) reported on the discovery of shocked quartz grains from a TR/J boundary location in northern Italy, although the identification of the PDFs has been questioned (e.g., Mossman et al. 1998). A search for shocked quartz at TR/J boundary locations in Nova Scotia (Canada) was negative (Mossman et al. 1998). The 100 km diameter Manicouagan impact structure in

Quebec has an age that is comparable to that of the Tr/J boundary, but currently available dates indicate that with an age of 214 Ma (Hodych and Dunning 1992) it slightly predates the Tr/J boundary. Spray et al. (1998) discussed new evidence for a multiple impact event in the Late Triassic, probably slightly predating that Tr/J boundary. On the other hand, Kent (1998) disputed the evidence cited by Spray et al. (1998). However, as past experience has shown, it is very difficult to correlate radiometric ages obtained from impact melt rocks with biostratigraphic ages obtained from the sedimentary record, as a correlation between the two records implies the use of the same time scale, which is basically never the case. Thus, confirmation of an impact signature from other Tr/J boundaries might be the next step in the investigation of the end-Triassic event.

3.8
Outlook

Distal impact ejecta layers are maybe more common in the geologic record than what was assumed before the K/T boundary was shown to be exactly such a layer. During the last 20 years the number of impact structures known on Earth has more than doubled, and it is likely that the ejecta of at least some of them are preserved in the stratigraphic record. Locations with a well-preserved stratigraphic record, such as the Umbria–Marque region in Italy, provide an ideal ground to search for (and study) possible impactoclastic layers (see Chap. 5). Table 3.8.1 gives a list of those distal ejecta layers that have been definitely, or with some degree of certainty linked to source craters.

Table 3.8.1. Distal ejecta layers and source craters

Name of Crater (Age in Ma)	Type of Distal Ejecta	Location
Bosumtwi (1.07)	tektites; microtektites	Ivory Coast; eastern equatorial Atlantic Ocean
Ries (15)	tektites	Central Europe (Czech Republic, Germany, Austria)
	limestone blocks	Switzerland
Chesapeake Bay (35)	tektites; microtektites	North America; southern Ocean; possibly worldwide?
Popigai (35)	microkrystites	Pacific and Indian Ocean, Europe; worldwide?
Chicxulub (65)	K/T boundary	worldwide
Manson (74)	Crow Creek Member	South Dakota, Nebraska
Acraman (590)	shocked minerals	Australia

An interesting aspect is the influence that impact events have had on the very early evolution of the Earth (e.g., Sleep et al. 1989). Planetary scientists assume that the Earth and other planets formed by accretion of smaller objects (planetesimals). Towards the end of the accretion of the Earth, at about 4.55 Ga, it may have been impacted by a Mars-sized body; from the debris of this giant impact event formed the Moon (e.g., Cameron and Benz 1991; Taylor 1993). The material remaining in orbit after accretion of the Moon would have continued to impact onto the Earth for quite some time. Due to later geological activity, no record of this very early bombardment may remain on the surface of the Earth. On the other hand, there is abundant evidence from Apollo rocks that indicates that the Moon was subjected to intense post-accretionary bombardment between about 4.5 and 3.9 billion years ago (e.g., Ryder 1990; Dalrymple and Ryder 1993, 1996). In addition, the lunar highlands show much evidence for isotopic resetting, consistent with a short and intense late heavy bombardment period, around 3.85 ± 0.05 Ga (e.g., Tera et al. 1974; Ryder 1990). However, the interpretation of these data in terms of a distinct period of heavy bombardment were disputed by others (e.g., Baldwin 1974; Hartmann 1975), who suggested that the resetting was caused by the tail-end of the decaying late accretionary impact flux. The reason for a peak in the impact flux is not clear, but could be related to the breakup of asteroids in the asteroid belt (e.g., Zappalà et al. 1998). This heavy bombardment may also have occurred on other terrestrial planets (cf. Wetherilll 1975). No matter what duration and form the bombardment took, at this time the Earth would have been subjected to a significantly larger number of impact events compared to the Moon, as it has a larger diameter and a much stronger gravitational attraction than the Moon.

Some Nd isotope data suggested that the Earth's upper mantle had already undergone some differentiation at the time of formation of the oldest rocks preserved on the Earth's surface (Greenland, Canada, Australia); these data could be interpreted to suggest that small amounts of crust had formed prior to 4 billion years ago, but had later been mixed back into the upper mantle. This view has been confirmed by Hf isotope studies on single zircons (Amelin et al. 1999). It is evident that any large-scale early impacts had some influence on the development of the continental crust (Goodwin 1976; Arndt and Chauvel 1991), and it is conceivable that the late heavy bombardment has been responsible for the rehomogenization of the minor amounts of pre-3.85 Ga crust with the upper mantle.

Large scale early impact events must have occurred, no matter if there was a significant peak in the impact flux at about 3.8 to 3.9 Ga, or if this was just the time of the end of an enhanced post-accretionary impact flux. The search for any evidence of such early impacts has important implications, as shown by papers in Gilmour and Koeberl (2000). It has been estimated (e.g., Grieve 1980; Frey 1980) that about 200 impact structures >1000 km in diameter formed on Earth between 4.6 and 3.8 billion years ago, which would have covered about 40% of the surface of the Earth. Using the minimum estimate for the cratering frequency, Grieve (1980) calculated that these impact events would have added a cumulative energy of about 10^{29} J to the Hadean Earth. It is of course possible that the high flux of large impacts led to large-scale melting of the Earth's crust (Grieve and Cintala 1992; Cintala and Grieve 1998), annealing any possible shock effects.

In an attempt to investigate if any record of such a late heavy bombardment period on the Earth has been preserved, Koeberl et al. (2000a) performed a petrographical and geochemical study of some of the oldest rocks on Earth, from Isua in Greenland, where there were earlier suggestions of a possible extraterrestrial component (Appel 1979). In this study, an attempt was made to identify any remnant evidence of shock metamorphism in these rocks by petrographical studies, and to use geochemical methods to detect the possible presence of an extraterrestrial component in these rocks, as described by Ryder et al. (2000) (see also Arnold et al. 1998). For the shock metamorphic study, zircon was thought to offer the best chances of success. Koeberl et al. (2000a) found that many of the studied zircon grains from Isua are strongly fractured, and single planar fractures do occur, but never as part of sets; none of the crystals studied shows any evidence of optically visible shock deformation. Several samples of Isua rocks were analyzed for their chemical composition, including the PGE abundances, by neutron activation analysis and ICP-MS. Only a few samples showed somewhat elevated Ir contents (up to about 0.2 ppb) compared to the detection limit, which is similar to the present-day crustal background content ($\bullet$0.02 ppb). However, the chondrite-normalized siderophile element abundance patterns are non-chondritic, which could (after subtraction of an indigenous component) indicate either the presence of a small extraterrestrial component that has been redistributed (as at, for example, the Acraman ejecta; Sect. 3.5), or indicate terrestrial (re)mobilization mechanisms. In absence of any evidence for shock metamorphism, and with ambiguous geochemical signals, no unequivocal conclusions regarding the presence of extraterrestrial matter (as a result of possible late heavy bombardment) in these Isua rocks can be reached. More details on this topic can be found in the review by Ryder et al. (2000) and in papers in Gilmour and Koeberl (2000).

The question that seems to arise more and more often is, how is it possible to identify "unusual" impact deposits as being of impact origin. By "unusual" we mean those that do not seem to show clear evidence of shock metamorphism (in the form of shocked minerals) or where any geochemical signal is not obvious. Many different causes for the absence of either of those two signs of impact could be conceived. For example, the search for shocked minerals in the stratigraphic record is similar to the famous search for the needle in the haystack (cf. Schmitz et al. 1994). Also, an impact into a wet target of mostly soft sediments will most likely not produce abundant shocked minerals (e.g., Buchanan et al. 1998). On the other hand, shock lithification of unconsolidated material is possible (Short 1966). The study of experimental craters may provide some background data (see, e.g., Short 1970; Cordier and Gratz 1995), as well as consideration of local shock formation and propagation (e.g., Spray 1999). Another problem is that microscopic shock features are best developed in crystalline rocks. It is still not clear how to verify shock effects in carbonates. There are only a few studies in that direction, for example by Skala and Rohovec (1998), who performed NMR and XRD measurements of shocked limestones. Employing techniques that are not traditionally used for shock studies, such as cathodoluminescence (e.g., Seyedolali et al. 1997), may help as well.

A recent controversy involving "unusual" deposits centered on the suggestion by Oberbeck et al. (1993) and Rampino (1994) that most tillite deposits in the geological record may not be glacial deposits, but impact-derived deposits. However, Reimold et al. (1997), in a detailed petrographical search for shock effects among South African tillites, and Huber et al. (2000), in a geochemical search for an extraterrestrial component in these samples, were not able to provide any confirming evidence for the hypothesis that tillites were produced in impact events. Of course this negative finding does not exclude the possibility that some tillite deposits, somewhere in the wold, could represent impact ejecta, as there is a distinct similarity in the lithological characteristics between the two types of deposits.

Oblique impacts at low impact angles may play a greater role in the distribution of distal ejecta than previously thought. We mentioned in Chap. 2 that it is very likely that the four known tektite strewn fields are the result of highly oblique impact events. Schultz et al. (1994) described impact glasses from the (highly elongated) Rio Cuarto impact craters, and Schultz et al. (1998) found some other significant (distal?) impact glass deposits in Argentina; so far no source craters are known. More details on the mechanism of oblique impacts are given by (for example) Schultz (1996) and Sugita et al. (1998).

Another open question, which will hopefully lead to intense future research, is, how many other geological boundaries and extinction events could be associated with impact events, as already envisioned by Urey (1973), who discussed a possible link between cometary collisions and geological periods. It is of course counterproductive to suddenly assume that all mass extinctions and catastrophes in the geological record are the result of impact events, just as it would be irrational to conclude that impact played no role whatsoever and that all extinctions resulted from sea level changes or continental flood basalt eruptions. The search for evidence of impact at some boundaries has just begun, including (for example), the Late Ordovician mass extinction (Wang et al. 1992, 1993b, Finney et al. 1999, c), the Devonian–Carboniferous boundary (Wang et al. 1993a), or the Cenomanian– Turonian boundary (Monteiro et al. 1998). In general, the connection between impact events and extinction mechanisms is still not clear (e.g., Keller 1986). What is the influence of the crater size and the target material? Some of these questions were discussed by Jansa et al. (1990), Raup (1992), Jansa (1993), and Poag (1997a). Of course it is conceivable that other extraterrestrial events may have influenced the biological evolution on Earth as well. In this respect it is interesting to note that for the first time it was possible to detect some physical indication (in the form of cosmogenic radionuclides) of a nearby supernova explosion in the geological record (Knie et al. 1999).

The possibility that there is some periodicity in the cratering record has also received a lot of attention (e.g., Stothers 1993; Alvarez and Muller 1984), based on the finding of a possible periodic signature in the extinction data (Raup and Sepkoski 1984, but see also Baksi 1990). Subsequent analyses failed to find any evidence for a periodicity among the impact crater ages (Grieve et al. 1985, 1988), partly because not enough craters have been dated with a high enough precision (Heisler and Tremaine 1989). If there would really be a periodic signature among the cratering record, the question regarding its cause arises. A variety of proposals have been made, ranging from comet showers (Davis et al. 1984; Hut et al. 1987), the orbital evolution of the sun in the galaxy (Rampino and Stothers 1984), to unknown companion stars of our sun (Alvarez and Muller 1984). It is interesting that recent helium isotope data (see Sects. 3.3 and 5.8.4) support the possibility of comet showers in the geologic past, but without any periodicity implications. The existence of crater chains on some of the large moons of Jupiter (Ganymede and Callisto; Melosh and Schenk 1993) and on the Moon (Melosh and Whittaker 1994) led Rampino and Volk (1996) to suggest that similar events could have happened on Earth. Despite support of this hypothesis by Spray et al. (1998), there may so far not be enough data to confirm or reject this proposal.

In a recent paper, Murray (1999) presented arguments based on celestial mechanics to explain the non-random clustering of aphelion distances of long-period comets (which are those that supposedly originate from the spherical Oort cloud around the solar system) by the possible presence of a so-far undiscovered distant planet with a mass several times that of Jupiter. Given the uncertainty in the ages of impact structures, the small number of precisely dated craters, and the apparent lack of a periodicity in the cratering record, it remains open if the existence of such a hypothetical planet or star can be confirmed by geological observations.

We have come a long way since Boon and Albritton (1938) presented their first list of possible meteorite craters and since Bucher (1963) disputed the existence of most impact structures on Earth. Today we have realized that impact events have played an important role in the geological evolution of the Earth, from the formation of the Moon (Cameron and Benz 1991; Taylor 1992, 1993) to the mass extinction at the end of the Cretaceous (Glen 1998), and that much can be learnt from the study of impact craters (Shoemaker 1977) and their distal ejecta, as described in the next chapters for Italian occurrences.

4 The Umbria–Marche Sequence

4.1
Tectonic Setting

The Umbria–Marche Apennines of northeastern Italy is a foreland fold-and-thrust belt, which was formed in the latest phase of the Alpine–Himalayan orogenesis (Fig. 4.1.1). These mountains are entirely made of marine sedimentary rocks of the so-called Umbria–Marche (U–M) Sequence, which represents a continuous record of the geotectonic evolution of an epeiric sea from the Late Triassic to the Pleistocene.

The Late Triassic–Early Jurassic rifting between Europe and Africa, at the time of the opening of the North Atlantic, led to the formation of new oceanic basins ancestral to the present Alpine mountain chains, including the Pennide–Liguride Ocean (Dercourt et al. 1993). This new ocean, perhaps connected with part of the ancient Tethyan Ocean to the east, outlined a northward-pointing promontory of African continental crust, commonly referred to as Adria or the Adriatic Promontory (Channel et al. 1979). This northward-pointing promontory was in many ways similar to the present south-pointing promontory of North America, which forms Florida and the Bahamas (D'Argenio 1970).

The Adriatic Promontory was isolated from the input of clastic sediments. As a large and nearly isolated continental passive margin, Adria underwent extensional faulting. Normal faults defined a complex of subsiding blocks leading to an irregular topography of structural highs (horsts) and adjacent depocenters (grabens and/or half-grabens). In cases where shallow water carbonate deposition could keep up with subsidence, and where the faulted blocks where large enough to support productive carbonate platform environments, very thick sequences of shallow water carbonates developed on the Adria's crust; a prominent example is the Central Apennines of the Abruzzo region. In other regions, subsidence and complex block faulting carried the sea floor down, below the photic zone and out of the zone of shallow-water carbonate production. Areas with this history became pelagic basins, such as the U–M basin, and they recorded, with remarkable continuity and completeness, the geologic, biologic and oceanographic evolution of this region (Fig. 4.1.2).

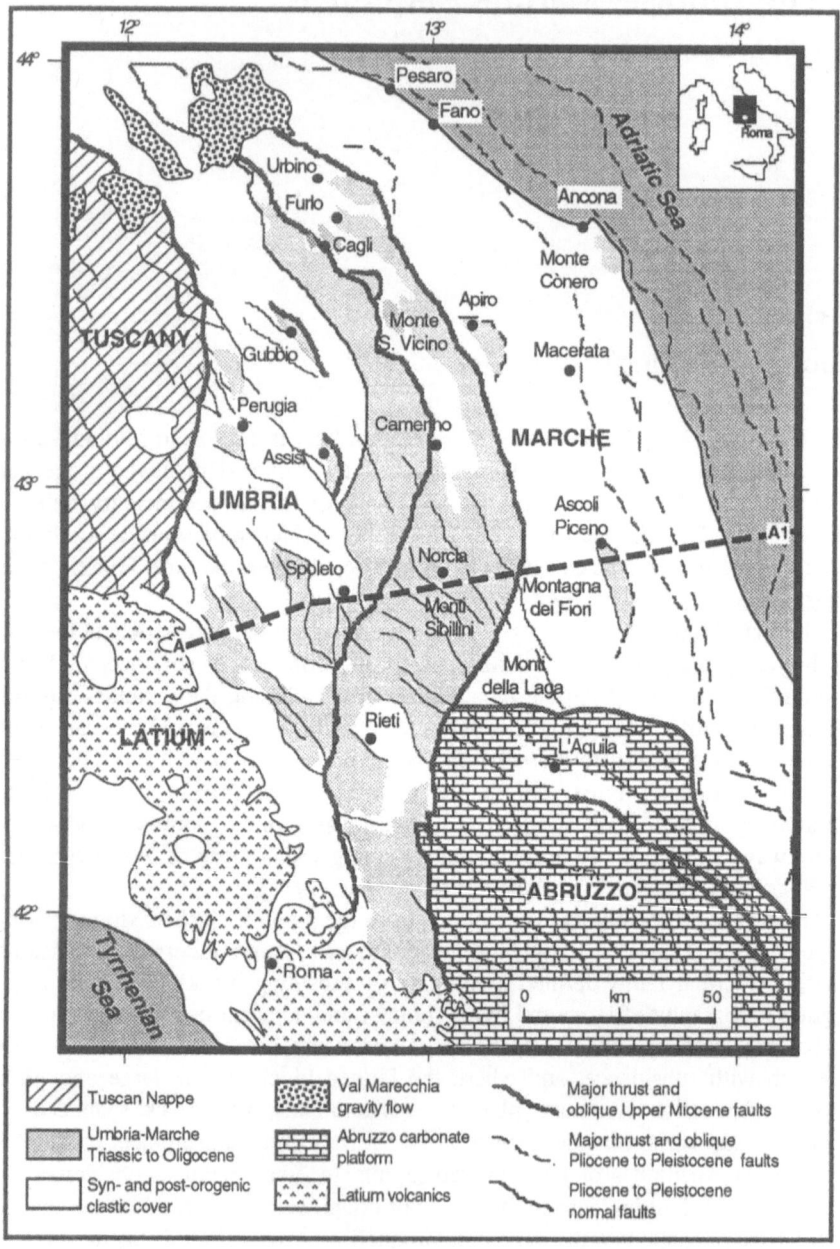

Fig. 4.1.1. Simplified geologic map of the northeastern Apennines. Fault pattern is from Lavecchia et al. (1994).

Following regional extensional tectonics from the Triassic to the Paleogene, the U–M Basin experienced an inversion of the tectonic regime in the Miocene, after which sedimentation was mainly controlled by compressional tectonics. At this time in the evolution of the paleobasin, the Apennine structures were formed in an active fold-and-thrust belt by the motion of thrust sheets from the southwest to the northeast (Fig. 4.1.3).

The broad, topographically even, sea floor was transformed in a variety of sub-basins at different scales. Loading of the crust by the advancing thrust sheets caused isostatic subsidence and formed a broad foredeep which gradually migrated toward the northeast; its position at the present time is marked by the foredeep basin that roughly corresponds to the Adriatic coastline of the Italian peninsula. Smaller synclinal basins were formed on the flanks of rising anticlines, and were carried to the northeast as piggyback basins. Meanwhile, the Miocene flysch facies reached the U–M region, spreading first as sheets of turbidite material across the region, and subsequently as narrow turbidite fingers reaching into the small synclinal basins left between rising anticlines (Centamore et al. 1978). Continuing thrust motion deformed the earlier turbidites, and the interplay of thrusting and turbidite deposition produced an extremely complex sedimentary geometry.

The U–M sedimentary sequence can be divided into two main parts: an upper Triassic to lower Miocene carbonate sequence, and a Middle Miocene to Pleistocene siliciclastic sequence (Fig. 4.1.2). The carbonate sequence was deposited during a long time of extensional tectonic regime and it is mainly represented by deep-water, pelagic limestone and marl formations. On the other hand, the U–M siliciclastic sequence is represented by syn- and post-orogenic clastic deposits, including the flysches of the Marnoso–Arenacea and Laga formations, the black shales and evaporites of the Messinian Gessoso–Solfifera Formation, the turbiditic silt and clay of the Colombacci Formation, and finally, by various sandy beach, and molasse deposits representing the ultimate emergence of the Apennine orogen.

In this book, we focus our attention on the carbonate sequence of the U–M Basin, which is mainly represented by deep water, fossiliferous, pelagic limestones and, thus may be considered as an excellent recorder of both local and global events for the past 200 Ma of Earth history. In fact, it has been in this sequence exposed near the Medieval city of Gubbio that Alvarez et al. (1980) first discovered the evidence for an extraterrestrial impact at the Cretaceous–Tertiary (K/T) boundary. The detection of a few part per billion (ppb) of the element iridium in the thin clay layer marking the K/T boundary opened the way to two decades of research activity, by a multitude of interdisciplinary earth and planetary scientists throughout the world, on the impact-induced mass extinction that ended the Mesozoic Era and the rule of the dinosaurs on Earth. And it was in this sequence, in the Global Stratotype Section and Point (GSSP) at Massignano, where Farley et al. (1998) discovered the first geochemical evidence for a comet shower through a stratigraphic interval (the terminal Eocene), in which multiple impacts were already reported by a number of researchers worldwide (see Montanari et al. 1993, and references therein).

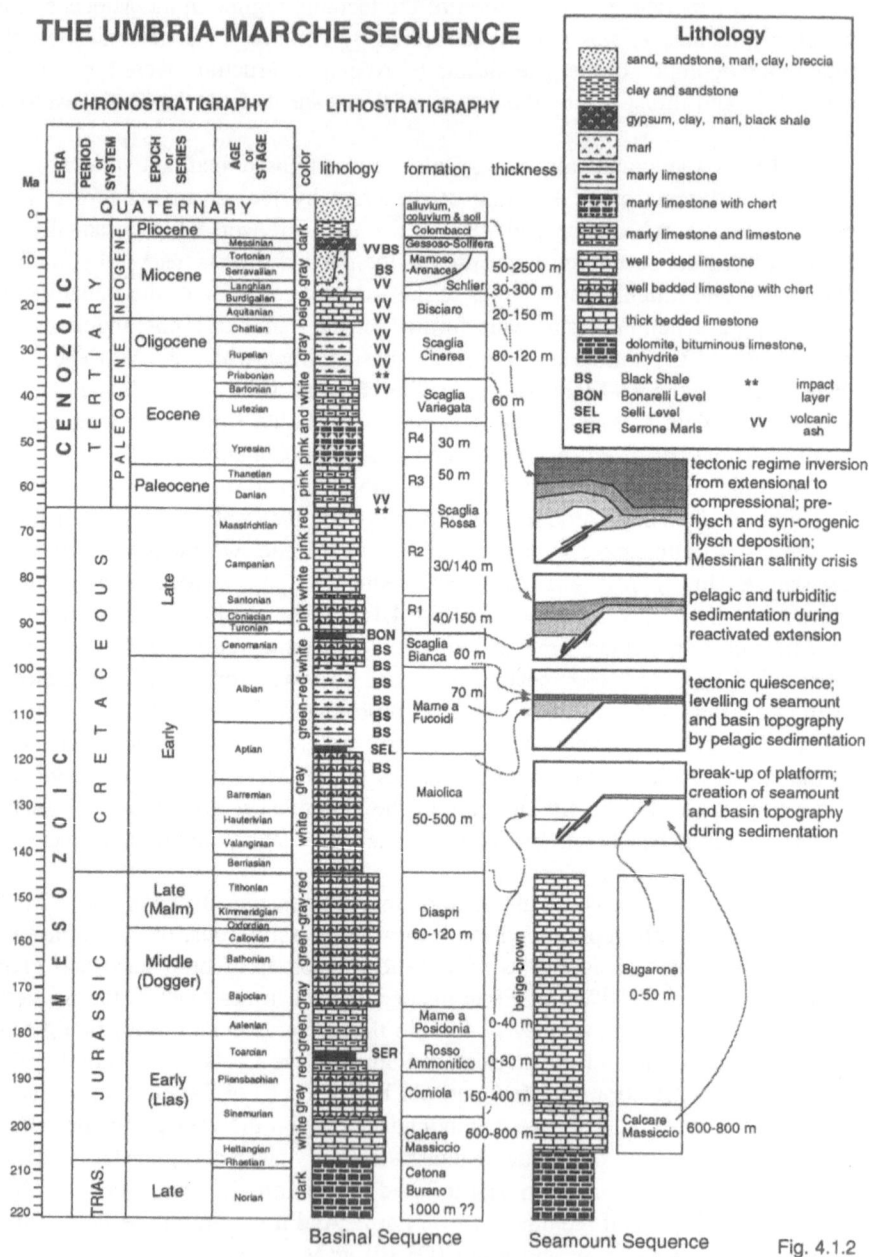

Fig. 4.1.2

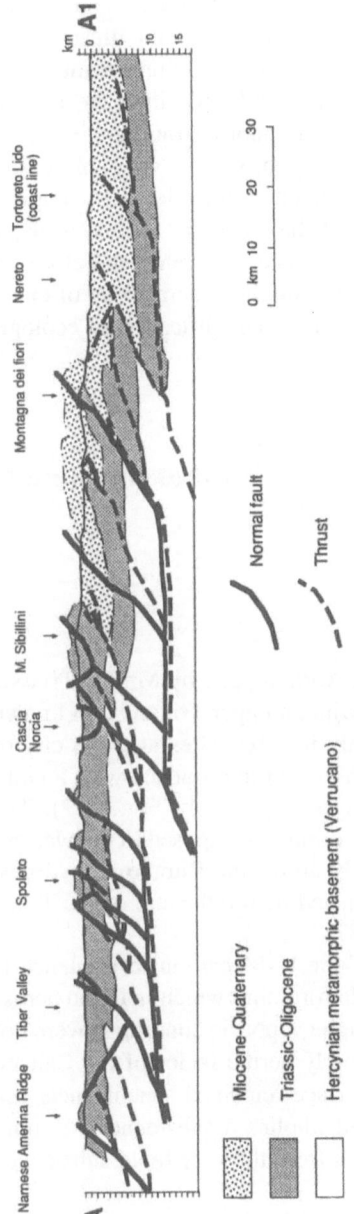

Fig. 4.1.3. Geologic section across the Umbria–Marche Apennines (see A–A1 in Fig. 4.1.1), redrawn and simplified after Lavecchia et al. (1994).

! **Fig. 4.1.2.** Stratigraphic and tectono-sedimentary synthesis of the Umbria–Marche Sequence.

We find it useful to give herein a brief summary on the sedimentological and stratigraphic characteristics of the U–M carbonate sequence, and illustrate the basic environmental conditions in which the signatures of distal impacts may have been preserved here. In fact, besides the well-studied K/T and the Late Eocene records, many other stretches of complete and continuous stratigraphic record exposed in several outcrops throughout the U–M region, were never explored for impact signatures, which, if present, may be correlated with evidence for impacts in coeval sedimentary sequences in other parts of the world, or with well dated impact craters elsewhere. The U–M carbonate sequence represents, therefore, an ideal ground to further the interdisciplinary research on the general issue of extraterrestrial impacts and the role they may have had in the modification of ecologic environments at local and global scales.

4.2
Stratigraphic Framework of the U–M Carbonate Sequence

4.2.1
Triassic–Lower Cretaceous

Above the Hercynian metamorphic basement of Adria topped by Middle Triassic continental deposits, the U–M sequence begins with an upper Triassic unit known as the Burano Anhydrite. This unit, exposed only in a few sites near the city of Perugia, and better known from boreholes, is an evaporite sequence with a complex facies geometry and locally it is as much as 1 km thick (Fig. 4.1.2). The *Rhaetavicula contorta* limestones of the Cetona Formation, exposed in the western part of the Umbrian region, lays stratigraphically above the Burano Anhydrites, and indicate that the evaporitic conditions terminated before the end of the Triassic (Martinis and Pieri, 1962; Ciarapica et al. 1987).

The oldest extensively exposed formation of the U–M carbonate sequence is the Hettangian Calcare Massiccio, a carbonate platform unit which is found across the entire area of the northern Apennines. Although evaporitic conditions were not present at this time throughout the basin, the clearly neritic facies of the Calcare Massiccio indicates that the sediment-water interface remained steadily near sea level. A typical thickness of 700 m for this unit implies a subsidence of about 100 m/Ma. Although rapid, this rate is a reduction from the very rapid subsidence in the Late Triassic, which averaged 170 m/Ma.

In the Calcare Massiccio there is a clear facies difference between areas that probably represent differential subsidence of faulted blocks. By the mid-Lias, the entire U–M basin had subsided to depths where neritic carbonate deposition was no longer possible, and pelagic sedimentation begun. Throughout the rest of the Jurassic, sediments deposited in the U–M basin show a clear facies difference between deeper basinal facies (grabens) and shallow water seamount facies (horst; see Fig. 4.1.2). These facies variations represent differential subsidence of fault-bounded crustal blocks inherited from the early Liassic extensional tectonic phase (e.g., Bice and Stewart 1985; Alvarez 1989a, b).

In basinal areas, the rest of the Jurassic is represented by the "Complete Sequence", which is about 300 m thick, overlaid by the Maiolica Formation, which may reach up to 500 m of thickness. The Complete Sequence (Còrniola, Rosso Ammonitico, Marne a Posidonia, and Diaspri formations) represents areas of greater subsidence rate, and is characterized by limestones often interbedded with radiolarian cherts. Frequent slumps, pebbly mudstones, megabreccias, and turbidites reflect the presence of steep slopes within the basin, which, in turn shows that extensional faulting was still active (e.g., Castellarin et al. 1978; Coltorti and Bosellini 1980; Lowrie and Alvarez 1984; Bice and Stewart 1986; Alvarez 1989a, b). During deposition of the thick, basinal Maiolica formation, these indications of sedimentary and tectonic instability are gradually reduced.

By the mid Cretaceous, the deposition of the Selli Level, a regional, 1 m thick, bituminous regional marker bed at the bottom of the Aptian–Albian Marne a Fucoidi Formation (Coccioni et al. 1987), and the uniformity of thickness and facies of these pelagic sediments throughout the U–M region (e.g., Fiet 1998), indicate that differential subsidence was largely terminated and that the sedimentary filling had almost completely levelled out the irregular topography of the paleobasin.

In the shallower, seamount areas (Fig. 4.1.2), the Jurassic, from late Pliensbachian time onward is represented by the very thin (50 m maximum) "Condensed Sequence" (Colacicchi et al. 1970; Centamore et al. 1971; Coltorti and Bosellini 1980; Farinacci et al. 1981). The entire Condensed Sequence at Monte Nerone (northwestern Marche) is represented by the Bugarone Formation, which has been divided into five members (Alvarez 1989 a, b). Cecca et al. (1990) demonstrated the presence of a 25 Ma hiatus within the Bugarone, representing the interval from the middle of the Bajocian through the early Kimmeridgian. Above the Bugarone, at Monte Nerone, there is a second major hiatus representing the whole Barriasian–Valanginian time. This is overlaid by the thin (less than 100 m) seamount version of the Maiolica Formation, lithologically similar to the basinal, up to 500 m thick, Maiolica.

In the seamount areas, as in the adjacent deep basins, the regionally uniform Aptian–Albian Marne a Fucoidi Formation demonstrates the end of active differential subsidence, and the levelling out of the pelagic sedimentary fill throughout the U–M paleobasin.

4.2.2
Upper Cretaceous–Oligocene

During this second major episode in the stratigraphic history of the U–M basin, the role of active extensional faulting and differential block subsidence was greatly reduced. There must have been local, minor episodes of normal fault reactivation, because we do find slump folds and some turbidites, but these are far less conspicuous than in the basinal Jurassic and lower Cretaceous. Moreover, throughout most of the U–M region, the late Cretaceous–Early Tertiary facies virtually lack extraformational reworked material. Evidently the sea floor was essentially levelled across the paleobasin, and, as a result, the facies of a particular stratigraphic unit is uniform throughout the entire region. Only near the margins of the basin, for example, in the Monte Cònero area near the coastal city of Ancona, or in proximity of the Abruzzo region (Central Apennines), where a broad shallow-water platform endured through this time, do the basinal facies change dramatically in this case to a turbidite and slump-rich slope facies, including debris directly derived from eroded carbonate platform sequences (Crescenti et al. 1969; Baldanza et al. 1982; Alvarez et al. 1985; Colacicchi and Baldanza 1986; Monaco et al. 1987).

This time of relatively mild synsedimentary tectonism that followed the rather quiet sedimentation of the Aptian–Albian Marne a Fucoidi is represented by the four formations composing the Scaglia sequence (Scaglia Bianca, Scaglia Rossa, Scaglia Variegata, and Scaglia Cinerea; see Fig. 4.1.2). The Italian term Scaglia means "scale" or "flake"; it refers to the thin, tabular stratification of this homogeneous pelagic limestone, often showing dense concoidal and penciling fracture systems, which give the typical scaly aspect to the tectonically disturbed and/or altered rock.

The Scaglia Bianca has a gradational contact with the underlying Marne a Fucoidi as the carbonate/clay ratio increases upwards, and interbedded black shales disappear. It is a mostly white biomicritic limestone containing nodular chert, with a reddish-purple member in the middle, and is about 60 m thick. In the upper 20 m of the unit, white limestone layers are interbedded with black, tabular, laminated chert containing recrystallized radiolarian skeletons and silicified planktonic foraminiferal tests. These black cherts are often associated with thin black shale layers. The analysis of the distances between successive black (shale and/or chert) levels shows that they are organized as multiples of 20 cm (20, 40, 60, ... n20), recording, in fact, a Milankovich-type cyclicity (20 ka, 40 ka, 100 ka; Beaudoin et al. 1996). The Scaglia Bianca represents the latest Albian and the Cenomanian, and the uniform thickness and correlative pattern of the black chert beds throughout the paleobasin (Beaudoin et al. 1996), indicate that this was still an essentially quite tectonic time with regionally uniform sedimentation.

The top of the Scaglia Bianca has been often considered coincident with the Bonarelli Level (e.g., Montanari et al. 1989), which is a prominent, carbonate-free, regional marker bed about one meter thick, very similar to the Aptian Selli Level, and representing an oceanic anoxic event close to the Cenomanian–Turonian boundary (e.g., Arthur and Premoli Silva 1982; Beaudoin et al. 1996).

The Scaglia Rossa Formation represents the time from the Turonian through the Early Eocene. A short distance above the Bonarelli Level, the pelagic limestones of this unit take on a pink color, and oxidizing conditions prevailed in the basin from then on through the rest of Scaglia Rossa time. The Scaglia Rossa can be divided into four members (Fig. 4.1.2). The lower (R1) member is characterized by the presence of red chert beds; it is overlaid by a chert-free, predominantly calcareous member (R2), followed by a Paleocene, chert-free, member (R3) characterized by marly limestones. The K/T boundary has been used to define the boundary between the R2 and R3 members of the Scaglia Rossa. It is recognized by the disappearance of Cretaceous planktonic Foraminifera and calcareous nannofossils, as well as by a white, bleached zone, about 20–30 cm thick, which commonly occurs just below the boundary (Luterbacher and Premoli Silva 1964; Lowrie et al. 1990; Montanari 1991). The upper member of the Scaglia Rossa (R4) is characterized by the renewed appearance of red, nodular chert beds. Compared to the very quiet Scaglia Bianca, the Scaglia Rossa shows an evident increase in syndepositional tectonism. Slump folding is noted in several levels within this unit (Baldanza et al. 1982; Alvarez and Lowrie 1984; Alvarez et al. 1985; Chan et al. 1985; Montanari et al. 1989), and in parts of the basin there are striking occurrences of white calcarenitic turbidites essentially made of reworked, intraformational planktonic foraminiferal tests (Montanari et al. 1989).

The Scaglia Variegata Formation represents the transition from the pink limestones and red cherts of the upper Scaglia Rossa to the gray marly limestones of the Scaglia Cinerea Formation. It is characterized by chert-free marly limestone beds of pink, white, and gray color. The lower boundary is marked by the last occurrence of nodular chert (Lowrie et al. 1982; Chan et al. 1985). The top of the Scaglia Variegata is conventionally located at the top of the uppermost reddish interval (e.g., Odin and Montanari 1988; Coccioni et al. 1988; Montanari et al. 1993). This unit represents the Middle Eocene and most of the Late Eocene.

Above the Scaglia Variegata, the thick, rather uniform, and apparently more marly Scaglia Cinerea Formation begins. "Cinerea" means ashy in Italian, but refers to the gray color rather than to the several biotite-rich volcanic ashes contained in this unit (e.g., Montanari et al. 1985, 1988b). Synsedimentary disturbances, such as calcareous turbidites made of extraformational, shallow-water debris, commonly occur in the southern Umbria region, in areas close to the margin of the Lazio–Abruzzo carbonate platform (Monaco et al. 1987). Throughout the rest of the U–M region, the Scaglia Cinerea testifies to rather quiet tectonic conditions in the paleobasin following the mild synsedimentary, extensional tectonic phase of the Scaglia Rossa described above, and preceding the onset of the Miocene synorogenic sedimentation. However, the relative abundance of clay in the Scaglia Cinerea is evidence of tectonic events not far away to the west, which were soon to overrun the area of the U–M Basin and bring to a close the long pelagic, carbonate phase of its evolution.

The Scaglia Cinerea represents the Late Eocene and the Oligocene, a time during which the siliciclastic turbidites of the Macigno flysch were being deposited in the Tuscan basin to the west (Abbate and Sagri 1970; Valloni and Zuffa 1983). These turbidites were derived from the erosion of the Alps, and their spread into the area of the future Apennines coincided with the beginning of tectonic deformation of the sedimentary cover in the western part of the Apennines. It seems likely that the clay of the Scaglia Cinerea was derived from the clay component of these turbidites in the area just to the southwest.

4.2.3
Miocene

Following the extensional tectonics of the Triassic–Early Cretaceous and Late Cretaceous–Early Eocene phases, and the relative tectonic quiescence of the Middle Eocene–Oligocene period, the third phase clearly reflects a sedimentation controlled by compressional tectonics. At the beginning of the Miocene, marly deposition of the Scaglia Cinerea was interrupted by a brief return to dominantly limestone deposition. The resulting unit is the Bisciaro, a formation whose geometry and significance are not fully understood despite its excellent integrated stratigraphic characteristics recently determined by a number of authors in several U–M sections (e.g., Amorosi et al. 1994; Montanari et al. 1991; 1997a, b; Coccioni et al. 1997; Deino et al. 1997).

The Bisciaro consists of rhythmic interbeds of shales, volcanic ashes, sandy (and sometimes glauconitic) limestones, and very hard, pure biomicritic limestones. The unit ranges in thickness from 20 to 150 m (Guerrera et al. 1986). It has been suggested (Montanari et al. 1991) that the Bisciaro near Gubbio (eastern Umbria) may represent the slight bulging of the basin floor caused by the advancing load of the northeast-moving Apennine thrust sheet.

The Bisciaro is overlaid by the Schlier Formation, consisting of a rhythmic sequence of pelagic and hemipelagic marls and marly limestones containing sporadic intercalations of volcanic ashes (Deino et al. 1997; Montanari et al. 1997b). In the western part of the U–M Basin (i.e., the area around Gubbio), the Schlier is a few tens of meters thick and its hemipelagic deposition is interrupted, in the Langhian, by the arrival of syn-orogenic siliciclastic turbidites of the Marnoso Arenacea flysch. On the other hand, in the easternmost part of the region (i.e., Monte Cònero area), the Schlier reaches a total thickness of over 300 m and covers the whole interval from the top of the Burdigalian to the Tortonian–Messinian boundary. This chronostratigraphic boundary roughly coincides to the base of the Gessoso–Solfifera Formation, which starts with the Euxinic Shale unit that marks the beginning of the Messinian salinity crisis of the proto-Mediterranean. The Apennine synorogenic siliciclastic facies arrived in the Cònero area only in the Pliocene. Thus, this area represents a unique situation in the region where the entire Miocene is still represented by an essentially continuous, and complete sequence of pelagic and hemipelagic carbonate rocks, and constitutes an excellent record of local and possibly global paleoecologic events (Montanari et al. 1997b).

5 Impact Stratigraphy in the U–M Sequence

5.1
Searching for Impact Signatures in the Stratigraphic Record

The distal ejecta of an impact found in a sedimentary layer, and a biologic event that may be caused by an impact, have to be dated in terms of numerical age (Sect. 1.9) in order to be correlated with a known impact crater. The most reliable way to assign a precise age to a crater is to date its impact melt rock with a radio-isotopic method. In fact, the majority of terrestrial impact craters are found on land and are buried or filled by incomplete sequences of continental sediments that do not permit a direct and accurate stratigraphic correlation with marine, fossili-ferous sediments. In some other cases, impact craters are filled with marine sedi-ments with a disturbed (i.e., mixed or reworked) paleontological record that is not conducive to precise and accurate numerical dating of the impact event.

Marine sedimentary sequences, on the other hand, are dated by interpolation of few radioisotopic ages, obtained from interbedded volcanic ashes in distant sec-tions around the world, and tied to paleontological, geochemical, and geophysical signatures which are assumed to be contemporaneous worldwide. Therefore, the strength of a correlation between an impact signature found in the sedimentary record and an apparently contemporaneous crater is solely based on the accuracy and precision of the geochronologic time scale, and the reliability of the geochro-nologic date of that crater.

Only a small proportion of the 160 or so impact craters known on Earth have precise ages, which allow an accurate correlation with the stratigraphic record. Moreover, the more we go back in geologic time, the poorer is the accuracy of the time scale, and the rarer are continuous and complete sedimentary sections with well defined bio-, magneto-, and chemostratigraphic characteristics. Conse-quently, for Paleozoic series or older, the geochronologic calibration of chrono-stratigraphic units is largely imprecise and probably inaccurate. The geochro-nologic calibration for Mesozoic series and younger is much more precise and accurate, and sedimentary sequences covering this latest span of Earth's history are much more frequently and extensively exposed than are Paleozoic or older sequences.

In the general field of research on the role of impacts in the evolution of the Earth's biota, a first-order approach would be to search for impact evidence (i.e., geochemical signatures and/or impact spherules and shocked minerals) in those stratigraphic intervals where major biotic crises or mass extinctions are known to occur. For the interval between the Permian–Triassic (P/TR) boundary to the Present, the most recently updated diagram of Sepkoski (1996), which reports the percentage of extinctions of marine fossil genera across interstage boundaries, constitutes a good start for this kind of approach (Fig. 5. 1.1). Three major peaks of extinctions occurred in this 250 Ma interval: at the P/TR boundary (68% extinction of the marine animal genera), in the latest Norian (TR/J boundary; 47% extinction), and at the K/T boundary (48% extinction). Then, there are six other relevant but smaller extinction peaks across the Pliensbachian–Toarcian boundary (P/T boundary; 27% extinction), the Tithonian–Barriasian (Jurassic–Cretaceous, J/K) boundary (32% extinction), the Aptian–Albian (A/A) boundary (18% extinction), the Cenomanian–Turonian (C/T) boundary (28% extinction), the Late Eocene (LE; 16% extinction), and, finally, the Middle/Late Miocene boundary (M/M boundary; 9% extinction). Relatively high levels of extinction percentages are reported also in the mid–upper Jurassic, but no clearly defined peak is present here. According to Sepkoski (1990), these extinction peaks are regularly spaced by intervals of about 26 Ma.

Fig. 5.1.1 also reports the temporal distribution of the known terrestrial impact structures with diameters larger than 5 km (Table 1.9.1). The temporal distribution of the impact structures in the past 150 Ma (i.e., from the J/K boundary to the Present) was studied in detail by Montanari et al. (1998), who did not find any significant periodicity by statistical means. The flux of total energy in megaton delivered to the Earth in the past 150 Ma by large impactors is plotted in Fig. 5.1.2. This figure shows that three peaks of extinction, the J/K, K/T, and LE, correlate precisely with major impact events, the almost coeval Morokweng structure (70–100 km diameter) and the Mjølnir crater (40 km diameter), the Chicxulub structure (about 200 km diameter), and the coeval Chesapeake and Popigai craters (about 100 km each), respectively. Correlations between impact structures and other extinction peaks in the J/K to Present interval are very week or absent.

As for the Permian through Jurassic time interval, only few impact structures are known, and most of them have large age uncertainties. Among these, only Manicouagan (100 km diameter) may possibly be correlated with the TR/J extinction event (Fig. 5.1.1).

Fig. 5.1.1. Impact structures with diameters larger than 5 km, impact signatures known in the stratigraphic record, and percentages of marine genera extinctions in the past 250 Ma. The marine genera extinction record is from Sepkoski (1996). →

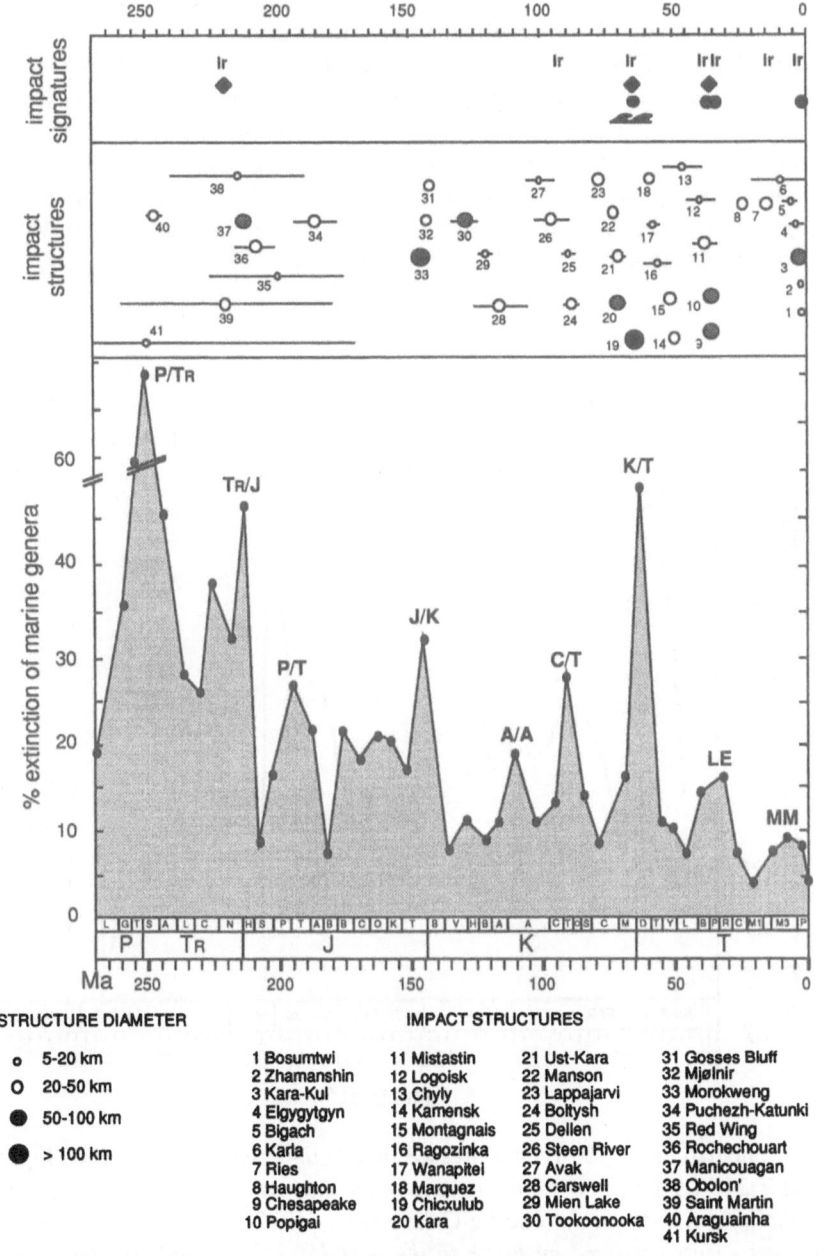

STRUCTURE DIAMETER

- ○ 5-20 km
- ○ 20-50 km
- ● 50-100 km
- ● > 100 km

IMPACT STRUCTURES

1 Bosumtwi	11 Mistastin	21 Ust-Kara	31 Gosses Bluff
2 Zhamanshin	12 Logoisk	22 Manson	32 Mjølnir
3 Kara-Kul	13 Chyly	23 Lappajarvi	33 Morokweng
4 Elgygytgyn	14 Kamensk	24 Boltysh	34 Puchezh-Katunki
5 Bigach	15 Montagnais	25 Dellen	35 Red Wing
6 Karla	16 Ragozinka	26 Steen River	36 Rochechouart
7 Ries	17 Wanapitei	27 Avak	37 Manicouagan
8 Haughton	18 Marquez	28 Carswell	38 Obolon'
9 Chesapeake	19 Chicxulub	29 Mien Lake	39 Saint Martin
10 Popigai	20 Kara	30 Tookoonooka	40 Araguainha
			41 Kursk

◆ shocked quartz   ● microspherules   ◀▰ megawave deposit   Ir iridium anomaly

Fig. 5.1.1

Montanari et al. (1998) pointed out that there is an obvious scarcity of impact signatures (iridium anomalies, spherules, shocked minerals, etc.) known at present in the stratigraphic record, compared to the number of large craters on the surface of the Earth (see Fig. 5.1.1). They explained that this "...*is mainly due to the fact that the search for signatures such as Ir anomalies, spherules, and shocked minerals, has been mainly focused across those few short stratigraphic intervals where major extinctions are recorded, or which were known to cover the time of major impacts. The work of the stratigrapher searching for a millimetric impact layer in a sedimentary sequence hundreds of meters thick is comparable to the proverbial search for the needle in the haystack*".

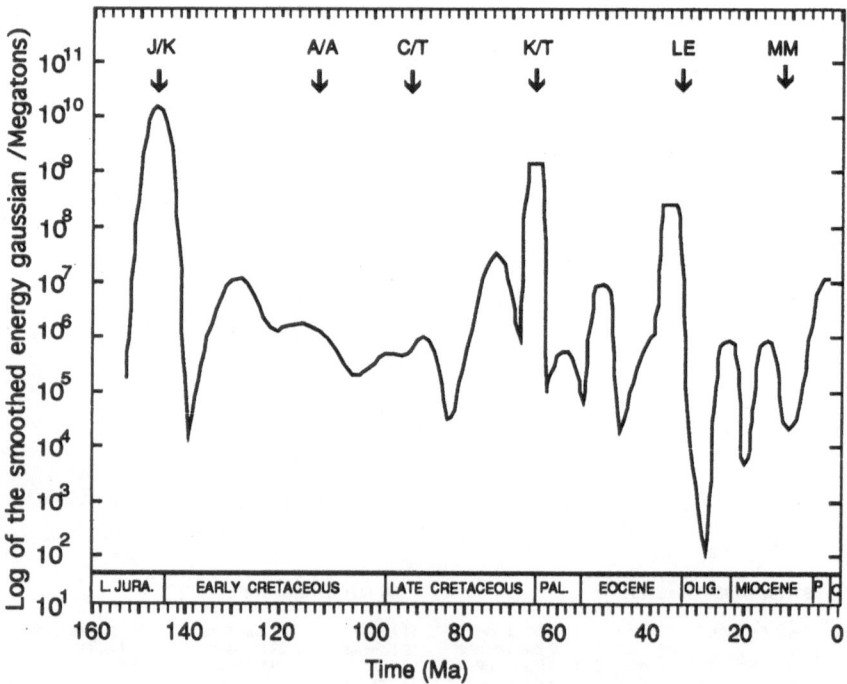

Montanari et al. (1998) also stressed that "...*the record shown*... (i.e., Fig. 5.1.1 and Table 1.9.1 in this work), *or any other that can be derived from published compilations such as the one by Grieve et al. (1995), does not represent all the impact events occurred on Earth in the past 150 Ma, but only the few ones that have left their signature on the Earth's surface and have been discovered up to now*". These authors, then, explained the reasons that limit the effectiveness of statistical work on the Earth's impact record, and pointed out the main problems that are encountered in studying the effect of impacts on the evolution of the Earth from the geological and stratigraphic records.

"There are at least three types of problems which limit the accuracy for a detailed statistical study of the impact record through Earth history.(...) A first source for analytical inaccuracy lies in the actual incompleteness of the cratering record due to the fact that the Earth's surface is extremely dynamic on a geologic time scale: many craters may have been completely obliterated by erosion or tectonic deformation, or buried under orogens, or subducted under tectonic plates. Moreover, the size vs. frequency relationship of impacts follows approximately a power law: small events (craters 20 km in diameter or less) are much more frequent than medium size events (20 to 100 km in diameter) which, in turn, are much more frequent than giant impacts producing craters larger than 100 km. On the other hand, the preservation probability of impact craters grows with their size: small structures may be erased from the record much more easily than large ones. Consequently, the record in Figure 1 and Table 1 [in Montanari and co-author's paper] shows an abnormal abundance of medium size craters (averaging 25 km in diameter), and an anomalous scarcity of small size craters which, according to a power law distribution, should be much more numerous.

This observation also suggests that the probability for a giant (~100 km) crater to be obliterated and, thus, unrecorded in the examined time interval is relatively low. Nevertheless, medium or large impact events are the ones that count more in the general study of impact flux and possible cause-and-effect relationship between impacts and biologic crises. Large impacts are those which may leave a world-wide signature in the geologic record, and may have climatic, environmental, and biologic effects on a global scale.

A second problem lies in the fact that many impact craters have been discovered recently, at a rate of approximately 1–2 new ones per year. This is due to the fact that on Earth there are large areas that have not yet been thoroughly explored for impact structures. Moreover, about 2/3 of the Earth's surface is made of oceanic crust, and at present all the known impact craters are located on continental crust...

Finally, there are still basic geochronologic uncertainties in the age estimate of impact craters.(...) For instance, for a long time the medium size Manson crater was believed to have an age close to the K/T boundary (i.e., about 66 Ma; Hartung et al. 1990). Recently, the melt rock of this crater has been re-dated yielding an age about 8 Ma older (Izett et al. 1993; Zeitler 1996). Similarly the Kamensk, another medium size crater (see Table 1 [in Montanari and co-author's paper]) was for long time considered a K/T boundary event and only recently was it re-dated at 49.2 Ma (i.e., 16 Ma younger; Izett et al. 1994).

Another relevant case is the Kara–Ust Kara double structure, which was dated by a Russian team using the traditional K/Ar technique at around 66 Ma (Kolesnikov et al. 1988), and in an American geochronology laboratory using the $^{40}Ar/^{39}Ar$ technique at about 74 Ma (Koeberl et al. 1990). Further geochronologic dating by Trieloff et al. (1992) led to an age estimate between 69 and 71 Ma. A compromise (conservative) age of 73 ±3 Ma was finally suggested by Grieve et al. (1995) and is the one we use in this paper.

(from Montanari et al. 1998)

Considering all these uncertainties and problems, the U–M carbonate sequence appears a rare opportunity to approach the search for impact signatures in a stratigraphic and sedimentary framework which is remarkably continuous and complete. In the following chapters of this book, we describe the stratigraphic details of outcrops in the U–M region that record major biologic crises that are possibly linked to extraterrestrial events. For those instances where impact signatures have been found, like the K/T and the LE events, we illustrate the impact signatures in detail (i.e., geochemical anomalies, spherules, shock metamorphosed minerals, etc.). For the other potentially interesting intervals, in which impact signatures have yet to be found, but which may correlate with extinction peaks and/or large known craters, we limit ourselves to a description of the stratigraphy commenting on the completeness and significance of the paleontological record.

The next chapter of this book is presented as a state of the knowledge on the impact stratigraphy in the U–M sequence, and at the same time as a sort of guide book for this special geologic region, which will hopefully encourage further research on the fascinating subject of extraterrestrial impacts as possible controlling factors in the evolution of life on Earth.

5.2
The TR/J Record

Three closely spaced shale levels containing probable shocked quartz were found by Bice et al. (1992) in the uppermost *Rhaetavicula contorta* unit near Corfino in northern Tuscany, just below the lithostratigraphic boundary with the overlying Calcare Massiccio Formation, which represents the Triassic–Jurassic (TR/J) boundary. Although no further paleoecologic, paleontologic, petrographic or geochemical research has been carried out at this particular outcrop after the discovery by Bice and co-workers, recent studies on the terminal Triassic record (see Sect. 3.7.4) indicate that the Earth may have been struck by several impacts (Spray et al. 1998), at a time coinciding with a prominent mass extinction peak of marine organisms comparable with that of the K/T boundary (Sepkoski 1996; see Fig. 5.1.1). Thus, the TR/J boundary is definitively an interesting interval for investigating the possible relationship between impacts and mass extinctions.

The sedimentology and biostratigraphy of the *Rhaetavicula contorta* limestone have been studied in several sections of the Northern Apennines, most of which are located in Tuscany (e.g., Ciarapica et al. 1987; Cirilli et al. 1994). Unfortunately, only a few exposures of this unit are known at present in the U–M Apennines. The best documented of them is the Monte Cerchio section, in the Monti Martani area near Spoleto (southern Umbria; see Fig. 4.1), which is exposed along a mountain dirt road between the villages of Terzo San Severo and Massa Martana (Fig. 5.2.1). The lithostratigraphy and the biostratigraphy based on palynomorphs and benthic foraminifers have been worked out, in this section, by Cirilli et al. (1994). However, the stratigraphic data available today from this and other Umbrian TR/J boundary exposures are still too scarce to attempt any inference on the response of the local marine fauna and flora to the extraterrestrial multiple impacts proposed by Spray et al. (1998). One of the main problems is the inaccuracy of biostratigraphic tools for drawing correlations, especially in a basin with marked lateral variations of sedimentary facies and ecologic environments.

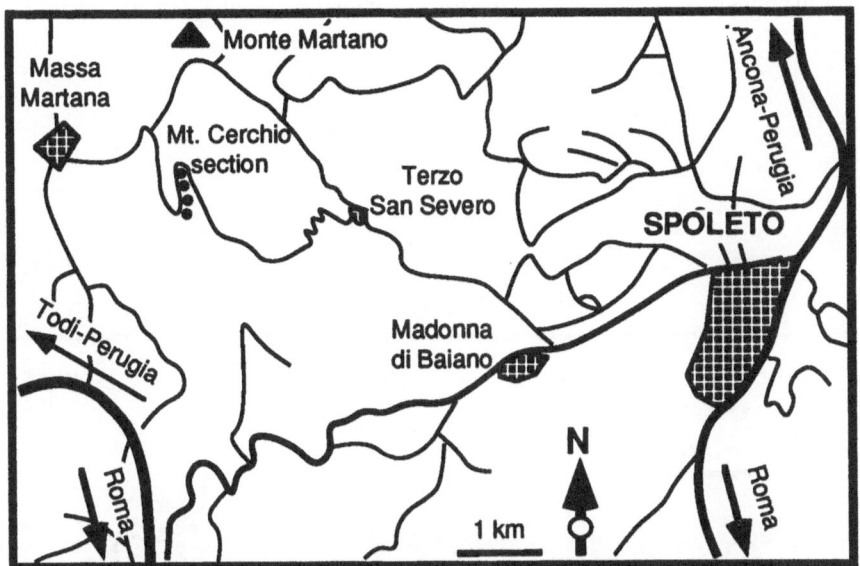

Fig. 5.2.1. Location map of the Monte Cerchio section covering the Triassic–Jurassic boundary.

One of us (A.M.) visited the Monte Cerchio section just before writing this chapter and realized that, despite the fair continuity of exposure, this outcrop is not quite suitable for the kind of high-resolution stratigraphical study necessary for pinpointing thin impactoclastic layers, such as those studied by Bice et al. (1992) in the Corfino section of Tuscany. The dolomitic limestones are fractured, and the bedding interfaces strongly sheared by tectonic deformation, especially near the contact between the *Rhaetavicula contorta* limestone and the overlying Calcare Massiccio Formation. Further research in the Triassic terrains of the Umbrian Apennines will hopefully yield, in the near future, better sections suitable for high-resolution impact stratigraphical studies.

5.3
The Pliensbachian–Toarcian boundary at Valdorbìa

According to Sepkoski (1996), an extinction peak of marine genera occurs at the boundary between the Pliensbachian and the Toarcian stages (see Fig. 5.1.1). In the time scale of Gradstein et al. (1994), this boundary is dated at 189.6 ± 4.0 Ma. The large uncertainty in the age assignment for this interstage boundary, which is about the same for all the interstage boundaries of the Early and Middle Jurassic, may be an impediment in the eventual correlation of geochemical and/or mineralogical signatures to a large impact. However, no large impact craters are known in this interval of geologic time and, thus, there is no hint in the geologic record that the Sepkoski (1996) peak across the P/T boundary is linked to an impact.

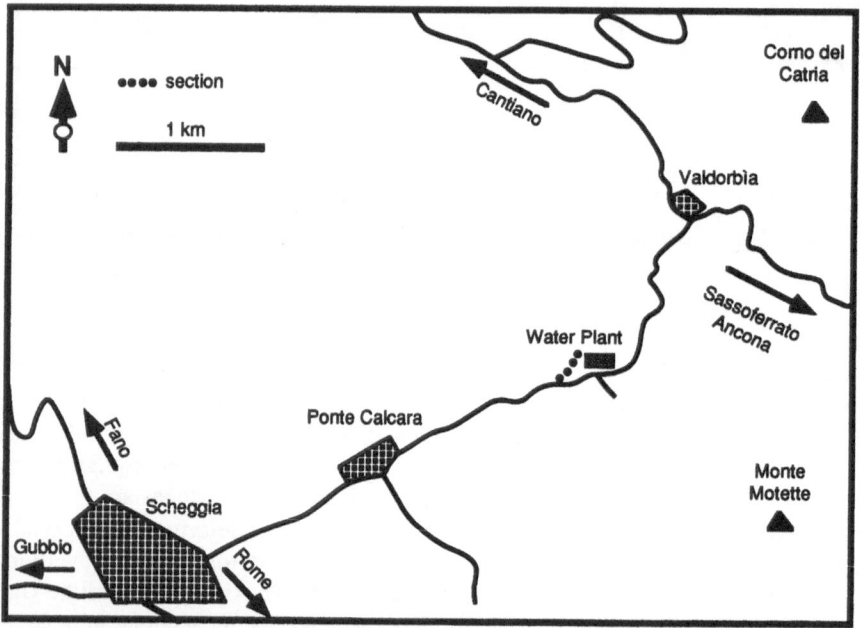

Fig. 5.3.1. Location map of the Pliensbachian–Toarcian section at Valdorbìa.

The Toarcian stage in the U–M Basin is almost entirely represented by a unit called the Marne del Serrone, locally found between the Sinemurian–Pliensbachian Còrniola Formation, and the upper Toarchian–lower Aalenian Rosso Ammonitico Formation (see Fig. 4.1.2). This unit is characterized by several horizons of black shales interbedded with clays and marls. This Toarcian anoxic event was recognized in other localities throughout the Tethyan domain (e.g., Pettinelli et al. 1995), as well as in Ethiopia (Canuti et al. 1983), and it has been attributed to a marine transgression over epicontinental areas (Hallam 1988; Haq et al. 1988). A biotic crisis associated with this event is probably the main factor making up Sepkoski's extinction peak, but at the current state of knowledge, this interesting event can only be attributed to global variations of sea level in a rifting geodynamic environment such as the U–M paleobasin.

One of the best sections to observe the P/T transition and the Marne del Serrone unit in the U–M region, besides the type locality at Monte Serrone, is located near the village of Valdorbìa, along the Arceviese road between the towns of Sassoferrato and Scheggia, and about 2 km east from the latter (Fig. 5.3.1). The main exposure is located within the yard of a mineral water plant (Fig. 5.3.2), but other stretches of the sequence are exposed in the immediate vicinity along the main road and the river, which flows at the bottom of the Gola del Corno valley.

Fig. 5.3.2. The exposure of the Marne del Serrone Formation at the Water Plant near Valdorbìa.

The stratigraphy of the Valdorbìa composite section, which can be correlated with other equivalent sections throughout the U–M basin (Bartolini et al. 1992), was worked out in detail by Monaco et al. (1994) and Parisi et al. (1996). A stratigraphic synthesis of this section is shown in Fig. 5.3.3.

176

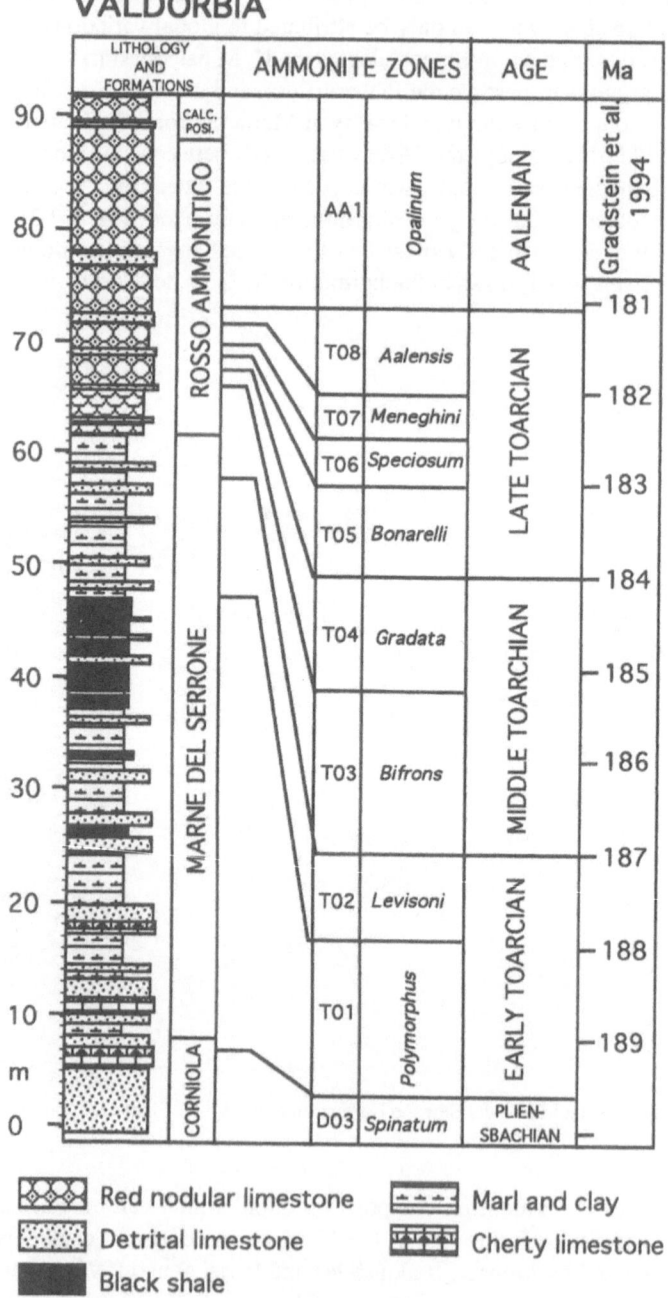

Red nodular limestone   Marl and clay
Detrital limestone   Cherty limestone
Black shale

Fig. 5.3.3. Stratigraphy of the Pliensbachian–Toarcian section at Valdorbìa (after Monaco et al. 1994, and Parisi et al. 1996).

5.4
The J/K Record

The fourth largest peak in Sepkoski (1996) diagram, with 32% extinction of marine fossil genera, is plotted at the Tithonian–Berriasian boundary (i.e., the J/K boundary). This event is associated with a minor decline in global diversity from one stage to the next although detailed quantitative micro- and macropaleontologic analyses are needed for a better local description of this event (Sepkoski 1996). Nevertheless, in the context of a possible cause and effect relationship between impacts and biotic crisis, the J/K transition is certainly an interesting one, because it covers the time of three large impacts (see Sect. 3.6): Morokweng (>70 km, maybe even up to 200 km in diameter), Mjølnir (40 km in diameter), and Gosses Bluff (22 km in diameter).

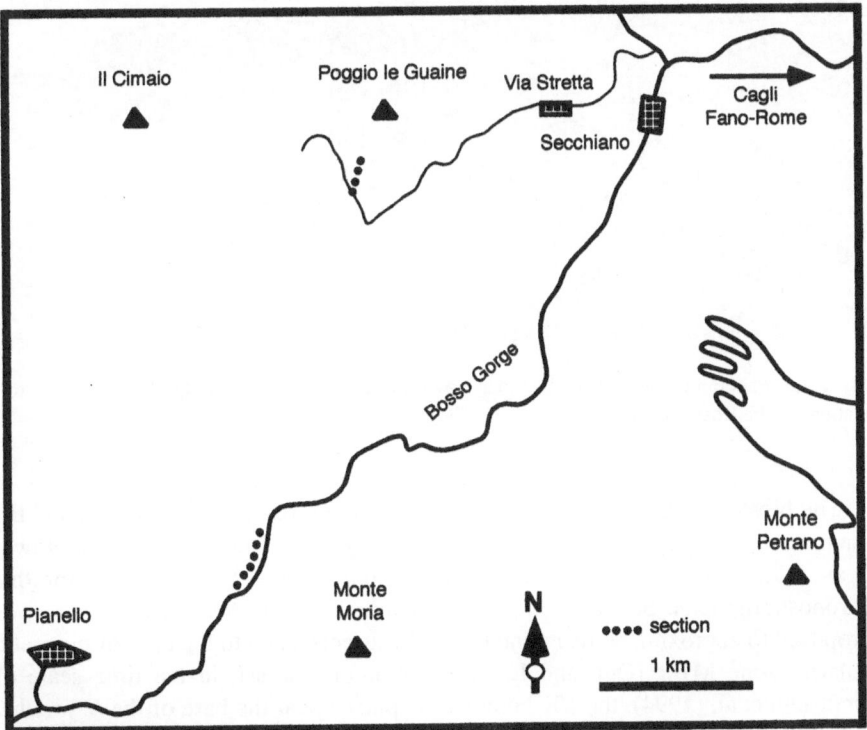

Fig. 5.4.1. Location map of the Jurassic–Cretaceous boundary in the Bosso section.

One of the best exposures of the J/K boundary in the U–M region is located along a road cut in the Bosso Gorge, near the town of Cagli (Fig. 5.4.1).

The boundary is found in the lowermost part of the Maiolica Formation, which has a short, transitional lower contact with the underlying Diaspri Formation. The Diaspri is characterized by thin bedded, variegated (green, red, gray), radiolarian-rich cherts and cherty limestones, whereas the Maiolica is a typically white, medium-thick bedded, biomicritic limestone interbedded with, or containing, gray and whitish nodular cherts (Fig. 5.4.2).

Fig. 5.4.2. Panoramic view of the boundary between the Diaspri (D) and the Maiolica (M) formations in the Bosso section.

The biostratigraphy based on calcareous nannofossils and calpionellids, and the magnetostratigraphy of this section (Fig. 5.4.3), were determined by Bralower et al. (1989). Although there is no internationally accepted definition for this chronostratigraphic boundary, the top of ammonite zone *Berriasella jacobi* was proposed to approximate the boundary, and was correlated to the base of magnetic polarity zone M18r (Ogg and Lowrie 1986). In contrast, in the time scale of Gradstein et al. (1994), the J/K boundary is placed near the base of the *B. jacobyi* Zone, in the upper part of magnetic polarity zone M19n. In the Bosso section, which contains no ammonites, this stratigraphic level corresponds to the top of calpionellid A3 Zone (i.e., top of *Calpionella brevis*), with an assigned age of 144.0 ± 2.6 Ma according to the time scale of Gradstein et al. (1995).

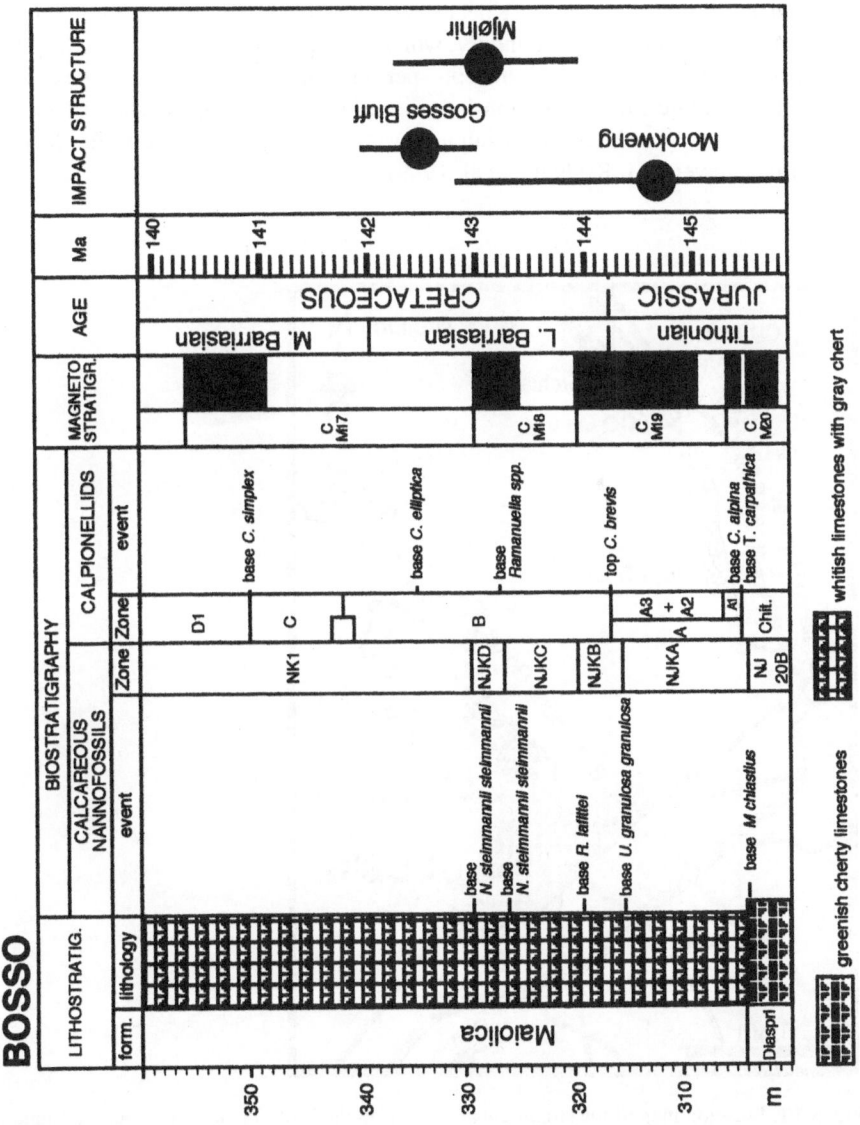

Fig. 5.4.3. Stratigraphic synthesis of the Jurassic–Cretaceous boundary in the Bosso section.

At the current state of knowledge, there seems to be no evidence for biotic crises in the calpionellid and calcareous nannofossil records through the lower part of the Maiolica Formation at Bosso. However, the lithofacies contrast between the Diaspri and the Maiolica formations is sharp enough to envision a major environmental change in this basin, which might correspond, indeed, to a significant explosion of the marine calcareous plankton. Despite the age uncertainties of this portion of the time scale (i.e., ± 2.6 Ma), and the age imprecision of the Morokweng giant impact (± 1.9 Ma; Koeberl et al. 1997c), the stratigraphic interval across the Diaspri–Maiolica boundary, which roughly corresponds to the transition between the Jurassic and the Cretaceous periods, may be regarded as an excellent situation for future high-resolution interdisciplinary stratigraphic studies, including a systematic search for distal impact ejecta possibly delivered by the Morokweng impact event (cf. Kudielka et al. 1999).

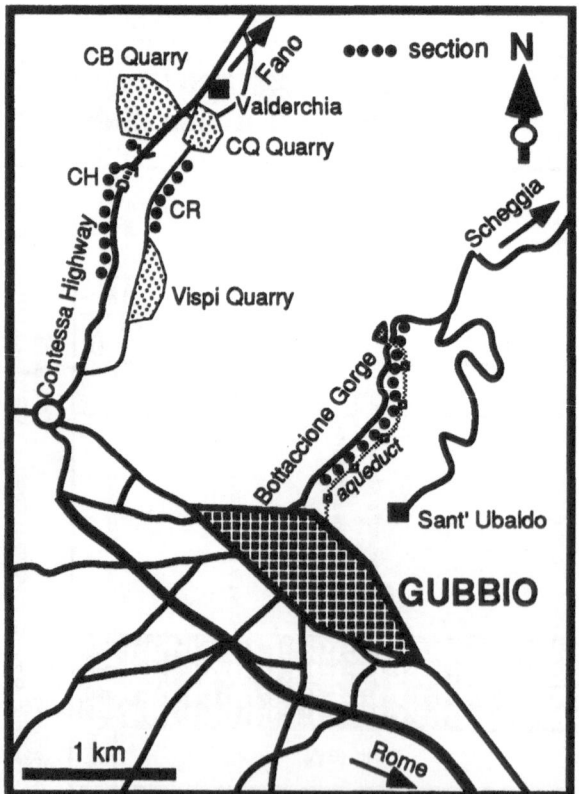

Fig. 5.5.1. Location map of the stratigraphic sections in the Bottaccione Gorge and the Contessa Valley near Gubbio. CH = Contessa Highway section; CB = Contessa Barbetti section; CQ = Contessa Quarry section; CR = Contessa Road section.

5.5
The Upper Aptian Event

An event of diversity decline in early planktonic Foraminifera during the latest Aptian–earliest Albian time was recognized by Leckie (1989). This event, which constitutes a minor extinction peak in Sepkoski's diagram (18% of marine genera; see Fig. 5.1.1), was attributed to widespread episodic production of warm, saline water masses in numerous, relatively isolated, marginal basins, and/or in subtropical oceanic areas (Leckie 1989). In the U–M sequence, this event is well recorded in the Brown Clay Member of the Marne a Fucoidi Formation, within the planktonic foraminiferal *Hedbergella planispira* Zone (Coccioni et al. 1987).

The Marne a Fucoidi Formation (see general description in Sect. 5.1) is extensively exposed throughout the U–M region. Its stratigraphy has been studied in detail by a number of researchers in several sections and in a core drilled near Piobbico (e.g., Arthur and Fischer 1977; Herbert and Fischer 1986; Coccioni et al. 1987; Premoli Silva et al. 1989; Erba 1988 and 1992; Fisher et al. 1991; Fiet 1998). Excellent and well documented outcrops are found in the classic area of Gubbio, namely in the Bottaccione Gorge and in the Contessa Valley (Fig. 5.5.1).

The most spectacular exposure of the stratigraphic interval covering the upper part of the Maiolica, the Marne a Fucoidi, the Scaglia Bianca, and the lowermost Scaglia Rossa, is located in the Vispi quarry, on the east slope of the Contessa Valley (see location map in Fig. 5.5.1, and panoramic view in Fig. 5.5.2). Here the contact between the Maiolica and the overlying Marne a Fucoidi, which corresponds roughly to the Barremian–Aptian boundary, is disturbed by a conspicuous soft-sediment slump representing the last major episode of instability in the Jurassic–Early Cretaceous extensional tectonic phase in this area. Similar slumps at the same stratigraphic level are recognized in numerous other sections throughout the U–M region (Coccioni et al. 1987). At Contessa, the bituminous-radiolaritic-ichthyiolitic Selli Level, which marks the boundary between the Maiolica and the Marne a Fucoidi (Coccioni et al. 1990), is clearly recognizable in the slump-deformed zone (Fig.5.5.2).

At Contessa, the Brown Clay Member of the Marne a Fucoidi Formation, which contains the Aptian–Albian boundary, is well exposed on the Vispi quarry face. However, the best and most studied exposure of this interval in the U–M region is located above a dirt road at the foot of the Poggio Le Guaine peak (see road map in Fig. 5.4.1 and panoramic view in Fig. 5.5.3). Although the detailed foraminiferal stratigraphy of this section by Coccioni et al. (1990) allows a precise location of the Aptian–Albian boundary (Fig. 5.5.4), and a fairly precise numerical age assignment using the Gradstein et al. (1994) time scale (i.e., 112.2 ± 1.1 Ma), no large craters are known with this age, and, to our knowledge, no detailed studies were ever carried out in this section, or in any other sections, that focussed on the search for impact signatures. Therefore, the A/A extinction peak remains a complete *incognita* in the general study of biotic crises possibly caused by impact.

Fig. 5.5.2. Panoramic view of the Vispi quarry in the Contessa Valley, exposing the U–M pelagic sequence from the Maiolica to the Scaglia Rossa. Note the conspicuous slump near the Maiolica–Marne a Fucoidi boundary. MA = Maiolica; SEL = Selli Level; MF = Marne a Fucoidi; SB = Scaglia Bianca; SR = Scaglia Rossa.

Fig. 5.5.3. Panoramic view of the Poggio Le Guaine section. The Brown Clay Member containing the Aptian–Albian boundary is exposed at the base of the outcrop.

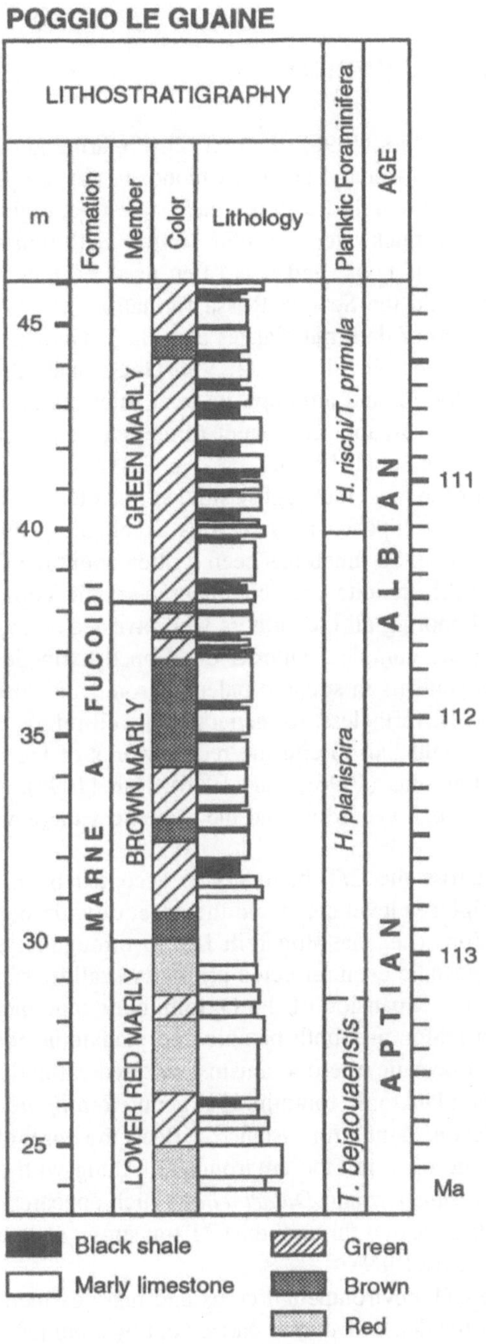

POGGIO LE GUAINE

Fig. 5.5.4. Stratigraphic synthesis of the Poggio le Guaine section (after Coccioni et al. 1990).

5.6
The Cenomanian–Turonian Boundary

A prominent extinction peak in Sepkoski's (1996) diagram (26% marine genera) is placed at the Cenomanian–Turonian boundary, and corresponds to the so-called Bonarelli Oceanic Anoxic Event. The Bonarelli Level is a carbonate-free, regional marker horizon, about 100 ± 40 cm thick, very similar to the Selli horizon described in the previous section of this book, and it is often used as boundary marker between the Scaglia Bianca and the Scaglia Rossa formations (Figs. 4.2 and 5.5.2). It is made of an alternation of dark radiolarites and black hychtiolitic, bituminous shales (e.g., Arthur and Premoli Silva 1982; Van Graas et al. 1983; Piergiovanni 1989; Marcucci et al. 1991), and although its total thickness varies laterally, its internal stratigraphy is remarkably consistent throughout the region (Montanari 1979, M'Ban 1994; Beaudoin et al. 1996).

The C/T boundary event is one of the most thoroughly studied mass extinctions of the Phanerozoic, and is extensively exposed in numerous excellent sections, especially in North America and Europe. So much has been written about the C/T boundary that it would be impossible to cite and comment here the copious literature on this subject, properly honoring all the authors who invested so much effort in studying and interpreting the complex patterns of stepped extinctions, survival, and recovery of marine organisms in stressed paleoenvironments during this critical interval of Earth history. Nevertheless, for general and well-referenced reviews of the C/T boundary, we would like to cite the recent works of Harries and Kaufmann (1990), Peryt and Lamolda (1996), Tur (1996), Hart (1996), and Fitzpatric (1996), as well as the thorough synthesis, and the references therein, in Hallam and Wignall (1997).

The cause of the extinctions across the C/T boundary is accepted by most authors to be related to a peak in high sea level stand and the onset of widespread dysoxic and locally anoxic conditions (i.e., the Bonarelli Level) throughout the world's oceans. Although such an eustatic event affected primarily shallow water environments and carbonate reefs, the expansion of the O_2 minimum zone had a serious toll also in the open ocean realm, and both benthic and planktonic foraminiferal communities, among other oceanic microorganisms, suffered extinctions and turnovers. The extinction of the planktonic foraminiferal genus *Rotalipora* in the latest Cenomanian (i.e., *R. cushmani* Zone), for instance, reflects the onset of a crisis in the mid-water column of the open marine environment. Along with *Ritalipora*, also the genera *Praeglobotruncana* and *Dicarinella*, which appeared in the late Cenomanian, temporarily disappeared through the C/T transition, and then reappeared in the early Turonian (Leckie 1989).

An extraterrestrial cause for the C/T environmental crisis and mass extinction did not find much support in any of the detailed studies carried out by a number of researchers throughout the world. First of all, no large craters are known with an age comparable to that of this stratigraphic boundary (i.e., 93.5 ± 0.2 Ma in the time scale of Gradstein et al. 1994), except for the Steen River structure in Canada which, in any case, has a modest diameter of 25 km, and an imprecise K/Ar age of 95 ± 7 Ma (Carrigy 1968).

Multiple impacts as a cause for a global climatic deterioration around the C/T boundary were proposed by Hut et al. (1987) on the basis of two small iridium anomalies (0.11 ppb over a background of about 0.02 ppb) detected by Orth et al. (1988) at the top of *R. cusmani* Zone in a section near Pueblo, in Colorado. Similar anomalies were then detected in other sections throughout the western interior of the United States (Orth et al. 1990). Although these authors did not preclude an (oceanic) impact source for these anomalies, the absence of impact spherules and shocked minerals in the Ir-rich layers, and the overall mafic proportions of other trace elements in the same layers, led them to suppose that these Ir enrichments were derived from a particularly active mid-oceanic ridge volcanism in the Atlantic at the end of the Cenomanian.

Needless to say, the C/T boundary is excellently exposed in numerous outcrops throughout the U–M Apennines. Once again, the Bottaccione Gorge at Gubbio is to be considered the classic site for this boundary and for the Bonarelli Level (Bonarelli 1891). The planktonic foraminiferal biostratigraphy across the boundary is documented in detail by Premoli Silva and Sliter (1995), and the sequence of biozones *Rotalipora cushmani–Whiteinella archeocretacea –Helvetoglobotruncana helvetica* was recognized by these authors upon closely-spaced sampling and quantitative analyses of thin sections. *The W. archeocretacea* Zone, which is about 10 m thick (i.e., 1.4 m.y. long), is the one that contains the C/T boundary and the Bonarelli Level.

As briefly mentioned in Sect. 4.2.2, the upper member of the Scaglia Bianca is characterized by a rhythmic sequence of white, biomicritic, pelagic limestones interbedded with black, laminated, tabular cherts, and expressing a Milankovich-type cyclicity. The sedimentology, quantitative biostratigraphy, and chemo-stratigraphy of the Scaglia Bianca has been studied in great detail by M'Ban (1994) in a number of outcrops throughout the U–M Apennines (see also Beaudoin et al. 1996). From an isopach study of the Bonarelli Level and the underlying black cherts of the Scaglia Bianca, Montanari (1979) and M'Ban (1994) reconstructed a slightly undulated sea floor, with relatively depressed areas in the center of the paleobasin (i.e., Furlo basin; see Fig. 4.1.2). Although bland, this slightly irregular topography was effective in controlling the development of pre–Bonarelli anoxic episodes at different times and in different parts of the paleobasin (Beaudoin et al. 1996), as well the first and last occurrences of index fossils, probably due to lateral variation of preservation factors.

Among the several C/T outcrops studied by M'Ban (1994) and Beaudoin et al. (1996), the Upper Road at Furlo (Fig. 5.6.1) can be considered as the reference section in this region, as it is the most complete and best exposed (Fig. 5.6.2). The stratigraphy of the Furlo Upper Road section across the C/T boundary is synthetized in Fig. 5.6.3.

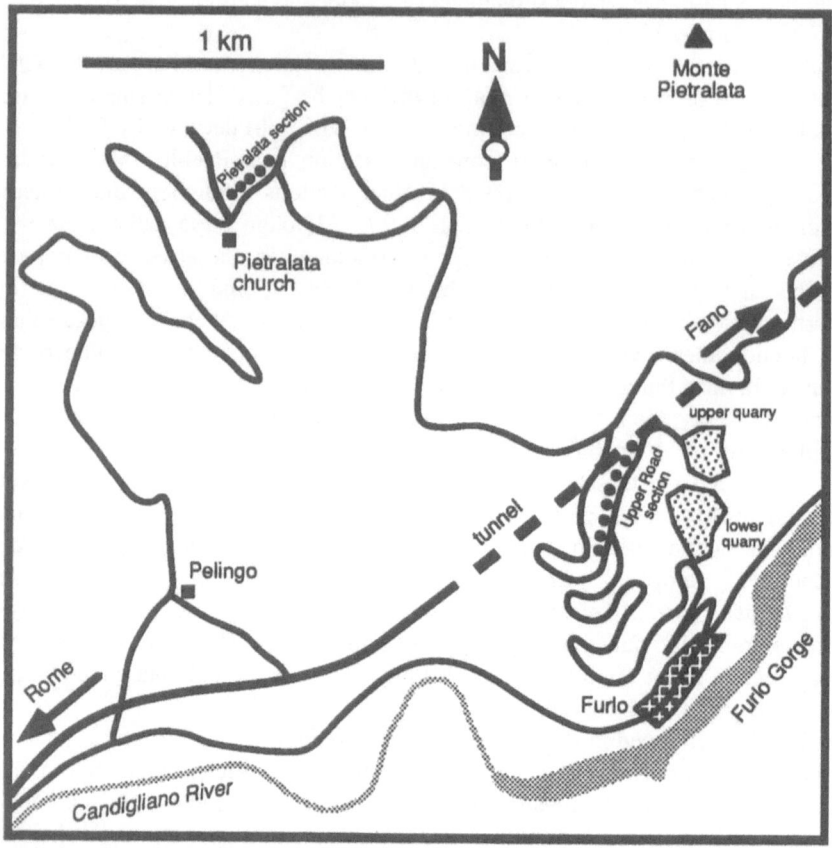

Fig. 5.6.1. Location map of the Pietralata and Upper Road sections at Furlo.

The search for impact signatures near the C/T boundary in the U–M Apennines dates back to the earliest development of the theory of impacts as a cause of mass extinctions. However, no evidence for impact has been produced so far. Wezel et al. (1981) reported iridium in samples from the Bonarelli Level at Burano (near Cagli) and Furlo, with anomalously high concentrations between 7.9 and 11.4 ppb. However, the reliability of the results of Wezel and co-workers was strongly questioned by Alvarez et al. (1984), who independently analyzed, by instrumental neutron activation analysis (INAA), a dozen samples covering the whole Bonarelli Level in the Furlo Upper Road section, and found no Ir higher than a best integrated value of 0.00 ± 0.03 ppb. Alvarez and co-workers concluded that the high Ir concentrations reported by Wezel et al. (1981) were due to sample and/or laboratory contamination.

Fig. 5.6.2. Panoramic view of the Furlo Upper Road outcrop showing the boundary between the Scaglia Bianca and the Scaglia Rossa formations marked by the Bonarelli Level.

In 1986, one of us (A.M.) collected some 500 one half inch core samples across the C/T boundary interval in the Furlo Upper Road section, completely covering an interval from 3 m below to 2 m above the Bonarelli Level. In addition, each alternating black shale and radiolarite of the Bonarelli was also collected. These samples were analyzed the following year at the Lawrence Berkeley Laboratory by Frank Asaro with the INAA Ir coincident spectrometer, but all proved to be practically bare of iridium. The highest Ir concentration of the whole sample collection was found in a thin black clay lamina at the base of the Bonarelli Level, which showed an insignificant Ir content of 0.05 ± 0.02 ppb.

These negative, unpublished results, along with the apparent lack of shocked minerals and impact spherules in several Bonarelli samples analyzed by one of us (A.M.), led the Berkeley team to withdraw from further costly and time-consuming analyses of the C/T boundary interval in Italy.

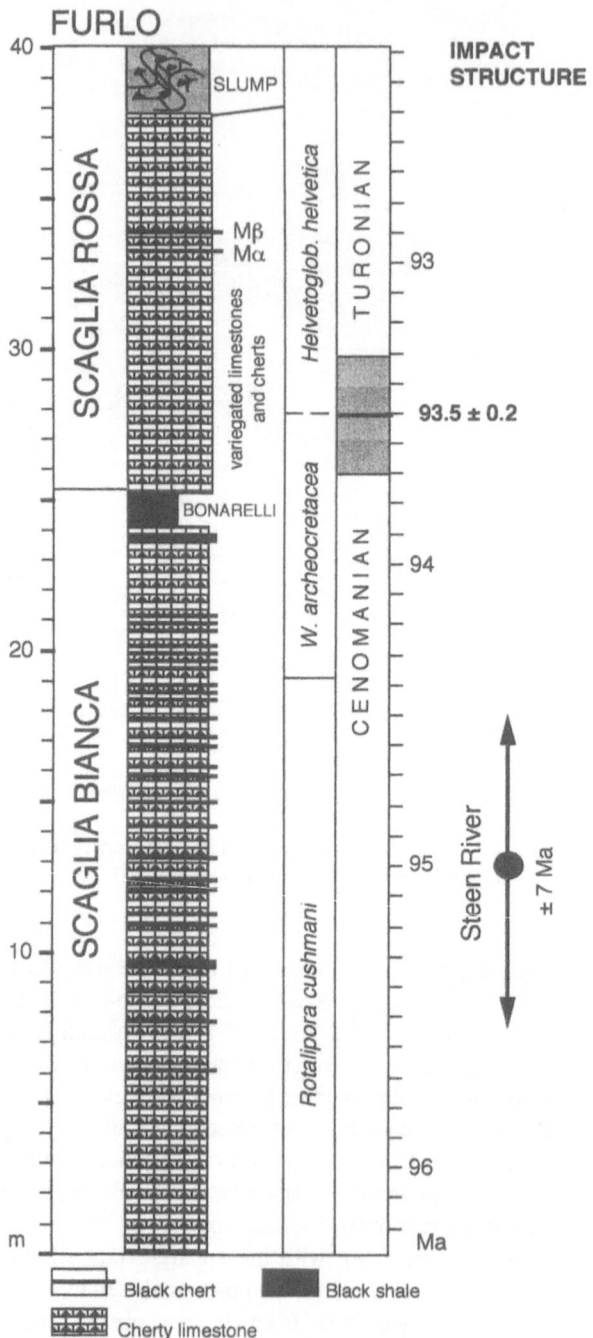

Fig. 5.6.3. Stratigraphic synthesis of the Scaglia Bianca–Scaglia Rossa transition (i.e., the C/T boundary) in the Furlo Upper Road section.

In summary, there is no evidence yet that extraterrestrial impacts had a role in the Cenomanian oceanic anoxic event and related mass extinction. There are no known large impact structures on Earth with a C/T age, and no impactoclastic layer was ever found in the numerous C/T boundary sections studied by a number of researchers in the United States and in Europe. The few minor iridium anomalies found in several North American sections are preferably interpreted as the result of oceanic volcanism. This is enough to discourage anyone in undertaking a focused impact-stratigraphical study on this interval.

Nevertheless, the Cenomanian anoxic crisis and the mass extinction documented in the fossil record at the end of this stage is still surrounded by an aura of mystery. That anoxic crises were caused by eustatic sea level changes, which were triggered by global climatic changes, is undoubtedly a reasonable hypothesis strongly supported by an enormous volume of data and observations. Complicated kill mechanisms can be devised or inferred from this hypothesis, such as reorganization and overturning of marine currents and water masses, changes in heat transfer patterns, and oxygen and nutrient redistribution in the oceans, with consequent alteration of delicate equilibria controlling the stability of various environments, and a chain effect of destructions of diverse, more or less interdependent ecosystems (see Hallam and Wignall 1997, for an excellently referenced review of these mechanisms). But the question remains: what caused the climatic change in the first place? This question can be asked for all of the mass extinctions boundary in the Earth history record.

Exceptionally active tectonism and related volcanism are certainly earthly factors capable of altering global climate, and constitute the basis for a working hypothesis that the vast majority of earth scientists have no difficulty in accepting as workable and rewarding. On the other hand, the extraterrestrial cause cannot be ruled out a priori, even if up to date there is no convincing evidence for impacts at the C/T boundary. In fact, a comet shower, such as the one predicted by Hut et al. (1987), not necessarily has to produce giant or large impacts that are capable of leaving an impactoclastic signature worldwide. On the other hand, the accretion of a huge volume of interplanetary dust particles during a comet shower, along with a myriad of small cometary impactors, for a period of 2–4 million years (the expected duration of a comet shower) may *per se* be an efficient agent for altering the Earth's atmosphere with consequent rapid climatic and biologic changes.

The geologic evidence of a comet shower has been documented, for the first time, by Farley et al. (1998) after the discovery of a significant enhancement of ^{3}He in a 2.5 million year interval across the uppermost Eocene in the Massignano section (see Sect. 5.8.2). Multiple impacts are known to have occurred in the terminal Eocene and at least one impactoclastic layer in the same time interval seems to have a global occurrence (e.g., Montanari et al. 1993), whereas several others have been found at different stratigraphic levels in upper Eocene sediments worldwide. To our knowledge, there has been no specific search for comet shower evidence (i.e., ^{3}He/^{4}He analysis) across the C/T boundary. This stratigraphic interval, with its Bonarelli Level so well exposed and documented in the U–M

Apennines, constitutes, therefore, an appealing invitation for further investigations on the possible extraterrestrial role in global environmental and ecologic crises.

5.7
The Terminal Cretaceous Record

The uppermost Cretaceous–lower Tertiary Scaglia Rossa Formation of the U–M basin has been the object of numerous interdisciplinary studies in the past decades, which yielded a remarkably complete and detailed picture of its sedimentological setting and tectonic evolution (e.g., Arthur and Fischer 1977; Baldanza et al. 1982; Alvarez et al. 1985; Montanari 1988; Montanari et al. 1989), calcareous plankton biostratigraphy (e.g., Luterbacher and Premoli Silva 1962; Premoli Silva 1977; Monechi 1977; Monechi and Thierstein 1985; Premoli Silva and Sliter 1994), magnetostratigraphy (Alvarez et al. 1977; Roggenthen and Napoleone 1977; Lowrie et al. 1982; Alvarez and Lowrie 1984; Chan et al. 1985), and stable isotope chemostratigraphy (e.g., Renard 1985; Corfield et al. 1991).

In this section, we review the state of knowledge of the terminal Cretaceous impact stratigraphy in the U–M sequence. This is a particularly interesting portion of the stratigraphic sequence, because it records the critical time of a major mass extinction (the K/T boundary event), which was preceded by a general deterioration of environmental conditions at a global scale (the Maastrichtian crisis; e.g., Barrera 1994; MacLeod and Huber 1996; Barrera and Savin 1999), especially in shallow marine environments (e.g., Swinburne 1982). Moreover, the end of the Cretaceous Period records three large impact events. Two of them, Kara–Ust Kara and the Manson, occurred in the upper Campanian and the lower Maastrichtian respectively, and the third one, Chicxulub, coincides with, and is very likely the primary cause of, the K/T boundary mass extinction.

For a better understanding of the chronostratigraphic framework of this critical interval of Earth history, we propose in Fig. 5.7.1 a stratigraphic synthesis of the upper Campanian to Danian interval in the classic sequence of Gubbio (Bottaccione and Contessa sections; see location maps in Figs. 4.1.1 and 5.5.1), along with the chronologic location of the three aforementioned impact events. We then proceed with the documentation of impact signatures in the Maastrichtian and the K/T boundary events separately.

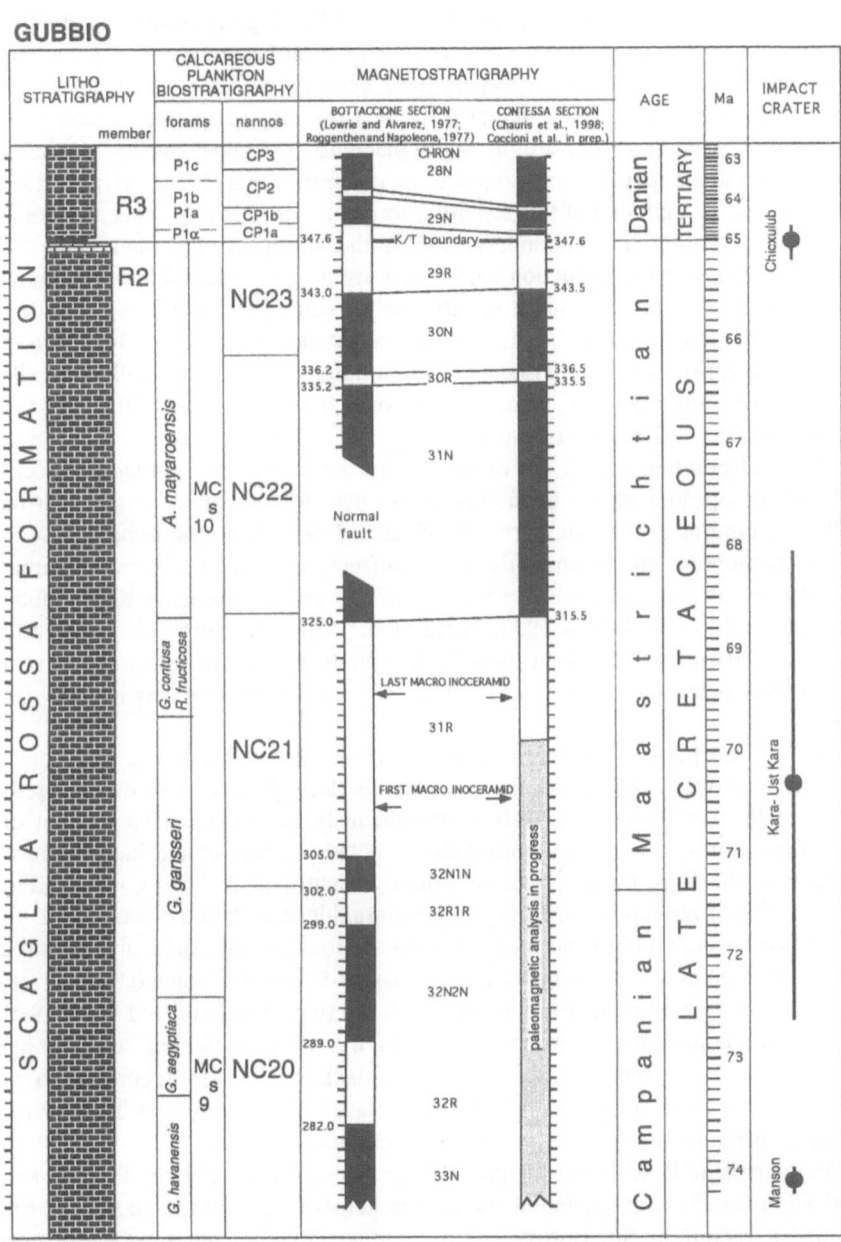

Fig. 5.7.1. Stratigraphic synthesis of the Campanian to Paleocene interval at Gubbio.

5.7.1
The Maastrichtian

During nearly two decades of scientific debate following the Alvarez et al. (1980) hypothesis of a terminal Cretaceous mass extinction caused by an impact event, some three thousand scientific papers have been published on the subject. This enormous volume of printed matter examined the most disparate aspects of this extraordinary event, from astrophysics to paleontology, geochemistry to stratigraphy, mineralogy to paleoecology, and so on (e.g., Glen 1994; Alvarez 1997). Since the very beginning of this scientific ferment, which is unprecedented for any other chronostratigraphic boundary, one of the strongest objections against the impact-induced mass extinction idea came from the paleontologic community. This critical reaction was, and it is still now for some people, not just a matter of principle or paradigm (i.e., gradualism vs. catastrophism), but it is based on evidence that various ecologic systems and taxonomic groups, especially in shallow water marine environments, were already going through a crisis millions of years before the fatal K/T boundary event.

The reef-building rudists, for example, after a radiation in the Late Cretaceous, all but disappeared at the beginning of the late Maastrichtian (e.g., Kauffman 1984; Swinburne 1990; Swinburne and Noacco 1992). A similar trend is observed in the ammonoids and belemnoids, which suffered an accelerated decline through the late Maastrichtian, starting some 4 million years before the K/T boundary (Hallam and Perch Nielsen 1990; Ward et al. 1991). This pre–K/T biotic crisis, which has particularly evident in carbonate platform and shallow water environments, has been attributed to an eustatic sea level drop (e.g., Swinburne 1990 1992).

As for the pelagic realm, the pre–K/T crisis seems much less pronounced. Among deep water organisms, the inoceramids (large bivalves) all disappeared in the mid Maastrichtian, apparently instantaneously in individual basins, but diachronically in distant basins worldwide. A detailed study of the inoceramid extinction in the U–M basin was carried out by Chauris et al. (1998), who, through precise lithostratigraphic and magnetostratigraphic correlations, established that these bivalves at Gubbio and in other U–M sections became suddenly and simultaneously extinct within the upper part of magnetochron 31r, some 0.9 Ma prior to the genus extinction in the Basque Country basin (e.g., MacLeod and Ward 1990).

Also in the planktonic foraminiferal realm, a turnover or decline seems to have started already in the mid Maastrichtian (MacLeod 1994). According to this author, the transition between the latest Cretaceous and the basal Tertiary took place at different rates in different basins throughout the world.

In the pelagic U–M basin, the pre–K/T decline is documented by Premoli Silva and Sliter (1995) with a detailed and comprehensive study of the planktonic foraminiferal record in the Bottaccione section at Gubbio, and by Chauris et al. (1998), who also studied the relation between the foraminiferal record and the inoceramid extinction in several well-exposed sections throughout the region,

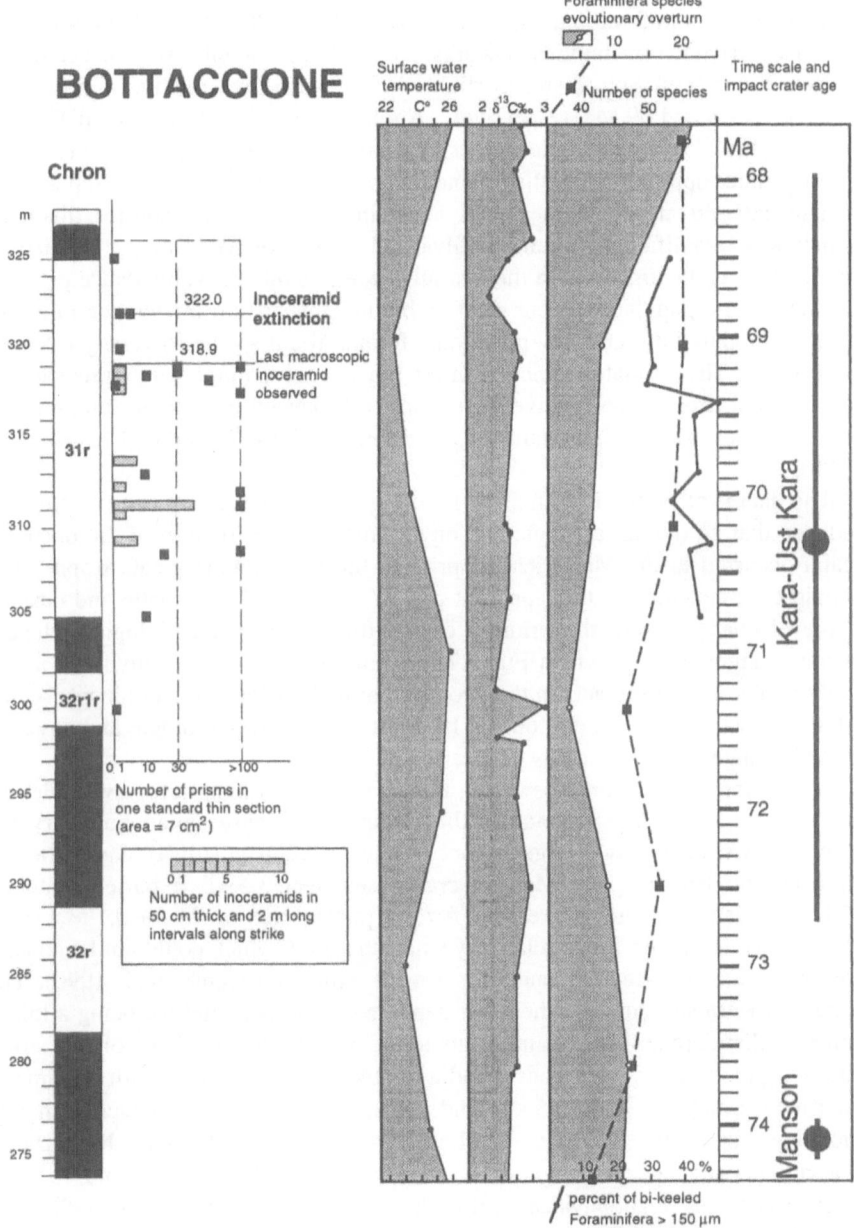

Fig. 5.7.1.1. Stratigraphic synthesis of the mid-Maastrichtian environmental crisis in the U–M pelagic basin.

including Bottaccione. In Fig. 5.7.1.1 we propose a synthesis of the inoceramid, planktonic foraminiferal, and stable isotope records of the classic section of Bottaccione derived from published works. The $\delta^{18}O$ paleotemperature curve of the surface water from Spicer and Corfield (1992) shows two relative minima centered at 73 Ma and 69 Ma, respectively. A relative negative shift of the $\delta^{13}C$ curve by Renard (1985), which could reflect a period of low primary productivity in this basin, mimics approximately the second interval of relative low paleotemperature.

This interval also coincides with a decline in planktonic species diversity, which was quantified by Premoli Silva and Sliter (1995) from the number of originations and extinctions in the overall species count. Two abundance peaks in the inoceramid population occur right within this interval of low temperature, low primary productivity, and low planktonic foraminiferal species diversity (Chauris et al. 1998). The ultimate extinction of the inoceramid is immediately preceded by a significant drop in the relative abundance of bi-keeled planktonic foraminifers, and coincides with a significant negative shift in the $\delta^{13}C$ curve (Chauris et al. 1998).

The data synthetized in Fig. 5.7.1.1 from various independent sources strongly indicate that an overall ecologic and environmental deterioration of the open sea realm occurred in the Maastrichtian prior to the K/T boundary catastrophe. The simplest explanation for this apparent pre–K/T crisis is that climatic and eustatic sea-level changes were the primary causes for the observed biologic and geochemical signatures shown in Fig. 5.7.1.1. For the few who are involved in the study on the role of impacts in the evolution of the atmosphere and biosphere, the question goes a step further: could it be that these pre–K/T biologic and environmental changes were caused by extraterrestrial events?

A first approach toward answering this question is to look at the available impact record. In the interval across the Campanian–Maastrichtian boundary the record shows, in fact, two large impact structures: Manson (Iowa), and Kara–Ust Kara (Arctic Russia). The Manson crater (see Sect. 3.4) is a buried structure 37 km in diameter, with a precise $^{40}Ar/^{39}Ar$ plateau age of 74.1 ± 0.1 Ma (Izett et al. 1998). However, it is a relatively small structure, which could not be considered *per se* as an effective cause for a mass extinction (Jansa et al. 1990). The Kara–Ust Kara structure, on the other hand, has some potential for being a killer, although there are still uncertainties about the actual size and shape of this structure. Originally it has been considered a double impact structure with a diameter for Kara estimated at about 65 km, and a diameter for Ust Kara, which is mostly submerged under the Kara Sea, of about 25 km. Geological studies by Nazarov et al. (1991) suggested a single crater with a diameter between 70 and 155 km, whereas analyses of geophysical profiles by Koeberl et al. (1990) led to a diameter estimate of 80 km for the Ust Kara crater. In our compilation in Table 1.9.1 we have adopted the smallest, most conservative estimates, but this may turn out to be wrong in the future, also considering that Nazarov et al. (1991) argued that there is not enough evidence for two Kara craters, and that it may well be a single, >120 km diameter structure. This latter estimate would make of the Kara the largest known impact structure in Russia.

As for the energy flux onto Earth delivered by the combined Kara–Ust Kara double impact, it defines the fourth largest peak in the past 150 Ma record (Montanari et al. 1998; see Fig. 5.1.2). In short, the Kara–Ust Kara structure may be a big enough event to contribute to a global ecologic and environmental change. Coincidentally, it's accurate $^{40}Ar/^{39}Ar$ plateau age of 70.3 ± 2.2 Ma (Trieloff et al. 1998) puts it right in the middle of the lower Maastrichtian critical interval marked by low temperature, low primary productivity, and low planktonic foraminiferal species diversity in the pelagic record (Fig. 5.7.1.1).

Recent high-resolution studies on the pelagic record of the U–M sequence are revealing new evidence that the Manson and the Kara–Ust Kara impacts may be part of a cosmic crisis that lasted several million years across the Campanian–Maastrichtian boundary. Mukhopadhyay et al. (1998) reported an ^{3}He increase in the Scaglia Rossa limestones of the Poggio San Vicino (see location map in Fig. 5.7.6.7) and the Gubbio sections at Bottaccione and Contessa (Fig. 5.5.1), spanning about 4 m.y., with a ^{3}He concentration peak a factor of five higher than average terrestrial values, approximately centered at 70 Ma.

A similar ^{3}He anomaly found in the upper Eocene section of Massignano (see Chap. 5.8.4 below) was interpreted by Farley et al. (1998) as evidence for a comet shower, which would have delivered onto Earth an enormous amount of ^{3}He-carrying interplanetary dust particles (IDPs). During this terminal Eocene time, several impact events are documented (i.e., the 90 km Chesapeake structure at 35.3 ± 0.2 Ma; the 100 km Popigai structure at 35.7 ± 0.8 Ma; the 28 km Mistastin crater at 38 ± 4 Ma; and the 17 km Logoisk crater at 40 ± 5 Ma; see Table 1.9.1). This whole cosmic crisis scenario leads to the hypothesis that a large number of small- and medium-size impacts, accompanied by an anomalously high influx of IDPs extending through a period of several million years, have the potential of causing global climatic (and, consequently, ecologic) changes by altering the transparency of the atmosphere, without the need for a giant killer crater the size of the K/T Chicxulub structure.

The next critical question in testing the hypothesis that extraterrestrial accretion had a role in the pre–K/T Maastrichtian crisis is, where we search for the distal impactoclastic layers produced by these events in our marine sedimentary sequence? In doing so, we may face the problem of the needle in the haystack. In fact, the Late Eocene Massignano experience, and the finding of an impactoclastic layer with Ir anomaly, impact spherules, shocked quartz, spinels, and ^{3}He (see Sect. 5.8.4), shows that a large impact event the size of Popigai or Chesapeake does not produce a significant lithologic change in a deep marine, carbonate sediment. Unlike the case of the K/T boundary, where a clear break of the carbonate production is marked by a thin but recognizable clay-rich layer worldwide, the upper Eocene impactoclastic level at Massignano is contained within a strongly bioturbated and homogenized marly limestone layer. Only a very subtle color change can be seen there, and the impact spherules are sparse throughout a 5–10 cm interval of indurated carbonate rock.

In the case of the Kara–Ust Kara impact, its age uncertainty of ± 2.2 Ma would correspond to about 40 m of homogenized, hard limestone layers at Gubbio. The impact signatures may be somewhere in this sequence, but certainly it will require a non-trivial effort to pinpoint their locations.

As a first approach in the search of the needle in the haystack, one of us (A.M.) collected three clay layers in the region of the inoceramid event at Contessa. These cm thick layers, located at 297.3 m, 297.8 m, and 301.0 m respectively, are not pressure–solution residual seams, which are the common limestone pseudobedding partings in the Scaglia Rossa Formation (Alvarez et al. 1985). Their continuity and thickness are suggestive of primary sedimentary clay layers, indicating brief interruptions of the biogenic carbonate sedimentation, just like the K/T boundary clay. The washed residues >63 μm of these sample contain numerous shiny, metallic and highly magnetic spherules with diameters between 80 and 150 μm (Fig. 5.7.1.2). Impact spherules? Unfortunately not.

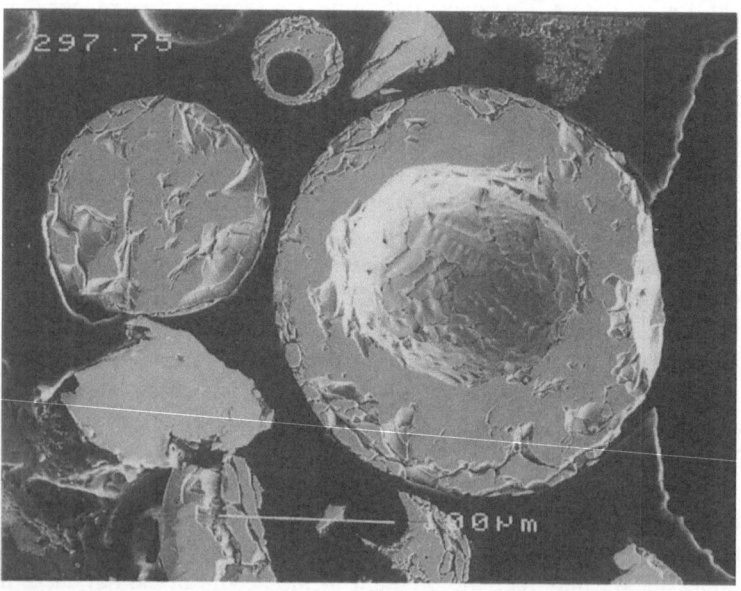

Fig. 5.7.1.2. Magnetic spherules from a clay layer at meter level 297.3 in the Contessa section. These objects, similar in shape to cosmic spherules, are made of pure iron and were probably produced by the exhaust of diesel engines. Thus, they represent a case of recent surficial outcrop contamination. Scale bar = 0.1 mm.

The spherules were analyzed by electron microscopy in Vienna by one of us (C.K.). These spherules turned out to be made of pure iron, which would indicate that they are not of extraterrestrial or impact origin, but rather the products of recent metallurgic industry. This is not the first time that metallic spherules are found as contaminants in soft rocks exposed near busy highways (Montanari 1986), or railroads (Byerly et al. 1990; Eustoquo Molina, personal communication, 1999). This type of spherules may have been produced by the friction of mechanical automotive parts, such as breaks and gears in trucks and locomotives, or, possibly, from the exhaust of diesel engines. The search for the needle continues.

5.7.2
Definition of the K/T Boundary in the U–M Basin

With the definition and proposal of the *Globigerina eugubina* planktonic foraminiferal biozone as the base of the Tertiary (i.e., Paleogene) in the type section of the Bottaccione Gorge, near Gubbio, Luterbacher and Premoli Silva (1964) described in detail a number of other K/T sections throughout the U–M Apennines. In doing so, they established a simple and precise criterion for recognizing the K/T boundary in any pelagic carbonate sequence in the world. In general, the boundary is marked by a thin (0.5–2.5 cm) clay layer sandwiched between the last Cretaceous limestone containing large (up to 1 mm) planktonic foraminifers, such as the Globotruncanidae, Rugoglobigenindae, and Heterohelicidae, which are recognizable with a hand lens, and the first Tertiary limestone containing tiny (less than 0.1 mm) globigerinids not visible with a hand lens (Fig. 5.7.2.1).

In the mid 1970s, a renewed interest in the nearly continuous and complete Upper Cretaceous–Paleogene sequence of Gubbio by a group of interdisciplinary researchers lead them to join into a perhaps unprecedented team effort for resolving the lithostratigraphy, sedimentology, planktonic foraminiferal biostratigraphy, and, above all, the magnetostratigraphy of this span of geologic time. The results of this coordinated research project were published in a series of papers in an issue of the Geological Society of America Bulletin (Arthur and Fischer 1977; Premoli Silva 1977; Lowrie and Alvarez 1977; Roggenthen and Napoleone 1977), along with the proposal of Gubbio as type section for the Late Cretaceous–Paleogene geomagnetic reversal time scale (Alvarez et al. 1977). For the first time it was possible to assign accurate biomagnetostratigraphic ages to the marine magnetic anomalies, and, thus, to date, with reliable geologic and biochronologic criteria, the expanding world's oceans.

A few years later, the discovery of an iridium anomaly at the K/T boundary of Gubbio, and in other U–M sections, and the proposal of the hypothesis that the impact on the Earth's surface of a large extraterrestrial object was the cause of the terminal Cretaceous mass extinction, was published by Alvarez et al. (1980), opening the way to two decades of extraordinary scientific debate and the development of a new theory on the evolution of terrestrial life.

Many people are still wondering today of how such a discovery came about: was is casual or was it the result of a methodic, scientific search? It would be fair to say that it was both. This fascinating story is told in detail by Alvarez (1997), and nobody better than the discoverer could have done it. Here is just a short account of one aspect of this discovery, which we think relevant and perhaps useful for anyone who, one day, will find her or himself wondering while looking at the original K/T boundary outcrop in the Bottaccione Gorge.

The distinct difference in fossil microfauna between the limestone layers above and below the K/T clay (see Fig. 5.7.2.1) was something more than a puzzling curiosity for the various interdisciplinary specialists, who worked together in the Gubbio team. The paleontologic answer to such a sharp change was that the thin clay layer was a hiatus, i.e., a long period of time represented by little or no sediment. How long was this unrepresented geologic time nobody could tell with confidence and precision, and that was the problem for the magnetostratigraphers in the team, who wanted to use the sequence of magnetic polarity reversals as a precise tool for calibrating the time scale. They needed to know how complete their Gubbio sequence really was.

It is so that Ir was suggested, by nuclear physicist Luis Alvarez, father of geologist Walter of the Gubbio team, as a tracer for estimating the sedimentation rate of these pelagic limestones. This was based on the assumption that Ir, along with other platinum group elements, which are very scarce in the Earth's crust, but much more abundant in undifferentiated meteoritic matter (see Sect. 1.8), is accreted onto the planet's surface at a more or less steady rate in the form of the so-called perennial fallout (i.e., micrometeorites).

Pelagic carbonate sediments are made essentially of biogenic calcite constituting the hard parts of phyto- and zooplankton. These organisms are produced in the water column, and deposited on the sea floor at variable rates through time. The biogenic carbonate ooze may contain a variable component of windblown siliciclastic dust (i.e., mostly clay and some silt). The Scaglia Rossa limestones may contain from 3% up to 15% of this siliciclastic component. In the case of a prolonged suppression of carbonate production, or as a consequence of carbonate dissolution, only windblown clay is deposited and preserved on the sea floor.

The basic idea of Luis Alvarez was to measure the concentration of the iridium tracer in the boundary clay and in the limestones below and above it through a known span of time, and by normalization, simply estimate the time represented by the K/T boundary clay. Following this reasoning, a higher than limestone background concentration of Ir was expected in the K/T clay, and that would have been proportional to the length of the alleged hiatus. While Ir turned out to be too scarce in the limestones to be used as a practical tracer for estimating the sedimentation rate, the boundary clay contained too much of it to be explained by a hiatus: the integrated amount of Ir in the boundary clay and in the basal 20 cm of the *eugubina* limestone above it would have represented a much longer time than the overall time span analyzed.

This apparent paradox was explained by Alvarez et al. (1980) with the hypothesis of a catastrophic, massive accretion of an Ir-rich extraterrestrial object, such as a chondritic asteroid or perhaps the rocky part of a comet with an estimated diameter of 10 ± 4 km. The impact explosion and its ejecta would have caused a sudden deterioration of the world's climate mainly by temporary obscuration of sunlight, with imaginable disastrous effects on the global biota and whole terrestrial and marine ecosystems (see Sect. 3.2.3).

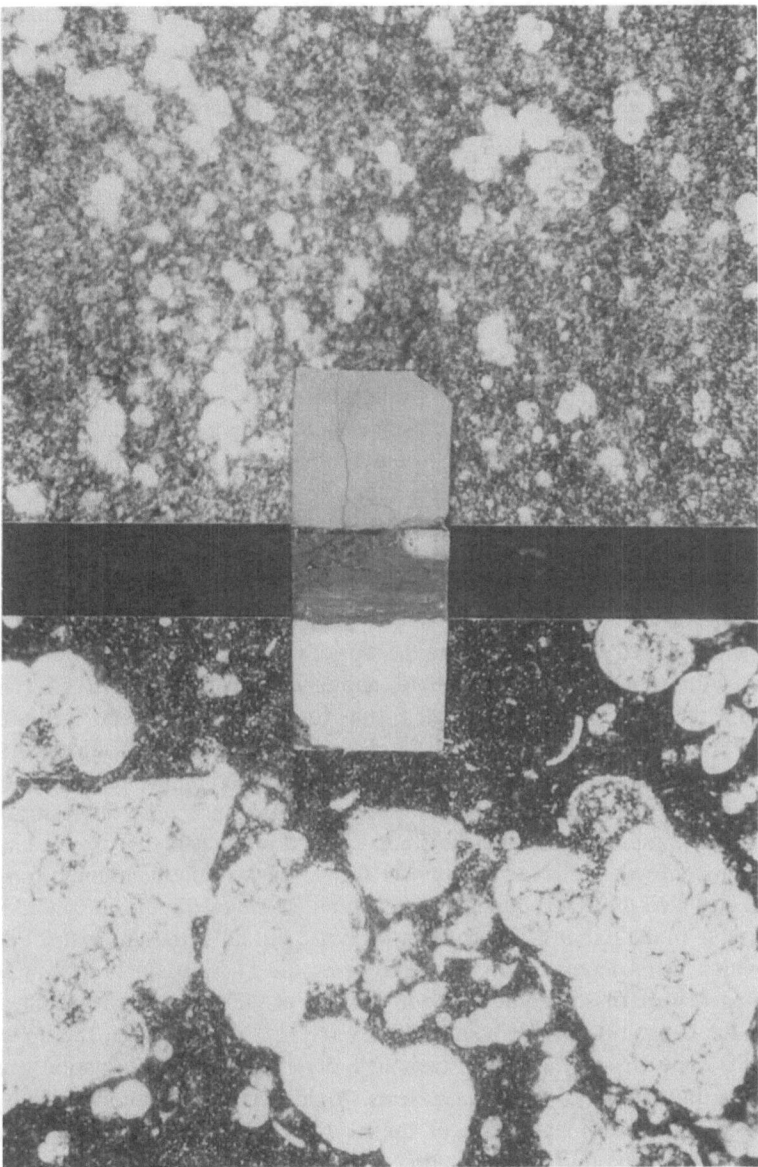

Fig. 5.7.2.1. Polished slab of the 25 mm thick K/T boundary clay layer from Petriccio sandwiched between the top, white limestone of the Cretaceous, and the basal, pink limestone of the Tertiary (the *eugubina* limestone). Thin section photomicrographs (same magnification) show the planktonic foraminiferal assemblages in the limestones below and above the boundary. The largest foram test in the lower part of the picture is about 0.6 mm across.

It is not in the intention of this book to review all the work done by innumerable researchers, especially paleontologists and biostratigraphers, on the K/T boundary in the two decades following the Alvarez et al. (1980) paper. To those who are interested, we suggest to consult the Special Papers of the Geological Society of America edited by Silver and Schultz (1982), Sharpton and Ward (1990), Ryder et al. (1996), and the book by Hallam and Wignall (1997), all with exhaustive reference lists. A discussion of the evidence for impact at the K/T boundary is given in Sect. 3.2.

As for the K/T boundary in our U–M Apennines, it must be recognized that no high-resolution biostratigraphic or micropaleontologic work has been carried out in the years immediately following the publication of the Alvarez hypothesis. One of the reasons for this neglect toward the classic Italian K/T boundary is that it may have seemed, in the eye of the stratigrapher, not so suitable for high-resolution stratigraphical studies. The limestones bracketing the K/T clay are thoroughly homogenized by bioturbation, strongly cemented, and were deposited at rather slow rates, varying from section to sections from about 6–10 mm/ka for the top Cretaceous limestone, representing the *Abatomphalus mayaroensis* Zone, to 2–5 mm/ka for the basal limestone of the Tertiary, representing the *Parvularuglobigerina eugubina* Zone (the former *Globigerina eugubina* of Luterbacher and Premoli Silva 1964).

In other K/T boundary sections around the world, where sedimentation rates are orders of magnitude higher than at Gubbio, and/or bioturbation less effective in homogenizing the soft marine sediment, a thin foraminiferal biozone is found between the *A. mayaroensis* Zone, and the *P. eugubina* Zone. This biozone is called P0 Zone and was first recognized by Smit (1982) based on the abundance of one single, dwarf foraminifer, the *Gumbelitria cretacea*, in a very short stratigraphic interval immediately above the sharp extinction level of the Cretaceous fauna. This creature was believed by Smit (1982) to be the sole planktonic foraminifer that survived the K/T boundary catastrophe. In short, the thin P0 Zone is not recognizable in the Italian K/T boundary sections. Possibly, the tiny tests of *G. cretacea*, which were deposited on the seafloor of the U–M basin immediately after the impact, were dissolved at a time of low pH at the water–sediment interface during the deposition of the boundary clay (see below). Perhaps, the upper part of the P0 Zone was formed when carbonate deposition and preservation was restored, but a thin lamina of *G. cretacea* tests would have been homogenized by bioturbation with the carbonate ooze of the overlying *P. eugubina* Zone. This would have rendered the thin P0 Zone unrecognizable. Moreover, the hardness of the *eugubina* limestone in the Italian sections is such that foraminiferal tests cannot be extracted and isolated, and species have to be recognized in thin section, which is a difficult, if not an impossible task, considering that these forams have dimensions around 40–50 µm.

The impossibility of recognizing a millimetric P0 Zone in the Gubbio and other Italian K/T sections may have discouraged high-resolution biostratigraphers in undertaking a detailed study in this unfavorable situation, and may have led some of them to believe that the K/T boundary in this area is a hiatus. As we will show in the following chapters, there is no real hiatus in most of the K/T sections in the U–M basin, and the absence of a P0 Zone cannot be considered *per se* the evidence of unrecorded sedimentation in the time immediately following the K/T impact.

In summary, following the aim of this book, we are now proceeding with a description of the high-resolution impact stratigraphy of the K/T boundary in the U–M Apennines, documenting the evidences of the impact that can be readily recognized in the field and with simple laboratory tools. Then, we will describe the most representative outcrops in the region, where the K/T boundary can be easily reached and observed.

5.7.3
Sedimentary Setting and K/T Boundary "facies"

Throughout the U–M basin, the K/T boundary is exposed in numerous accessible sections, and most of them have been studied to some extent. Even though the K/T boundary is always found in a genuine pelagic environment, it exhibits different lithologic characteristics from sections to section. Lateral variations in sedimentary facies of the Scaglia Rossa Formation reflect an irregular topography of the paleo-sea floor in an active synsedimentary tectonic setting (e.g., Chan et al. 1985; Alvarez et al. 1985; Montanari et al. 1989; see also Fig. 4.1.2).

In the areas west of the Cagli and Camerino (i.e., Gubbio, see Fig. 4.1.1), the Scaglia Rossa is free of turbidites, although small scale soft sediment slumping is found locally (see, for instance, the K/T boundary section at Rocca Leonella, near Piobbico, described by Alvarez et al. 1985). On the other hand, the areas around Furlo and Monte San Vicino, in the central part of the U–M anticlinorium, were occupied by intrabasinal depocenters which were collecting frequent earthquake-triggered slump masses, and turbidites essentially made of reworked planktonic foraminiferal tests. Finally, in the areas near the margins of the Abruzzo and Adriatic carbonate platforms (i.e., Monti Sibillini, Montagna dei Fiori, and Monte Cònero), the Campanian to lower Eocene Scaglia Rossa pelagites are interbedded with thick and coarse calcarenitic turbidites continuing shallow water bioclasts.

On the basis of the color of the K/T boundary clay and the enclosing limestones, and differences in the type and relative abundances of authigenic and pseudomorphic minerals within altered impact spherules, Montanari (1991) recognized at least four different U–M K/T boundary "facies" (Fig. 5.7.3.1). Apart from the authigenic mineralogy of the altered impact spherules, which is described below, rock colors are sufficient for recognizing these different "facies" in the field.

OUTCROP	SPHEROIDS RELATIVE ABUNDANCE	K/T CLAY THICKNESS AND COLOR	GLAUCONY COLOR	CRETACEOUS LIMESTONE COLOR	TERTIARY LIMESTONE COLOR	BULK K/T CLAY Ir (ppb)
CONTESSA (Gubbio)		10 mm red / 4 mm green	bright green	white	dark pink	2.94 ± 0.16
PETRICCIO (Acqualagna)		21 mm red / 5 mm green	pale green	white	dark pink	8.30 ± 0.51
PIETRALATA (Furlo)		8 mm red	pale green	light pink	dark pink	2.14 ± 0.18
FORNACI QUARRY C. (Monte Cònero)		16 mm ochre / 3 mm green	pale green to ochre	white	beige	10.29 ± 0.48
MONTAGNA DEI FIORI		15 mm green	gray-green to gray	white	whitish	8.30 ± 0.50

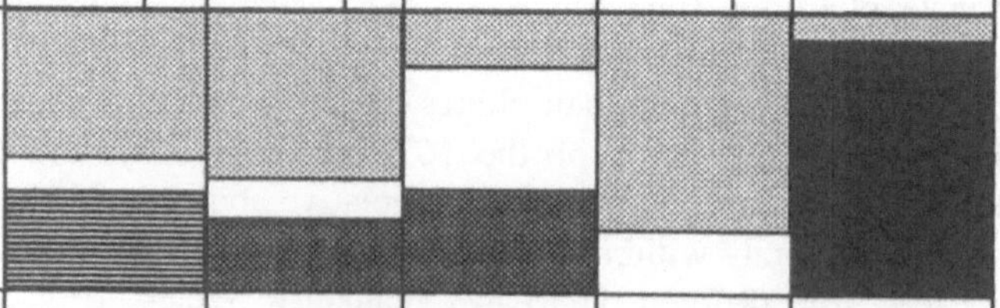

Legend: pyrite · hematite · goethite · glaucony · K-feldspar

Fig. 5.7.3.1. K/T boundary "facies" of the U–M basin from five representative sections (from Montanari 1990).

At Gubbio (i.e., Bottaccione, and Contessa sections), as in every other outcrop of Scaglia Rossa in the U–M Apennines exhibiting a typical turbidite-free facies (see the descriptions of these outcrops and those of the Petriccio and Frontale below), the uppermost layer of the Cretaceous is white, which contrasts with the characteristic pink color of underlying Cretaceous limestones, and the slightly darker pink limestones of the basal Paleocene. The K/T boundary clay layer of the Gubbio "facies" is dark red, with a thin gray-green sole at the base (Fig. 5.7.2.1). The white color of the top Cretaceous limestone has been attributed to iron reduction and leaching, and consequent bleaching of the oxidized, pink pelagic ooze, following a brief episode of anoxia at the water–sediment interface (Lowrie et al. 1990; Montanari 1991). This low-Eh condition was probably caused by anoxic bottom waters, rich in organic matter, which formed on the abyssal sea floor immediately after the mass killing of the marine plankton.

In the Furlo turbiditic basin, the K/T boundary "facies" is characterized by a thin, entirely red clay layer (no green sole is present here), and a light pink color of the top Cretaceous, unbleached limestone.

The Monte Cònero K/T boundary "facies", which is well exposed in several outcrops in the Cònero promontory south of Ancona, is characterized by a light ochre color of the boundary clay, which exhibits a thin greenish-gray sole. The Cretaceous pelagic limestones, which are interbedded with thick and coarse calcarenitic turbidites, are here whitish all the way up to the boundary, whereas the basal Paleocene limestones (i.e., the *P. eugubina* Zone) are of a light yellowish-beige color. Therefore, in this area the bleached top Cretaceous limestone is not recognized.

Finally, in the Montagna dei Fiori section, where the sedimentary facies of the uppermost Cretaceous and lowermost Tertiary Scaglia Rossa is similar to the one of Monte Cònero (i.e., thick and coarse calcarenitic, white turbidites interbedded with whitish pelagic limestones), the K/T boundary clay is entirely green. This is the only locality known in the U–M basin where pyrite is found in the K/T boundary clay, and neither K-feldspar nor goethite or hematite are present (Fig. 5.7.3.1).

5.7.4
Impact and Catastrophe Signatures in the U–M K/T Clay

In this section, we review briefly the impactoclastic evidence that can be found in the K/T boundary clay in the U–M basin. Besides iridium (and other trace elements that are related to the impact fallout but can be detected only with sophisticated laboratory analysis), microspherules and shocked quartz grains can be easily picked and observed after simple washing and concentration procedures, and with standard optical microscopes (see Sect. 6.2.2). In addition to this mineralogic evidence, we found it helpful to discuss also the benthic foraminifers and bioturbation, which constitute clear biologic signatures of the impact-triggered crisis in this deep marine environment. An explanation of the high-resolution event stratigraphy across the Italian K/T boundary will be proposed in Sect. 5.7.5 and it is based mainly on the geochemical, mineralogical, and paleontological indicators of the unusual and short-lived local paleoenvironmental conditions related to the global catastrophe. More detailed and specific information for representative K/T boundary outcrops in the U–M region will be given in Sect. 5.7.6.

Iridium. In addition to INAA by Alvarez et al. (1980), which was carried out on samples from the U–M sections of Bottaccione, Contessa, Acqualagna, Petriccio, and Gorgo a Cerbara, the bulk K/T boundary clays from thirteen other U–M sections were also analyzed with the coincidence spectrometer by Frank Asaro at the Lawrence Berkeley Laboratory, as reported in Montanari (1991; see Table 5.7.4.1). No significant Ir concentrations were detected in the U–M sections at Genga and Fossombrone. In both cases, the analyzed clay samples were not from the "standard" K/T boundary, but, rather, from clay-rich levels marking stratigraphic hiati between limestones with Cretaceous foraminifers below, and Paleocene foraminifers above. No spherules or shocked quartz grains were found in these clay-rich layers. These hiati resulted from the removal, probably by bottom current winnowing and/or slumping, of unconsolidated pelagic sediment. Several basal Paleocene biozones are, in fact, missing in these sections (Luterbacher and Premoli Silva 1964; Montanari 1991).

In every other K/T section analyzed in the region, anomalous concentrations of iridium were found in the boundary clay. However, the elemental concentrations vary significantly from section to section, from a minimum of 1.19 ± 0.16 ppb in the Furlo Via Flaminia section, to a maximum of 10.29 ± 0.48 ppb in the Fornaci Quarry C. section.

This variability is probably due to local preservation factors, such as bioturbation, alteration and/or tectonization of the outcrop, which may have caused remobilization and loss of Ir. An exception may be the two sections analyzed at Furlo (Pietralata and Via Flaminia), where in both cases the concentration of iridium is relatively low (Montanari 1991). In this area of the paleobasin, winnowing by bottom currents effectively remobilized the fine fraction of the pelagic sediment (Montanari et al. 1989), and may have remobilized also some of the Ir-rich impact fallout component during, or soon after, its deposition.

Fig. 5.7.4.1. a) Altered microkrystites from the Italian K/T boundary clay: the black spherules are made of goethite (Go), the green ones of glaucony (Gl), and the white ones of K-feldspar (Kf). Notes some dumbbell-shaped K-feldspar spherules (db). b) thin section microphotograph (plain light) of a K-feldspar spherule from the Petriccio section. Note the dendritic (D) and fibrous (F) structures reminiscent of quenched clinopyroxene and calcic feldspar crystals. c) SEM image of an altered impact spherule made of glaucony (Petriccio section) showing dissolution pits identical to those found on Late Eocene microtektites (for comparison see Fig. 2a8 in Sanfilippo et al. 1985), and on impact glass spherules from Mexican K/T boundary sections (see Smit et al. 1992). d) SEM image of a polished glaucony spheroid from the Fornaci central quarry section of Monte Cònero, exhibiting clear dendritic structures identical to those found in Late Eocene clinopyroxene microkrystites (compare with Fig. 5g in Glass et al. 1985). e) extra thin section (1 μm thick) microphotograph (transmitted, plain light) of goethite impact spherules from the Petriccio section, showing various arrangements of spinel microcrystals (black spots), and internal fibrous structures reminiscent of a quenched silicate melt. f) back-scattered SEM image of a magnesioferrite spinel from a goethite spheroid from the K/T boundary clay of Petriccio exhibiting skeletal structure.   →

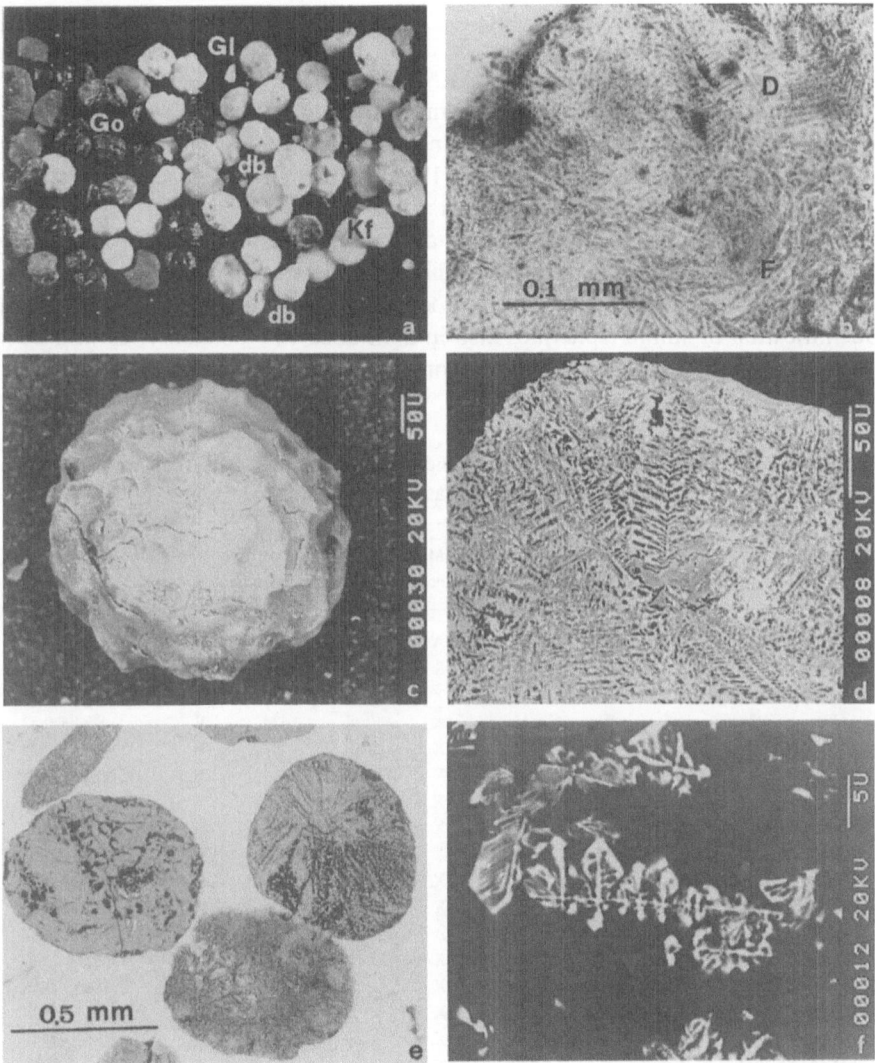

Detailed Ir measurements across the K/T boundary interval, and through the actual K/T boundary clay layer, were carried out at Bottaccione and Petriccio, respectively, as will be discussed in Sect. 5.7.6.

Impact microspherules. Microspherules made of K-feldspar, with a diameter of 0.1–0.8 mm, were first discovered in the K/T boundary clay of the Spanish Caravaca section by Smit and Klaver (1981). Although an impact origin for the spherules was inferred by these authors, their mineralogy similar to (volcanic) high-sanidine remained somewhat more difficult to explain, until $\delta^{18}O$ analyses by DePaolo et al. (1983) showed the sanidine to be low-temperature K-feldspar replacement. Soon after this discovery, cooperative work with Jan Smit led to the identification of similar spherules also in the Italian K/T boundary, as well as in the Central Pacific DSDP Site 465A (Montanari et al. 1983).

Table 5.7.4.1. Major components of the sand-size fraction, and bulk-rock iridium content in the K/T boundary clay layer of the northeastern Apennines

outcrop	sparry calcite	Cretac. plankton	Tertiary plankton	Miocene contamin.	calcar. benthos	arenac. benthos	spheroids	Ir (ppb)*
1 Bottaccione	A	A	A	A	A	A	P	6.06 ± 0.13
2 Contessa	A	S	S	S	S	A	P	2.94 ± 0.16
3 Pontedazzo	A	N	N	X	S	A	P	3.13 ± 0.35
4 Gorgo a Cerbara	A	N	N	X	S	A	P	8.74 ± 0.57
5 Maestà Confibio	A	N	N	X	S	S	P	6.01 ± 0.71
6 Casacce	A	N	N	X	N	S	P	5.30 ± 0.40
7 Colcanino	S	N	N	X	S	A	P	6.28 ± 0.48
8 Acqualagna	A	N	N	X	S	A	P	5.10 ± 0.30
9 Petriccio	N	N	N	X	S	A	P	8.30 ± 0.51**
10 Furlo V. Flaminia	A	S	S	X	S	A	P	1.19 ± 0.16
11 Furlo Pietralata	S	S	S	X	S	A	P	2.14 ± 0.18
12 Fossombrone	A	A	A	A	X	A	X	< 0.38
13 Genga	A	A	A	X	S	S	X	< 0.23
14 Frontale	N	A	A	X	S	A	P	4.51 ± 0.58
15 Cingoli	S	A	A	X	S	A	P	7.11 ± 0.44
16 Ceselli	A	N	N	X	S	A	P	2.69 ± 0.32
17 M.gna dei Fiori	A	N	N	X	X	N	P	8.30 ± 0.50
18 Fornaci Quarry C.	S	S	S	X	S	A	P	10.29 ± 0.48

* Ir data are from Montanari (1991).
** Iridium contained in the central part of the layer.
A = abundant; S = scarce; N = very scarce or neglectable; X = absent: P = present.
Note: relative abundances were estimated visually with a binocular microscope on >64 μm washed residues.

The Italian K/T clay is made of various clay minerals, including illite, smectite, kaolinite, and mixed layered illite-smectite clay. This composition is essentially identical to the detrital clay component in the pelagic limestones below and above the boundary (e.g., Johnsson and Reynolds 1986; Montanari 1991). In addition, the K/T boundary clay contains variable amounts of secondary calcite crystals, which were precipitated within the clay during tectonic bedding-plane parallel flexural-slip folding. The insoluble residue >63 μm constitutes 0.01 wt.% of the bulk rock, and is made up of altered impact microspherules, arenaceous benthic foraminiferal tests, and very rare detrital quartz and (even rarer) feldspar grains.

Up to two hundred impact microspherules per cubic centimeter of bulk rock can be found in the Italian K/T boundary clay (Montanari et al. 1983). These are made of different authigenic minerals reflecting a complex process of diagenetic alteration. Moreover, ratios between these minerals differ from section to section, indicating lateral variations of redox conditions in the early diagenesis of the K/T boundary clay throughout the basin. In addition to whitish spherules made of a K-feldspar, structurally similar to high-sanidine, and, thus, almost identical to the K-feldspar of the Spanish spherules of Smit and Klaver (1981), the Italian K/T clay contains also spherules of bright green to gray glaucony, and black, strongly magnetic goethite (Fig. 5.7.4.1a). These spherules, especially the glauconitic ones, are flattened to various degrees, with axial ratios as large as 10:1, and for this they were originally called "spheroids" by Montanari and co-workers.

The K-feldspar spherules are often pitted on the surface and sometimes they exhibit dumbbell shapes (Fig. 5.7.4.1a). Internally, as seen in petrographic thin sections, they are characterized by fibro-radial or fan-shaped arrangements of K-feldspar crystals. More rarely, ghosts of skeletal or dendritic crystals are present (Fig. 5.7.4.1b). These crystal shapes, along with the composition of the K-feldspar, which is a common low temperature alteration product of oceanic pillow basalts, lead Montanari et al. (1983) to the inference that the K/T spherules were originally made of a quenched basaltic melt, similar to impact clinopyroxene spherules described by Glass et al. (1985) in upper Eocene marine sediments, and, thus, they represented altered microkrystites (see also DePaolo et al. 1983, and Smit et al. 1992). The Italian K-feldspar spherules contain 1.02 ± 0.11 ppb Ir (Montanari et al. 1983), indicating that they were contaminated by the impactor.

Glauconitic spherules are present in all the K/T boundary clay samples analyzed by Montanari (1991) in the U–M basin. Their color, from bright green to gray, and maturity as indicated by their relative K_2O content (Odin and Fullagar 1988), vary from section to section, reflecting variable redox conditions throughout the paleobasin's sea floor. In thin section, these flattened spherules exhibit the typical pinpoint texture of the glaucony, and no particular structure reminiscent of their precursor mineralogy is usually visible. However, in a few rare cases, claucony spherules exhibit a peculiar pitted surface very similar to that of Late Eocene Caribbean microtektites (Sanfilippo et al. 1985), and of K/T impact glass spherules from Mexican sections (Smit et al. 1992) (Fig. 5.7.4.1c). Only in some cases elaborate dendritic internal structures in glaucony spherules are emphasized by back-scattering SEM imaging of polished surfaces (Fig. 5.7.4.1d), suggesting that, in such cases, quenched clinopyroxene was the precursor mineral prior to diagenetic alteration. This, in fact, is the common mineralogy of Late Eocene micro-krystites described by Glass et al. (1985), which show internal dendritic structures virtually identical to those of the altered K/T spherules, as well as of some unaltered microkristites found by Smit et al. (1988) in the K/T boundary of an Atlantic deep sea core.

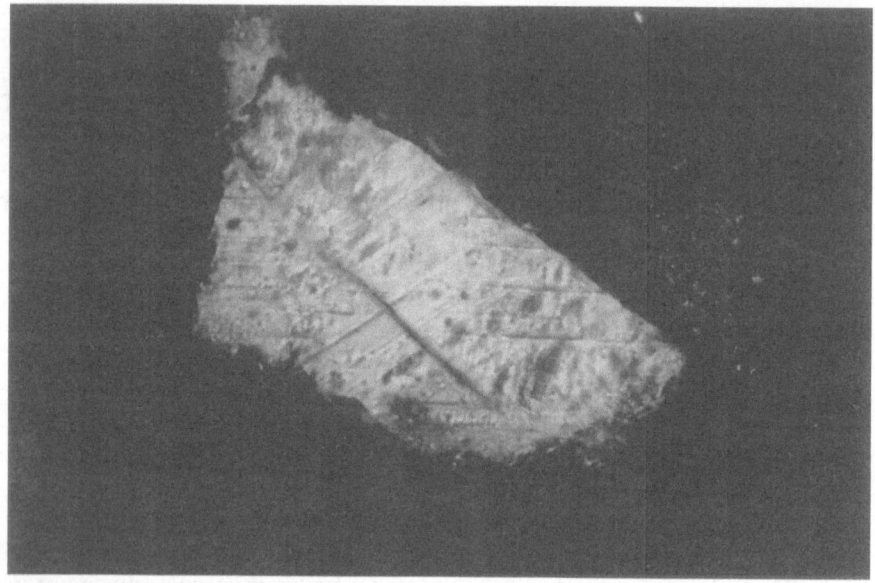

Fig. 5.7.4.2. Thin section microphotograph (crossed polars) of a shocked quartz grain, about 0.1 mm across, from the K/T boundary clay of Petriccio, exhibiting two clear sets of PDFs.

Black magnetic spherules in the Italian K/T boundary clay were fist described by Montanari et al. (1983) who, on the basis of early XRD and INAA data, misinterpreted their mineralogy: they thought they were made of magnetite crystallites suspended in an impure microcrystalline quartz matrix. Instead, specific geochemical and mineralogic studies (Smit and Kyte 1984; Kyte and Smit 1986; Montanari 1991) showed that these spherules are made of tiny crystallites of magnesioferrite-magnetite spinel suspended in a translucent, rust color goethite matrix. Flow-textural and fibrous structures, as seen in extra thin sections ($\sim$1 µm thick), characterize the goethite matrix, whereas the spinel crystallites are often aligned along these matrix structures (Fig. 5.7.4.1d). In other cases, spinel crystallites are sparse and randomly distributed throughout the spherules, or distributed in concentric rings. SEM imaging of the spinel reveals dendritic and skeletal habits indicating quenched crystallization (Fig. 5.7.4.1e). Further detailed mineralogic and geochemical analyses led Robin et al. (1992) and Gayraud et al. (1996) to propose that these K/T spinels were formed by ablation in the lower atmosphere of large projectile debris possibly generated in a low angle asteroid impact (see Sect. 3.2.1.8).

Spinel-bearing goethite spherules are absent in the relatively reduced K/T boundary clays of the Monte Cònero and Montagna dei Fiori sections. In the former, only more or less ferroan glaucony and K-feldspar spherules are present, whereas in the latter section, glaucony is associated with euhedral pyrite.

In addition to these dominant types of altered spherules, the Italian K/T clay contains rare spherules with mixed mineralogies (Montanari 1991). Most commonly, glaucony is associated in the same spherule with K-feldspar or with goethite, but rarely, if at all, the three mineralogies are present together in the same spherule.

Shocked Quartz. Shocked quartz grains (see Sect. 1.6) in the U–M K/T boundary clay were first reported by Bohor and Izett (1986), confirming the worldwide distribution of this unmistakable indicator of impact shock metamorphism originally discovered by Bohor et al. (1984) in North American K/T boundary sections (see Sect. 3.2.1.5). In the Italian K/T boundary, these grains are angular and rarely larger than 0.1 mm (Fig. 5.7.4.2). In the Petriccio outcrop (see Sect. 5.7.6), quartz grains are concentrated in the basal 5 mm thick green sole of the boundary clay. Out of ~230 quartz grains per 100 g of bulk rock, 20 grains show single and multiple sets of PDFs (Montanari 1991). In contrast, the red part of the boundary clay contains no more than 50 grains of quartz per 100 g of bulk rock, and about 10–15% of them show lamellar features. The concentration of quartz grains at the base of the boundary clay was confirmed by one of us (A.M. unpublished data) in a test in the Monte Cònero section of Fonte d'Olio (see Sect. 5.7.6).

Benthonic Foraminifera. In addition to spherules, variable amounts of secondary calcite crystals, very rare quartz grains (and even fewer feldspar and other non-carbonate minerals), and rare fish teeth, the >63 μm washed residue of the Italian K/T boundary clay, contains a well diversified benthic foraminiferal assemblage dominated by arenaceous tests of agglutinating *Textulariina* species. Calcareous tests of *Osangularidae* and *Gavellinellidae* are rare and often show clear signs of dissolution, whereas Cretaceous and/or Tertiary planktonic foraminiferal tests are usually absent or very rare, except in a few sections, where they were probably introduced into the clay layer by later bioturbation (Montanari 1991; see also the description of the Frontale outcrop in Sect. 5.7.6).

Textulariina are very rare in the limestones of the Scaglia Rossa formation. Although many species of this Suborder are well adapted to deep water environments, they favor siliciclastic sediments, which provide silt and fine sand grains of hard silicates, such as quartz and feldspar, that they selectively collect and agglutinate to make up their resistant tests. Arenaceous foraminifers are found also in clay layers above and below the K/T boundary, and throughout the whole U–M carbonate sequence, but the rich and diversified assemblage in the K/T boundary clay, which is represented by some 16 genera and 35 species for a total of up to 40 specimens per gram of bulk rock (Fig. 5.7.4.3), led Coccioni and Savelli (1983) to conclude that the K/T boundary clay represents a sudden change from normal to unusual lower bathyal paleoenvironmental conditions.

A recent detailed study of the benthic foraminiferal assemblage in several sections of the U–M basin by Coccioni and Galeotti (1998) confirmed these early observations. They concluded that during the sedimentation of the K/T boundary clay in the lower bathyal sea floor of this basin (i.e., 1500–1800 m below sea level), benthic foraminiferal communities show an increase in species richness, and a "bloom" of low-oxygen tolerant, infaunal, detritus-feeding species (e.g., *Spiroplectammina* ss.pp.), which indicate an increase in organic flux to the sea floor, and a concurrent decrease in oxygenation at the water–sediment interface.

Slight variations in reciprocal proportions among *Textulariina* genera and species from section to section throughout the basin indicate lateral variations of redox conditions, probably controlled by topography, bathymetry, and bottom current activity. In the relatively reduced, more proximal and perhaps shallower Monte Cònero area, for instance, *Spiroplectammina* tests (the most common genus in the Italian K/T boundary clay) are more abundant and larger than those found in deeper and better oxygenated central areas of the paleobasin (i.e., Petriccio, Furlo, Gubbio). In the pyrite-bearing, green K/T boundary clay at the Montagna dei Fiori section, benthic foraminifers are practically absent, except for a very few tests of *Ammodiscus*.

In summary, although in more proximal and shallower settings throughout the K/T world's oceans the benthic foraminiferal community testify to drastically stressed sea floor environments, and even mass extinction (see Coccioni and Galeotti 1998, and references therein), in the lower bathyal U–M basin arenaceous *Textulariina* were less affected by the catastrophic event in terms of extinction, and actually they were somewhat favored by the sudden low oxygen and high organic nutrients conditions produced by the planktonic mass killing.

Bioturbation. One of the most peculiar macroscopic aspects of the K/T boundary interval in the U–M basin is the high concentration of trace fossils within the top white layer of the Cretaceous. This anomalous occurrence can be better appreciated if compared with the normal sedimentological setting of the Scaglia Rossa limestones. In fact, these limestones are usually homogenized by bioturbation activity occurring during a slow (3 to 10 mm/ka) pelagic carbonate sedimentation on a well-oxygenated sea floor. Changes in sedimentation rate and/or brief interruptions of the carbonate production are probably controlled by cyclic climatic changes affecting the primary productivity of the surface sea water. The result is an obvious rhythmic layering of the Scaglia Rossa, emphasized by pressure-solution pseudobedding (stylolitization) produced by sediment loading (Alvarez et al. 1985).

Fig. 5.7.4.3. Some common *Textulariina* arenaceous benthic Foraminifera from the K/T boundary of Frontale. S = *Spiroplectammina* ; D = *Dorothia*; R = *Rhyzammina*; B = *Bathysiphon*; T = *Tritaxia*; V = *Verneulina*; Re = *Reophax*; So = *Sorosphaera*.

The abundant bottom dweller's activity on the very top sediment of the Scaglia Rossa sea floor is excellently documented at the interface between calcarenitic turbidites and the underlying pelagic limestones (i.e., Furlo and Monte San Vicino areas). Here, the sudden arrival of fine grained turbidites essentially made of reworked planktonic foraminifers, have stopped the ongoing bioturbation activity, and a well diversified ichnofauna is preserved (Fig. 5.7.4.4.). This is mainly represented by traces of *Planolites* and large, flattened *Zoophycos*, and more rare *Helmintoidea*, *Paleodychtion*, *Scolicia*, and *Subphyllochorda* (Montanari 1979). In the normal sedimentological setting of this formation, some well-preserved traces of *Zoophycos* can be seen in cross-section below prominent clay layers (Figs. 5.7.4.5a, b).

In summary, continuous reworking of the top sea floor sediment is carried out by bottom dwelling organisms. However, sudden interruption of planktonic productivity allows the sedimentation of only windblown clay and silt, and a concurrent cessation of bioturbation activity with the consequent preservation of a thin, unhomogenized clay-rich layer. As soon as normal nutrient-rich, but slow carbonate sedimentation is restored, bottom dwellers activity restarts, and the new surface sediment gets bioturbated while it accumulates. During this process, however, some organisms, while feeding at the surface, burrow deeper than others in the subbottom leaving patterned traces of *Zoophycos* (Fig. 5.7.4.5). A recent study by Le Rousseau et al. (1996) showed that, in the basal 10 m of the Tertiary at Gubbio, *Zoophycos* "acmes" have a rhythmic occurrence, possibly reflecting Earth's orbit eccentricity cycles (i.e., ~100 ka).

Fig. 5.7.4.4. Bottom surface of a calcarenitic turbidite of the Scaglia Rossa (Furlo Upper Road section), showing a complex network of *Planolites* trace fossils.

The ichnofossil record at the K/T boundary can be considered a magnification of the normal process of bioturbation versus sedimentation described above. The difference is that the interruption of planktonic carbonate production represented by the K/T boundary clay was more drastic and lasted a much longer time than any of the cyclic clay episodes recorded above and below the boundary.

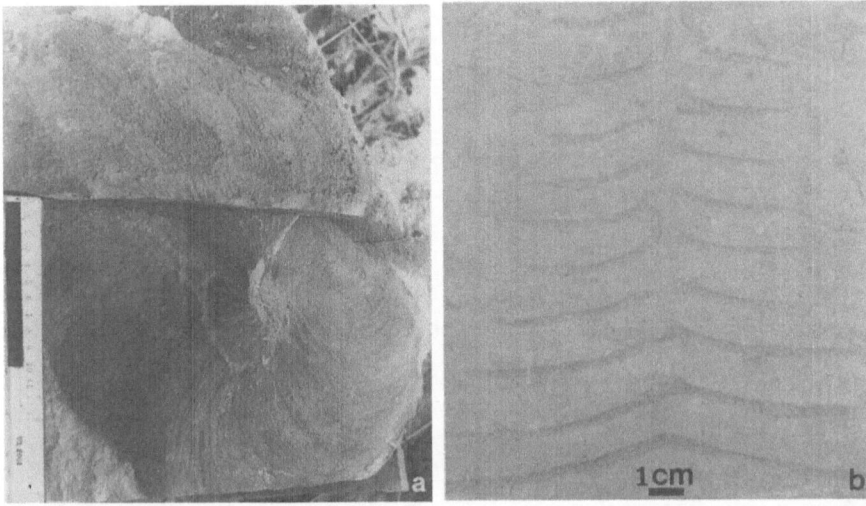

Fig. 5.7.4.5. Fossil traces of *Zoophycos* in the pink pelagic limestone of the Scaglia Rossa at Furlo. a) plane view (from the top surface of layer; b) vertical section on a polished slab of a limestone layer.

Zoophycos traces are the most conspicuous and are immediately visible on the outcrop. They are distributed along the white top limestone layer of the Cretaceous, and may reach 30–40 cm in diameter, 15–20 cm in height, and as many as two or three traces can be counted in a meter of exposed section along strike. The apex of these traces rarely reach the base of the overlying K/T clay layer. These ichnofossils are associated with randomly distributed *Planolites* traces. The number, size, and density of fossil traces vary from section to section. In the Petriccio section, for instance, as well as at Gubbio, *Zoophycos* in the range of 20 cm in diameter seem to be the predominant ichnofossils. In the Furlo area, where the light pink uppermost Cretaceous limestone is not completely bleached to a white color, and the K/T boundary clay is entirely red, bioturbation appears much more intense and diversified than in any other outcrop in the region. In the whitish K/T boundary transition of Monte Cònero, trace fossils are rare and small compared to Gubbio and Furlo, whereas in the pyrite-bearing Montagna dei Fiori section trace fossils are absent.

214

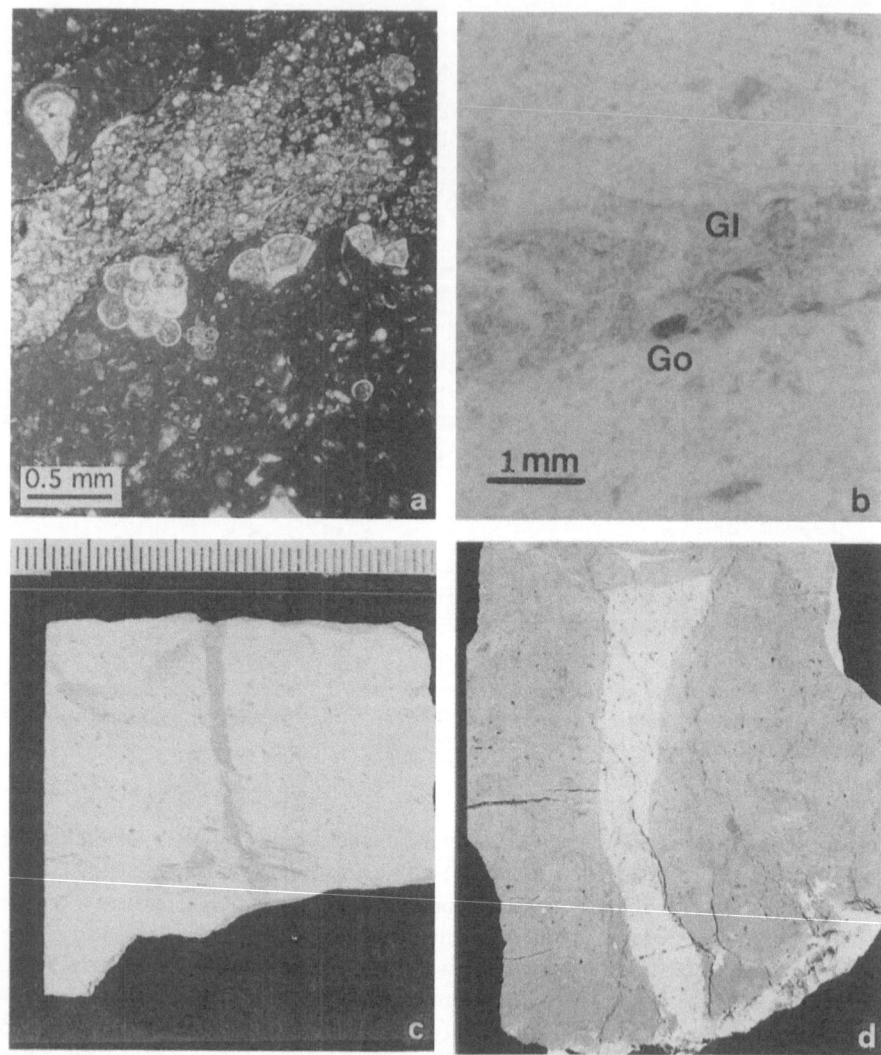

Fig. 5.7.4.6. a) Thin section microphotograph (plain light) of a *Planolites* burrow in the top Cretaceous limestone from the Pietralata section filled with tiny planktonic foraminiferal tests derived from the overlying Tertiary *eugubina* limestone; b) photograph of a polished slab of the top Cretaceous limestone from the Pietralata section showing the typical chevron structure of a *Zoopycos* burrow. Note a goethite (Go) and three glaucony (Gl) spherules contained in the burrow; c) reddish *Planolite* burrow penetrating the white top Cretaceous limestone immediately below the base of the K/T boundary clay at Petriccio. Small divisions in the scale at the top of the picture are millimeters; d) *Planolites* burrow filled with white top Cretaceous sediment in the red basal Tertiary limestone from the Torricella section (Furlo anticline). Tiny dark specks dispersed in the homogenized red biomicrite are altered microkrystites derived from the underlying K/T boundary clay. Same scale as in Fig. 5.7.4.6c.

Planolites and *Zoophycos* burrows penetrating the top Cretaceous sediment are filled with the tiny Tertiary foraminifers that make up the overlying *eugubina* limestone (Fig. 5.7.4.6a). Altered microkrystites of K-feldspar, glaucony, and goethite are often found in these burrows (Fig. 5.7.4.6b), along with fragments of green and red boundary clay. The color of these traces is usually dark or red, strongly contrasting with the bleached white color of the enclosing terminal Cretaceous limestone (Fig. 5.7.4.6c). Conversely, "return" burrows penetrating the dark pink or red *eugubina* limestone from below, are white, and filled with uppermost Cretaceous biomicrite (Fig. 5.7.4.6d). These color contrasts and the spatially confined location of these burrows clearly indicate that, after a brief pause of bioturbation, bottom dwellers restored their burrowing activity at the very base of the *eugubina* layer, penetrating downward across the K/T boundary layer into the still unconsolidated top Cretaceous sediment. While this renewed biologic activity was in progress, the top Cretaceous sediment was already reduced and bleached, the low-Eh authigenesis of glaucony was accomplished, and the environment at the water–sediment interface was oxidizing.

5.7.5
High-resolution Impact Stratigraphy of the K/T Event

Despite an elusive P0 Zone, which is unanimously accepted as the first biozone (i.e., the first biostratigraphic record) of the Tertiary, it cannot be said that the K/T boundary in the U–M basin is incomplete or that it bears a long hiatus. On the contrary, the combined geochemical, sedimentological, mineralogical, and paleontological data gathered in this deep water, pelagic basin, are sufficient for reconstructing with confidence and good resolution the sequence of exceptional events following the terminal Cretaceous catastrophe. It is just a matter of scale: in the U–M sections the appropriate tool for sampling and studying the stratigraphy of the K/T boundary are tweezers, instead of the classical hammer that is used in more expanded sections throughout the world.

On the basis of what we have documented in the preceding chapter, we describe now the various interpretative steps of environmental changes across the K/T event, starting in the uppermost Cretaceous, and ending in the lowermost Tertiary.

The Days Before. The general environmental crisis that characterized the lower-mid Maastrichtian with an extended period of low productivity, water cooling, extinctions, and biologic turnovers (and also impacts; see Chap. 5.7.1) was over, and the situation in our U–M deep-water basin at the very end of the Cretaceous was, oceanographically speaking, florid. Variations in productivity and diversity of the planktonic foraminiferal population was mainly controlled by Milankovich cyclicity, which imparted to the Scaglia Rossa its typical rhythmic bedding. Chauris and Le Rousseau (1996) carried out a detailed (~5 cm intervals) cyclostratigraphic analysis of the top 110 cm of the Cretaceous in the classic Contessa highway section, using carbonate abundance and relative counting in thin section of important foraminiferal genera and species. They came up with the identification of two peaks in carbonate abundance coinciding with relative abundance peaks of selected foraminiferal taxa, which they interpreted, on the basis of the duration and mean sedimentation rate for the C29r interval (~6–6.2 m/m.y.), as reflections of the ~100 ka orbital excentricity cycle (Fig. 5.7.5.1). The last peak in relative abundance of six foraminiferal taxa (*Rosita contusa, Globotruncanita* sp., *Racemiguemelina fructicosa, Rugoglobigerina* sp., *Globotruncana* sp., and *Hedbergella* sp.), which coincides with a peak in $CaCO_3$, is located about 45 cm below the K/T boundary clay. The foraminifers in the uppermost sample of the Cretaceous seem to suggest the beginning of a new positive cycle, which was sharply interrupted by the K/T event.

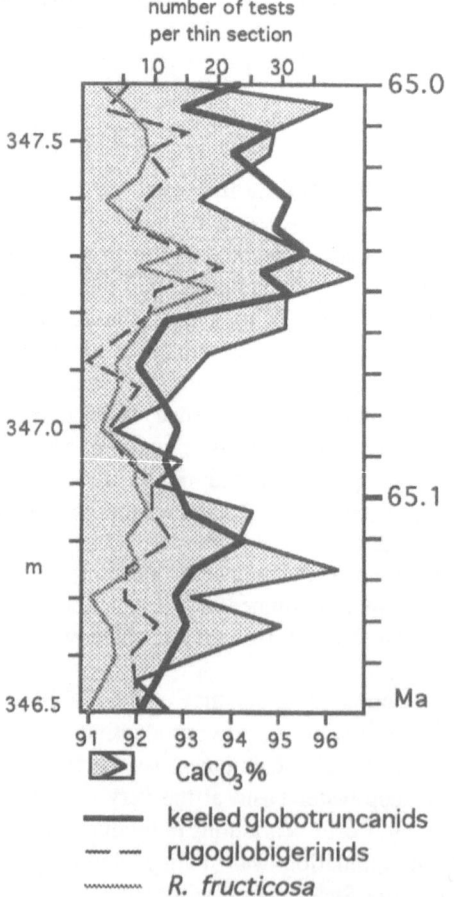

Fig. 5.7.5.1. Variations of index fossils abundance and whole rock calcium carbonate content in the last 1.1 m of the Cretaceous at Gubbio. The two main peaks are probably related to the excentricity of the Earth's orbit (from Chauris and LeRousseau 1996).

Renon (1997) extended for several meters down section, at Gubbio and elsewhere in the U–M basin (i.e., Furlo and Monte Cònero), the investigation of Chauris and Le Rousseau (1996), using detailed stratigraphic, calcimetric, and quantitative biostratigraphic analyses, and reached the same conclusions as his French colleagues. In summary, at the very end of the Cretaceous, the planktonic foraminiferal population in the U–M pelagic basin went over a minimum in productivity, reflecting environmental stress, as they did so many other times, cycle after cycle, in the preceding millions of years. Unfortunately, these marine organisms could not make a profit of a newly restored positive trend, as they were suddenly and radically exterminated by the bolide impact at the Yucatan Peninsula in Mexico.

The Days After. It must have been a difficult awakening for the U–M plankton the day after the armageddonian impact on the Campeche platform in Yucatan. Gaseous and particulate matter produced by the expanding impact cloud may have completely blocked the sun light, and a deadly blanket of darkness and cold enveloped the whole planet in a matter of hours or days. The once fertile photic zone of the marine realm became quickly a field of death.

It is difficult to say how long the agony of the world's oceans lasted. Darkness and cold may have gone on for months before the most opaque fraction of the impact cloud settled on the Earth's surface, but the marine plankton would have probably survived less than that. The phytoplankton, including the *Coccolitophorida*, which were at the very bottom of the oceanic food chain, and provided most of the calcium carbonate making up the Scaglia Rossa pelagic mud, was probably the first to have died out. The sudden darkness would have interrupted their vital photosynthesis inexorably. Soon after that (perhaps a matter of days to months) was the turn of the zooplankton, including the Foraminifera, which provided 10 to 30% of the calcium carbonate of the Scaglia Rossa limestone. Some species, such as those which were adapted to dwelling also below the photic zone, may have had the chance to see a feeble sunlight once more after the gloomy impact winter was over; but the atmosphere, now free of most of the opaque impactoclastic matter, still contained gases such as CO_2, water vapor, and $SO_2 + SO_3$ released from the instantaneous vaporization of the limestones and evaporites in the shallow sea of the Campeche platform, along with nitrous monoxide formed by the impact-generated superheating of the atmosphere. This formed a deadly atmospheric cocktail, which would be reprecipitated as torrential acid rains, causing long lasting greenhouse conditions at a global scale.

All this had an immediate effect on the pelagic sedimentation, in the U–M basin as everywhere else around the world. Carbonate production was suddenly interrupted, and only windblown silt and clay reached the sea floor, along with the fine ejecta of the distant impact. In the months after the catastrophe, a huge amount of dead organic matter, plankton and nekton alike, with or without hard parts, slowly settled in the deep of the oceans. At first it could have seemed a manna for the benthic inhabitants of the abyss, but soon this excess of quickly decaying organic matter turned the already precarious environment of the bathyal sea floor into an anoxic, acidic place, where life could barely be supported.

Worms and other bottom dwellers, which in past times were meticulously gardening the well-oxygenated sea floor with an incessant bioturbation activity, came to a sudden rest. Only detritus feeding, low-oxygen tolerant benthic microorganisms managed to populate this silent, barren environment.

After this, the K/T boundary clay started to accumulate on the Scaglia Rossa sea floor. A few millimeters of the very top Cretaceous carbonate ooze, as well as the lamina of dead calcareous plankton killed by the impact, were probably dissolved in the film of stagnating bottom water acidified by the decaying organic matter. This is suggested by the corroded top surface of the uppermost Cretaceous limestone, and the thin green clay lamina at the base of the K/T clay layer perfectly adhering on top of it.

Reduction progressed downward in the water-saturated, well-oxidized terminal Cretaceous ooze causing dissolution and remobilization of now reduced iron particles, and bleaching of the once pink carbonate sediment (Lowrie et al. 1990). This process of dissolution and bleaching did not happen in the Furlo basin, where bottom currents freed the sea floor from the stagnating film of low-pH, low-Eh water. Here, the top limestone layer of the Cretaceous exhibits a light pink color, and it does not seem to have a corroded upper surface. Moreover, the overlying K/T clay lacks the green sole, and its Ir content is significantly lower than the average Italian K/T clay (Montanari 1991). In summary, while throughout most of the deep sea floor of the Scaglia Rossa basin the thin carbonate layer of the "day after" hecatomb was dissolved away by the acidic "nutrient soup" (Coccioni and Galeotti 1994), at Furlo it was swept away by local bottom winnowing (Montanari et al. 1989).

How long did this situation last before almost normal conditions in the surface water restored carbonate sedimentation? The answer to this intriguing question is contained within the 0.8 to 2.6 mm thick boundary clay: way too thin a page of Earth history, anyone would say, to read what happened in the oceans during the millenia after the impact. But cautious tweezers may find, in fact, interesting hints leading to a satisfactory answer.

The Millennia After. Accepting that *Gumbelitria cretacea* was, in the planktonic foraminiferal population, the only survivor of the K/T catastrophe (the P0 Zone), and that in more expanded sections such as the GSSP for the K/T boundary at El Kef, Tunisia, a high-resolution biostratigraphy can be drawn from the base of the P0 Zone up through the various subsequent phases of the repopulation of the planktonic realm (Smit et al. 1997; Canudo 1997; Masters 1997; Olsson 1997; Orue-etxebarria 1997; Keller 1997; Smit and Nederbragt 1997), the question arises: why isn't this lucky planktic foraminifer present in the Italian K/T clay?

One reasonable answer to this question is that the tiny tests of this biserial globigerinid, along with the micrometric platelets of surviving *Coccolitophorida*, were dissolved in the low pH-low Eh post-impact environment of the Scaglia Rossa sea floor. It could well be that *G. cretacica* and coccoliths were forming their thin P0 Zone also after the K/T clay was deposited, when the low-pH conditions at the water–sediment interface were buffered by the arrival of conspicuous new biogenic carbonate, but at that point, restored bioturbation activity would have mixed up the new Tertiary sediment, yielding what is today recognizable as a homogenized *eugubina* limestone. Thus, because *G. cretacea* cannot help answering any question, either because it was dissolved (so it is no longer there) or because it was vertically mixed by bioturbation and confused with its descendants, the question should be addressed to the inorganic world and their numbers: how long did the crisis last in this basin?

A first estimate can be derived from the thickness of the K/T clay layer. The clay itself is practically identical to the clay making up the fine windblown, siliciclastic fraction of the enclosing Cretaceous and Tertiary limestones. At Gubbio, the Cretaceous part of Chron 29r is about 4.1 m thick and represents about 500 ka (using the time scale of Gradstein et al. 1995; see Fig. 5.7.1). Thus, the average sedimentation rate for this pre–K/T interval is 8.2 mm/ka. An average of 5% of the pre–K/T limestones is made of clay. From this, we can derive a sedimentation rate for the clay of about 0.04 mm/ka. In short, a 20 mm boundary clay would correlate roughly to 50 ka. One problem here is that the K/T clay thickness vary laterally due to local tectonic shearing and compaction. A thickness of 20 ± 10 mm can be considered a realistic estimate representing the U–M K/T clay layer, and, thus, a time of 50 ± 25 ka is the best estimate we can derive for its time duration. This estimate would become even smaller, roughly 30 ± 15 ka, if we use the sedimentation rate of about 5.6 mm/ka for the lowermost Tertiary limestones in the Bottaccione section, derived from the time scale of Berggren et al. (1995).

The imprecision of the time estimate for the duration of the K/T boundary clay episode in the U–M basin is mainly due to the inaccuracy of the actual stratigraphic record (completeness, compaction, and consistent rate of sedimentation), and possibly the inaccuracy of the numerical time scale(s). In fact, a recent high-resolution analysis of extraterrestrial ^{3}He through the K/T boundary at Gubbio by Mukhopadhyay et al. (1988) shows equal concentrations in the uppermost Cretaceous and lowermost Tertiary limestones, indicating practically identical sedimentation rates. In summary, on a simple sedimentological basis, we can infer a duration of the K/T clay episode longer than 15 ka and shorter than 75 ka, with a preferred, possibly more realistic, figure of 50 ka.

A confirmation of this time estimate came from considerations about the authigenesis of glaucony. The glauconitic spherules, in fact, are the ones that best reveal the unusual early diagenetic environment of the K/T boundary clay. Glaucony is virtually absent in the well-oxidized, deep-water carbonate sediments of the Scaglia Rossa Formation, and was never described in any of the Jurassic to Oligocene pelagic carbonate formations of the U–M basin. At Gubbio, the first occurrence of glauconitic pellets have been reported in a few lower Miocene sandy limestones layers at the base of the Bisciaro Formation (e.g., Montanari et al. 1991). Therefore, glaucony in the thin K/T boundary clay clearly indicates an unusual diagenetic environment.

Authigenic growth of glaucony at the expense of a mineral precursor, such as volcanic glass or a silicate crystal, occurs in a slightly reducing marine environment at or immediately below the water–sediment interface. Circulation of sea water in such an environment is necessary for ion exchange and glauconitization (e.g., Odin and Dodson 1982). The evolution of glaucony can progress only if these conditions are met, and the growing mineral acquires from sea water more and more K_2O: excessive burial and restricted water circulation or a change in the Eh would stop the process. A more or less mature glaucony cannot regress to a less evolute phase but will be preserved as is or can be altered into another authigenic mineral (i.e., pyrite, goethite).

On these premises, and on the basis of empirical observation of glauconitization in recent sediments, Odin and Fullagar (1988) constructed a chart of glaucony evolution (i.e., K_2O from 2 to 9 wt.%) versus time of formation. This relation practically indicates how long the particular low-Eh and/or ion exchange conditions in the diagenetic environment lasted before the glauconitization process is stopped by a change in water chemistry or burial isolation. In the case of the K/T glaucony, its K_2O content ranging between 5 wt.% and 6.5 wt.% would indicate that the necessary low-Eh conditions for glauconitization lasted between 7 and 70 ka (Fig. 5.7.5.2). Only the bright green glaucony at Contessa, among others analyzed by Montanari (1991) in the U–M basin, with a K_2O content of about 7.3 wt.%, would indicate a longer duration of authigenic evolution around 100–200 ka (Fig. 5.7.5.2).

The interpretation by Montanari (1991) of the Italian K/T glaucony is that it was formed on the sea floor during the short lived low-Eh event immediately following the mass kill of the marine plankton. Iron-rich microkrystites would have been transformed authigenically into glaucony, whereas silica-rich ones (i.e., mostly made of quenched calcic feldspar and perhaps clinopyroxene), would have been altered into K-feldspar. The return to normal oxygenation on the sea floor would have definitively stopped the process of glauconitization.

The fact that the glaucony at different levels within the K/T clay in the well-preserved section of Petriccio (see Sect. 5.7.6) have all the same maturity led Montanari (1991) to infer that the impact spherules reached the sea floor at once, and were spread upward later while the clay was forming, perhaps by a tenuous microbioturbation activity. The end result is a peak of spherule abundance in the central part of the K/T clay layer, which is mimicked by an abundance peak of Ir.

The problem with this high-resolution stratigraphic observation is that quartz grains, including shocked quartz, which are rarely larger than 100 μm, are concentrated in the basal green sole of the K/T clay layer.

The decoupling of quartz grains from the peak concentration of Ir and spherules in the middle of the red part of the layer, lead Montanari (1991) to suggest that at least two impacts occurred at the K/T boundary, one on continental crust producing shocked quartz grains, and the other in the ocean yielding Ir and spherules. This strange microstratigraphic distribution, however, is different from the one found in terrestrial K/T boundary sections of the North America interior, where Bohor et al. (1987b) reported that Ir is associated with shocked quartz at the top of the boundary layer, whereas altered impact spherules are found just below the Ir-shocked quartz layer.

Although the multiple impact scenario could still be considered an interesting, workable hypothesis, the oddities in the high-resolution stratigraphy of the K/T boundary clay may be explained also with a combination of local diagenetic dissolution, vertical mixing, and remobilization. The dissolution event at the beginning of the K/T clay deposition may have concentrated the quartz and remobilized (removed) the Ir, which would explain at least half of the puzzle.

More difficult to explain is the concentration of spherules in the middle part of the clay and their paucity in the green sole: the process of dissolution of the carbonate (i.e., a low pH environment at the water–sediment interface) would not affect the silicate microkrystites. Vertical mixing of the impactoclastic material by post-impact reworking may explain, at least in part, the concentration of spherules and Ir in the middle part of the boundary clay. On the other hand, it remains difficult to explain how such a reworking would have removed Ir and spherules from the basal clay, but not much of the quartz grains. In conclusion, the high-resolution stratigraphy in Italy still bears unsolved problems, which could be approached by further detailed work in other well preserved K/T boundary sections throughout the U–M basin.

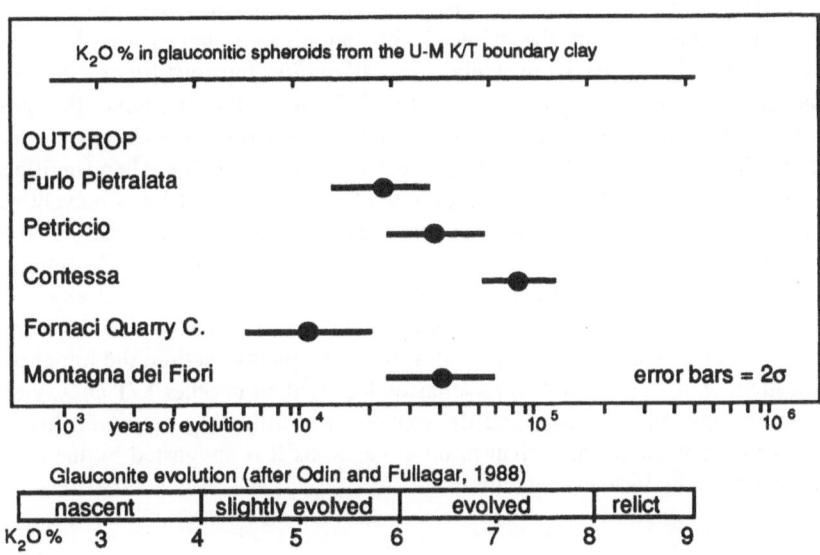

Fig. 5.7.5.2. Evolution of glaucony vs. time in the U–M K/T boundary.

The return to carbonate sedimentation. The *eugubina* limestone layer represents the restoration of conspicuous biogenic carbonate production in the surface water of the U–M basin, and/or the preservation of it after renormalization of pH conditions on the sea floor. This event occurred several thousands to several tens of thousands of years after the impact, and, judging from the sharp and smooth surface contact between the *eugubina* limestone and the underlying K/T clay, it seems that the return to normal carbonate production and sedimentation happened rather suddenly.

The *eugubina* limestone, which in the U–M basin is about 40–50 cm thick, is well homogenized and usually red, indicating a significant bioturbation activity on a well-oxygenated sea floor. A subtle increase in the size of the tiny (0.06 to 0.15 mm) globigerinid tests from the bottom to the top of the layer (Luterbacher and Premoli Silva 1964), suggests that some of the original stratigraphic order has been preserved within the layer despite bioturbation and vertical mixing.

Vertical mixing is also indicated by altered spherules and tests of Cretaceous planktonic foraminifers scattered within the basal 2–5 cm of the *G. eugubina* limestone. However, this bioturbation activity, although efficient in mixing up the slowly sedimenting biogenic carbonate, was not effective in obliterating or mixing the thin K/T boundary clay.

Some large detritus-feeding organisms managed to penetrate the subbottom sediment, burrowing through the K/T clay layer, and bringing down into the white bleached top-Cretaceous limestone the red, biogenic carbonate sediment of the basal-Tertiary *eugubina* limestone. These organisms left numerous traces of *Zoophycos* and *Planolites* in the white layer immediately below the K/T clay (Figs. 5.7.4.6 a, b, and d). Conversely, some *Planolites* burrows filled with white Cretaceous sediment are found in the basal 5 cm of the *eugubina* limestone (Fig. 5.7.4.5b). Often these burrows contain altered spherules and chunks of green and red K/T clay (Fig. 5.7.4.6b). These observations indicate that when significant bioturbation restarted on the basal Tertiary sea floor, the low pH-low Eh event was already over: the top Cretaceous limestone was already bleached, the spherules authigenically altered, and the K/T clay oxidized to a red color. Thus, the return of conspicuous planktonic carbonate production and bottom dweller activity coincided with a return of oxidizing conditions, which would have caused the formation of goethite in some Fe-rich spherules. In the Contessa section, the initial process of glauconitization seems to have lasted longer than in other K/T sites, yielding a more evolved glaucony, and the oxygenation following the initial low-Eh event was somewhat stronger than in other cases, as it is suggested by the authigenesis of hematite instead of goethite.

5.7.6
Representative K/T Boundary Outcrops in the U–M Basin

In areas where the Scaglia Rossa exhibits its most typical turbidite-free facies, an exposed section with the K/T boundary, either in a quarry face, or a cliff, or along a road cut, can be confidently recognized from a distance, even from a moving car. The boundary closely coincides with the lithologic contact between the pink limestones of the R2 member of the Scaglia Rossa, and the slightly darker, more densely bedded marly limestones of the overlying R3 member. In this colored interval, the white-bleached limestone layer of the top of the Cretaceous sticks out as a distinct marker for pointing out the K/T boundary. In areas like the Furlo basin or at Monte Cònero, where the white layer is absent, or where the whole interval is whitish, the lithologic difference between the R2 and R3 members, from more to less calcareous, can still be used to spot the boundary from a distance. In any case, the basic resources of a field geologist (*mente et melleo*) are sufficient to find the boundary in any other situation: a hand sample exhibiting angular tests of globotruncanids means that the boundary is somewhere up section, whereas a piece of limestone with small rounded tests of globigerinids is an arrow indicating that the K/T boundary is down section. A few tries, and the fatal boundary will be narrowed down to a hand span of outcropping rock.

However, the boundary hunt may end up with some disappointment. The magic layer can be covered by dirt or vegetation, or protected by a rock-fall retention steel net, or hanging too high on the cliff or quarry face, or is sheared and tectonized, as often happens, beyond recognition. For this, we would like to end this chapter on the Italian K/T boundary by guiding a potential visitor to those outcrops in the U–M region that best represent the different "facies" described above in Sect. 5.7.3. For a matter of historical respect, we will start with the classic sections near Gubbio, where the whole story started some twenty years ago. Then we will proceed to Petriccio, where the boundary has been studied in some detail, to Frontale (at walking distance from the Geological Observatory of Coldigioco), to Furlo (no bleached white layer), and finally to Monte Cònero and Montagna dei Fiori, where the K/T clay was never oxidized to its typical red color.

Gubbio. The original K/T boundary outcrop of Gubbio is located in the Bottaccione Gorge, along the Gubbio–Scheggia provincial road, just before the crossing of a Medieval aqueduct (see location map in Fig. 5.5.1, and outcrop views in Fig. 5.7.6.1a, b). Here, all the main characteristics of the typical Italian K/T boundary are well expressed, such as the white-bleached top-K limestone, the *Zoophycos* "acme" zone just below the boundary, the bicolor green and red boundary clay, the altered impact spherules, and the dark pink, porcellaneous *eugubina* limestone. A summary of the stratigraphy of the K/T interval in this section is shown in Fig. 5.7.6.2.

The Bottaccione outcrop is the only one in the U–M basin that has been scanned for Ir using a closely-spaced collection of samples spanning some 10 m.y. (i.e., about 60 m of stratigraphic thickness) from the uppermost Cretaceous to the lowermost Tertiary (Alvarez et al. 1990). Several Ir peaks of 0.06–0.08 ppb over a background of 0.02 ppb have been found in a 2 m interval across the K/T boundary. These peaks are two orders of magnitude smaller than the main peak in the boundary clay, and they may be attributed to vertical mixing and reworking. In addition, this is the only section where detailed geochemical and rock magnetism analyses have been carried out in the limestones bracketing the boundary (Lowrie et al. 1990), yielding evidence that the top Cretaceous white limestone has been bleached as a consequence of reduction and dissolution of its iron mineral component. The Fe content is, in fact, lower in this layer in respect to the pink limestones above and below it, and significantly lower Fe/Si ratios indicate that this is not attributable to a decrease in detrital input, but rather to removal of iron (Fig. 5.7.6.2).

Despite its fame of being the representative K/T boundary in Italy, the Bottaccione K/T clay is one of the worst known in the U–M region in terms of preservation and suitability for a high-resolution microstratigraphic study. In fact, in addition to strong shearing caused by flexural slip folding, the K/T clay is here intimately contaminated by the surrounding soil. A washed residue of the clay would yield abundant secondary calcite, a conspicuous amount of sandstone grains derived from the Langhian Marnoso–Arenacea flysch, and well-preserved Miocene microfossils (Fig. 6.7.6.3). Rootlets, insect parts, seeds, shiny spherules of HF-insoluble organic matter, and other recent contaminants are also found in the clay. All this material represents practically the sand-size fraction of the soil surrounding the outcrop, and constitutes the residue of the erosion and dismantling by the Bottaccione creek of the soft Miocene formations exposed upstream along the valley (Montanari 1986). They were introduced into clay seams and fractures of the outcropping Scaglia Rossa by runoff.

Abundant biotite (>0.1‰ of the whole rock) and ~0.1 mm nuggets of titaniferous magnetite are found in a 2 cm thick red clay layer about 2 m above the K/T boundary (Fig. 5.7.6.2). Biotite flakes and booklets are often euhedral and well preserved showing no mineral alteration in X–ray diffractograms. In addition to biotite and titano-magnetite, this thin layer contains also small crystals of apatite and zircon. Rare white mica flakes suggest that some contamination from the Marnoso–Arenacea sandstone may also have occurred here as in the K/T clay.

Nevertheless, this layer, which is also present at the same stratigraphic height in the Contessa section, is clearly derived from some volcanic rock and may represent, in fact, a synsedimentary volcaniclastic event. However, preliminary fission track dating from 18 apatite grains by Odin et al. (1992) yielded a wide scatter of dates, with a mean crystal age of 93.2 ± 2.8 Ma, which is much older than an age of about 64.5 Ma expected for this stratigraphic level. No attempts have been made yet to date the biotite with K/Ar techniques.

In the Contessa Valley, not far from the Bottaccione Gorge (see location of the Contessa Highway – CH – section in Fig. 5.5.1), the boundary is well exposed on both sides of the highway, right after the northern exit of a short rock-fall protection tunnel (Fig. 5.7.6.1b). The K/T clay here has the same problem of tectonic shearing and outcrop contamination as at Bottaccione, although somewhat less severe. The only significant difference from its twin section is that the K/T clay contains a more evolute glaucony which, as explained in the previous section of this book, would indicate that early diagenetic low-Eh conditions in this site lasted a bit longer than at Bottaccione. Moreover, strongly flattened, red hematite magnetic spheroids, which are absent in the Bottaccione and in other known K/T clay outcrops in the region, suggest that at Contessa the re-oxidation process following the initial low-Eh event was more intense than elsewhere in the U–M basin.

Fig. 5.7.6.1. Outcrop views of the K/T boundary in the Gubbio sections: a) Bottaccione Gorge; b) Contessa Highway (= CH in Fig. 5.5.1).

BOTTACCIONE

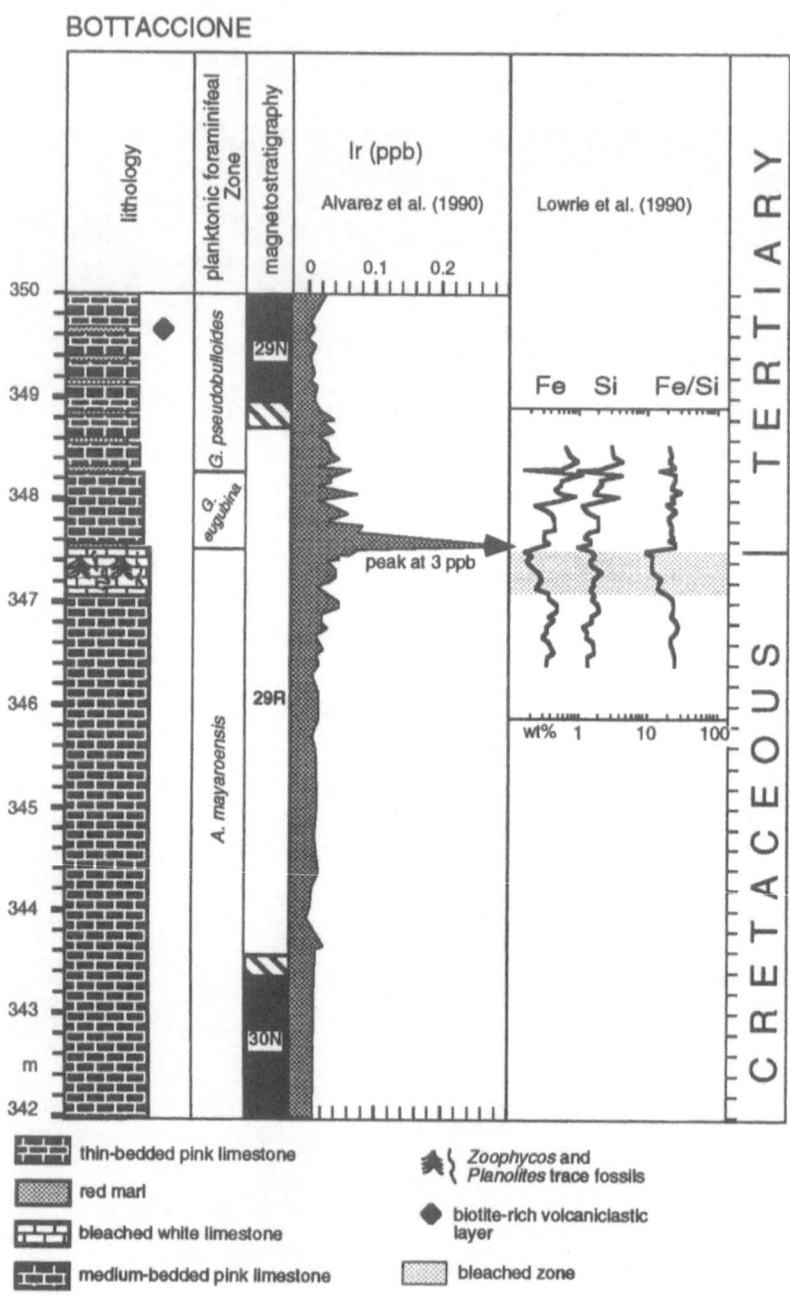

Fig. 5.7.6.2. Stratigraphic synthesis of the K/T boundary interval at Bottaccione.

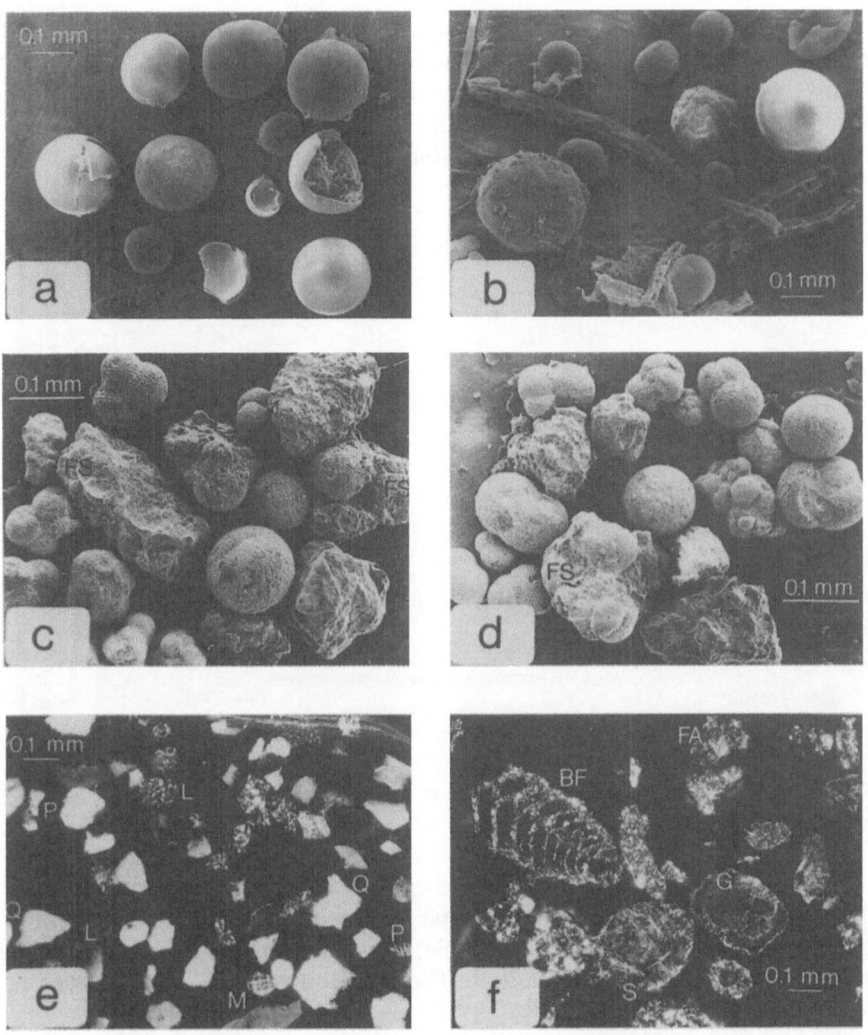

Fig. 5.7.6.3. The problem of outcrop contamination in the K/T section of Gubbio (from Montanari 1986). SEM images of shiny spherules made of recent organic matter from a) the K/T boundary clay in the Bottaccione section, and b) the soil surrounding the same outcrop; c) and d) are SEM images of Miocene planktonic foraminifers and fragments of fossiliferous (FS) Langhian Marnoso–Arenacea sandstone contained in the K/T boundary clay at Bottaccione, and in the surrounding soil, respectively; e) thin section microphotograph (crossed polarizers) of loose silicate sand grains from the contaminated K/T boundary clay in the Bottaccione section obtained by dissolving in HCl the K/T bulk layer fraction >64 μm: L = lithic fragments; Q = quartz; M = microcline; P = plagioclase; f) thin section microphotograph (crossed polarizers) of the HCl-insoluble residue >64 μm from the uncontaminated K/T boundary clay layer of Petriccio; BF arenaceous benthic foraminifer (*Spiroplectammina* sp.); S = sanidine-altered microkrystite; G = glauconitized microkrystite; FA = fragments of arenaceous benthic foraminifers.

Petriccio. The best preserved K/T boundary exposure in the U–M region with a typical "Gubbio facies", that is with a clear white-bleached top-K limestone, a relatively expanded bicolored boundary clay, and the dark pink *eugubina* limestone, is located on the right bank of the Candigliano River, near the village of Petriccio, in the vicinity of Acqualagna (Fig. 5.7.6.4).

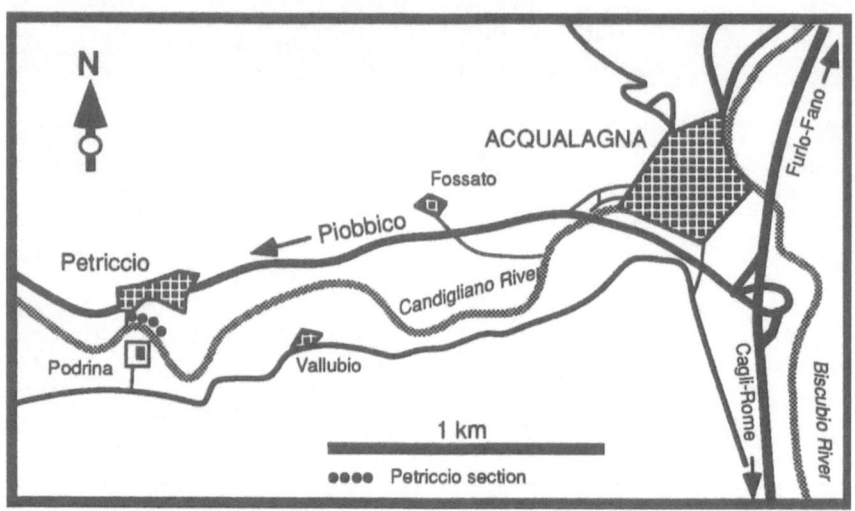

Fig. 5.7.6.4. Location map of the Petriccio section.

For many years, this outcrop has been easily accessible by wading through the shallow river from Petriccio (the decrepit suspension bridge described by Montanari et al. 1983, is no longer practicable), or descending the right bank along a trail from the once abandoned farm of Podrina. However, the farm has been recently reclaimed and fenced in.

The passage outside and around the fence is still accessible in theory, but the property is now (1999–2000) guarded by a couple of large Rottweiler dogs, Nala and Pumba. So, in practice, given that these dogs often get loose by biting holes through the metal chickenwire fence, and that they are extremely vicious, the Petriccio section no longer has easy, safe access. Our strong recommendation for anyone who wants to visit it is to call ahead the owners of the farm, asking the courtesy of holding the dogs. The farm's masters, Mr. and Mrs. Mascarucci (Phone 0347–1192826) are kind, cooperative persons (despite their arguable canine preferences), and they both speak fluent English.

Fig. 5.7.6.5. Outcrop views of the K/T boundary at Petriccio. a) The K/T clay is located at the base of the overhanging *eugubina* limestone; b) close-up of the K/T clay layer. Note the dark trace fossils in the upper part of the white, top Cretaceous limestone.

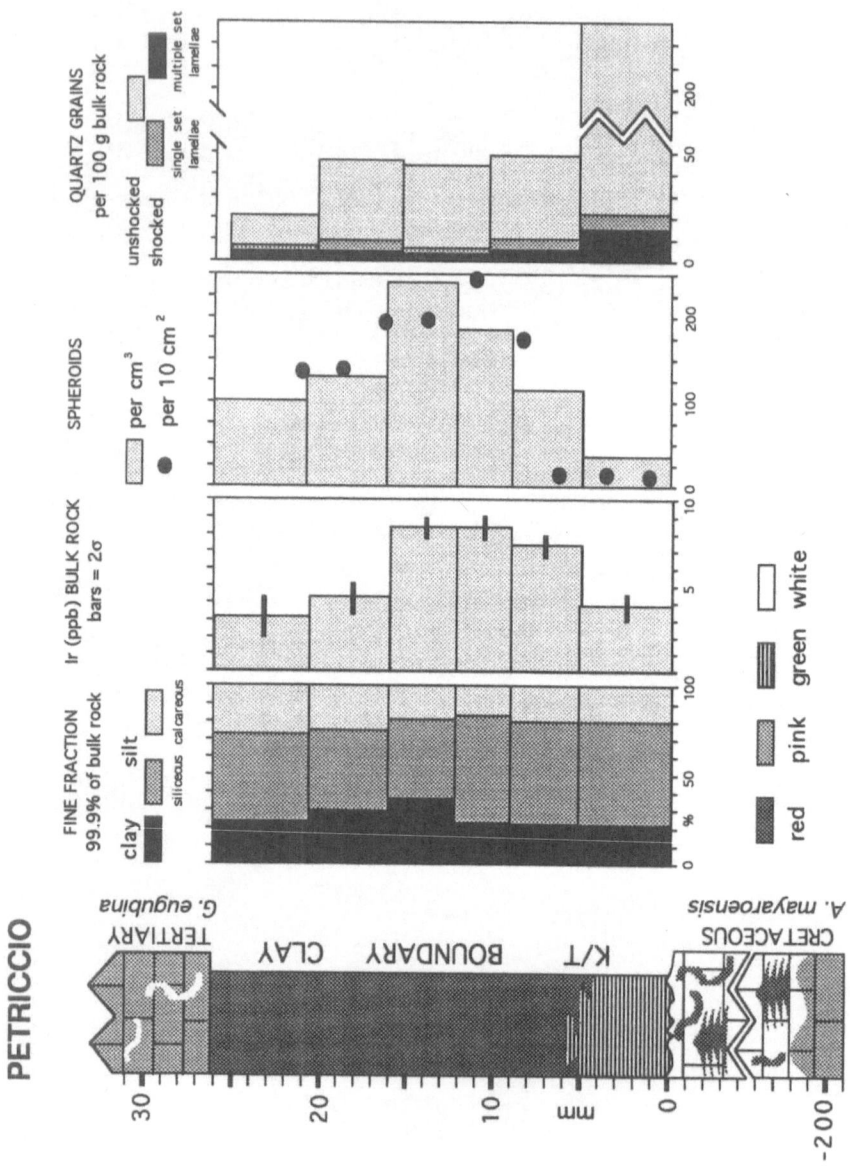

Fig. 5.7.6.6. High-resolution stratigraphic synthesis of the K/T boundary of Petriccio (from Montanari 1991).

The Petriccio outcrop extends along the collapsed river bank for about 15 m. Bedding is subhorizontal and just slightly disturbed in a few spots by small displacement conjugate faults. The white-bleached top Cretaceous layer is clearly visible and contains numerous *Zoophycos* traces (Fig. 5.7.6.5a). The K/T clay layer is 20–30 mm thick, uncontaminated and relatively unsheared (Fig. 5.7.6.5b). A few Tertiary globigerinid tests less than 80 μm in size, probably introduced into the clay layer by bioturbation from the *eugubina* ooze, are recognizable in smear slides. The overall good preservation of the Petriccio K/T boundary outcrop allowed Montanari et al. (1983) and Montanari (1990) to perform a high-resolution microstratigraphic study of the clay, which is summarized in Fig. 5.7.6.6, and the interpretation of which is described in detail in the preceding Sect. 5.7.4.

Frontale. An extended section of the Scaglia Rossa is exposed along the mountain road connecting the village of Frontale to the resort area of Pian dell'Elmo, at the foot of Monte San Vicino (Fig. 5.7.6.7).

The magnetostratigraphy of this section, along with the identification of some important planktonic foraminiferal events, was determined by Chan et al. (1985; see Fig. 5.7.6.8), whereas the general sedimentology of this area was studied by Montanari al. (1989) in the context of a regional basin analysis. Fig. 5.7.6.8 reports also the stratigraphy of the close-by Poggio San Vicino section (see map in Fig. 5.7.6.7) where the inoceramid extinction event has been studied by Chauris et al. (1997), and where a significant enhancement of extraterrestrial ³He was reported by Mukhopadhyay et al. (1998) in the lower Maastrichtian (see also Sect. 5.7.1). Unfortunately, in the Poggio San Vicino section the K/T boundary was involved in a soft sediment slump and it is not resolvable. We would like to point out, however, that lithified clasts of the *eugubina* limestone are found at the base of a 10 cm thick turbidite resting on top of this slump. The turbidite itself is partitioned with a Ta Bauma interval at the base, which is made of large Cretaceous planktonic foraminiferal tests, and a Tb made of small Tertiary globigerinids. This would indicate that the fine-grained, porcellaneous *eugubina* limestone experienced an early diagenetic cementation in contrast with the enclosing Cretaceous and Tertiary pelagic muds, which were remobilized as loose sediment and sorted out during a slumping event (Alvarez et al. 1985; Montanari et al. 1989).

At Frontale, the K/T boundary is exposed on a large quarry face, which offers a panoramic view of the R2–R3 interval of the Scaglia Rossa (Fig. 5.7.6.9). The white top Cretaceous limestone layer is present in this section. However, the white layer visible on the quarry face is not the top Cretaceous marker, but rather the *eugubina* limestone. In fact, unlike the typical "Gubbio facies", at Frontale the *eugubina* limestone is pink only for its basal three or four centimeters, and for the rest it is whitish. Just above the *eugubina* limestone, the limestones and marls of the R3 member are visibly deformed by soft-sediment slump structures which may be correlated with the slump in the nearby section of Poggio San Vicino. Above the slump interval, a few thin layers of white limestone represent very fine calcarenitic turbidites made of reworked foraminiferal tests. These turbidites are exposed also along the Frontale road cut, where the conspicuous slump structure in the lowermost Tertiary is apparently missing.

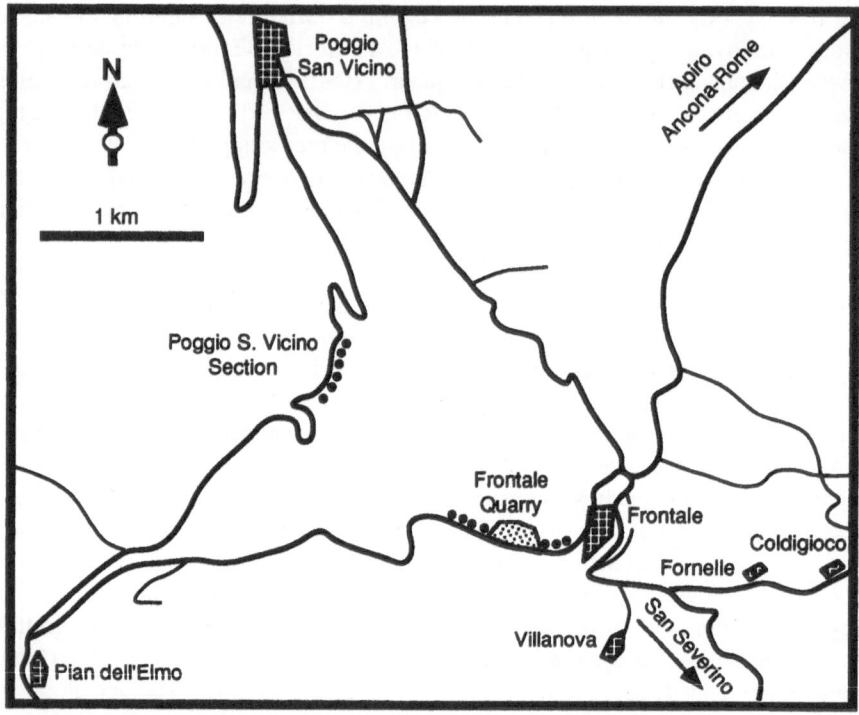

Fig. 5.7.6.7. Location map of the Frontale and Poggio San Vicino sections.

Approaching the K/T boundary in the quarry may be dangerous due to frequent rock falls. Moreover, a normal, synsedimentary fault in the eastern corner of the quarry makes it difficult to reach the boundary, which remains exposed some 3–4 meters above the quarry floor. On the other hand, the same boundary, lowered down by the fault, is well exposed on a small outcrop at the end of the road cut section, right at the eastern entrance of the quarry (Fig. 5.7.6.9). This outcrop has been recently cleaned and reclaimed by the Mountain Community of Monte San Vicino, which installed an illustrated explanatory panel about the K/T event.

In the Frontale quarry, the limestones immediately below and above the K/T boundary clay layer are tectonically undisturbed, and free of closely spaced bedding-parallel pressure-solution stylolites, which usually disturb other K/T sites in the U–M region. On a cut surface, the gradual transition between the pink Cretaceous limestone layer, which has a total thickness of 30 cm, and the white-bleached band is located about 15 cm below the K/T boundary (Fig. 5.7.6.10a). Abundant trace fossils of *Zoophycos* and *Planolites* have diameters in the range of few millimeters, and are filled with reddish or reddish-brown basal Tertiary sediment. Occasionally, altered impact spherules are found in the burrows.

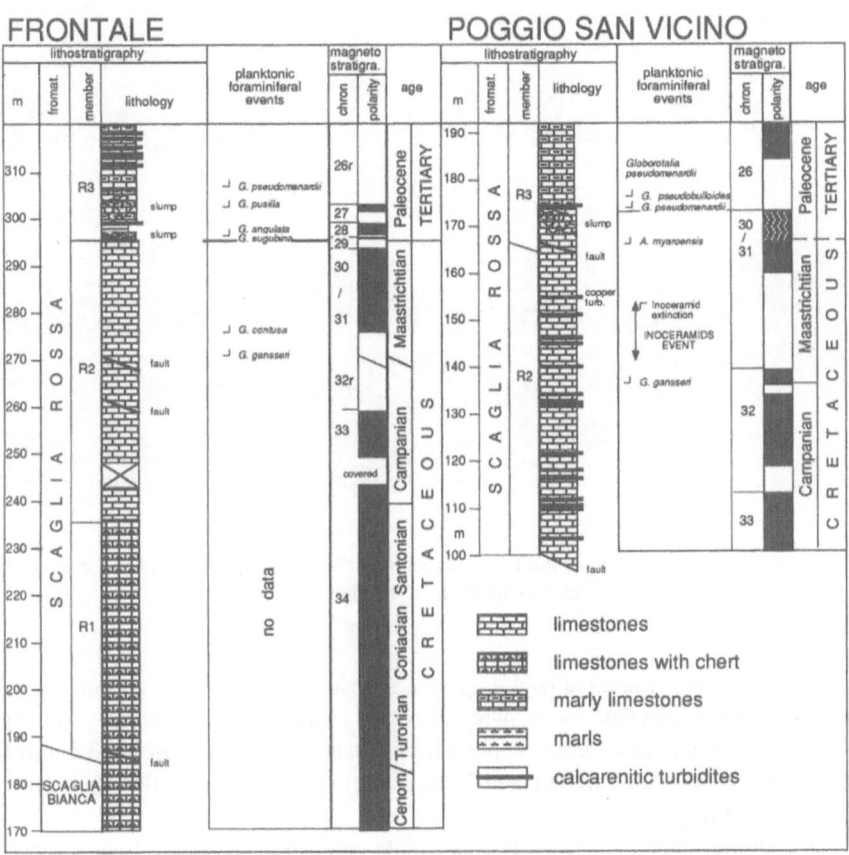

Fig. 5.7.6.8. Stratigraphic syntheses of the Frontale and Poggio San Vicino sections.

Fig. 5.7.6.9. Panoramic view of the Frontale Quarry (left side the photo), and of the small K/T boundary outcrop along the road cut (right side of the photo).

The basal, well-homogenized *eugubina* limestone layer exhibits a pink bottom, which grades upward to a whitish color some 3–4 cm above the base (Fig. 5.7.6.10b). Altered impact spherules are scattered in the lower 2–3 cm of the layer. Few *Planolites* burrows with diameters up to 5 mm, are filled with grayish Tertiary sediment.

Occasional large *Planolites* burrows with diameters of up to 2 cm are found laying parallel to bedding in the upper part of the layer, and they are filled with white-bleached Cretaceous sediment, including some chunks of whitish and gray Tertiary sediment (Fig. 5.7.6.10b).

Fig. 5.7.6.10. a) Polished slab of the top, bleached Cretaceous layer at Frontale. Note the worm burrows filled with dark *eugubina* ooze. **b)** Polished slab of the basal *eugubina* limestone. Note the color transition from pink to white at the base of the layer, and the large worm burrow filled with white ooze derived from the bleached top Cretaceous limestone layer. The small divisions of the rulers in pictures a) and b) are millimeters.

The 15–20 mm thick K/T clay exhibits its thin basal green sole, as at Gubbio and Petriccio, and it is fairly well preserved being almost free of tectonic shearing and secondary calcite. The most peculiar aspect of this boundary in contrast with other K/T clays studied in the region, is that it contains frequent, moderately well-preserved tests of Cretaceous and lowermost Tertiary planktonic foraminifers (Fig. 5.7.6.11a, b). These fossils were probably introduced into the clay by post-P0 bioturbation. This is suggested by the fact that the burrows in the top Cretaceous limestone are never exclusively filled with *G. cretacea* tests, but they contain tests of *G. eugubina*, *G. fringa*, *G. minutula*, and *W. hornestowensis*, which would indicate a P1 Zone source (Montanari 1990).

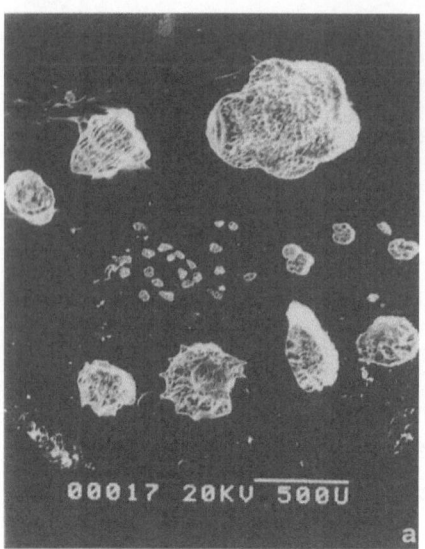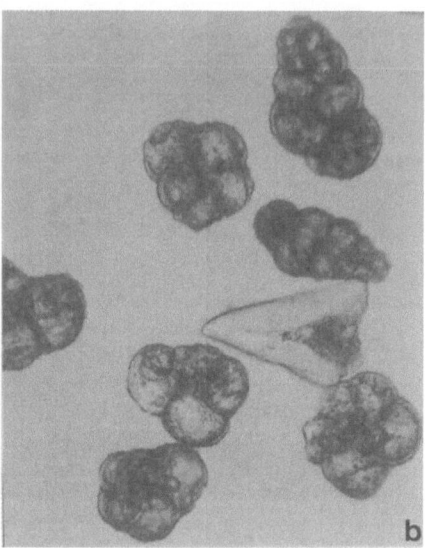

Fig. 5.7.6.11. a) SEM image of large Cretaceous planktonic foraminiferal tests (upper part of the image) and small basal Tertiary globigerinids (center of the image) extracted from the K/T boundary clay of Frontale. The good preservation of these reworked planktonic foraminiferal tests contrasts with the poor preservation of authochthon benthic foraminifers which exhibit clear signs of corrosion (see the four tests in the lower part of the image). Scale bar = 0.5 mm. **b)** Smear slide microphotograph (transmitted, plain light) of basal Tertiary globigerinids from the Frontale K/T clay. Note also the well-preserved fish tooth in the center of the photo. Tooth length approximately 0.05 mm.

The reasons why these microfossils are abundant in the Frontale K/T clay (and in a few other sections such as Cingoli and Ceselli; see Luterbacher and Premoli Silva 1964, and Montanari 1990), and not in other strongly bioturbated sections such as Furlo, remains unclear. The green sole adhering over a corroded top-Cretaceous, white-bleached limestone, and the clearly corroded tests of robust calcareous benthic foraminifers (Fig. 5.7.6.11a) indicate that also at Frontale the brief episode of low-pH and low-Eh conditions occurred during the deposition of the K/T clay. On the other hand, the burrows filled with red lowermost Tertiary sediment penetrating the top-Cretaceous limestone, along with the red base of the *G. eugubina* limestone, indicate that a return to normal oxidizing conditions coincided with the restoration of copious carbonate production and sedimentation.

Perhaps at other U–M sites the dissolution of delicate planktonic foraminiferal tests occurred at a later diagenetic phase.

Furlo. The Scaglia Rossa Formation in the anticlinal structure of Monte Paganuccio–Monte Pietralata, which is cut by the picturesque Furlo Gorge, exhibits a sedimentary facies different from that of the classic area of Gubbio. Here, the typically pink pelagic biomicrites, starting in the upper Santonian R1 member all the way up to the Paleocene R3 member, are interbedded with white foraminiferal turbidites. The magnetic stratigraphy of the upper Cretaceous–lower Tertiary Furlo sequence was determined by Alvarez and Lowrie (1984) and Chan et al. (1985), while a detailed basin and paleocurrent analysis of the Scaglia Rossa was carried out by Montanari et al. (1989). A stratigraphic synthesis of the Furlo sequence is shown in Fig. 5.7.6.12.

The Furlo area represented an intrabasinal depocenter, which started to subside in the Turonian, following a rejuvenated regional extensional tectonic phase (see Chap. 4 and Fig. 4.2). The white turbidites, probably triggered by earthquakes, are typically partitioned with Tb and Tc Bauma intervals, and are essentially made of reworked planktonic foraminiferal tests, which were sorted in turbidity currents of remobilized intraformational pelagic sediment. Through the R2 and lower R3 members of the Scaglia Rossa, these turbidites are arranged into two main thickening-up cycles, the lower culminating with a conspicuous turbidite marker made of three (or sometimes two, like at Pietralata) amalgamated units called the "triple turbidite" (TripleT), and the upper cycle ending with a 1 m thick turbidite called the "mega turbidite" (MegaT). The inoceramid abundance peak and extinction event, described in Sect. 5.7.1 of this book, is located just below the TripleT marker, and it can be easily found in several outcrops and in abandoned quarries along the road that from Furlo climbs up toward the top of the mountain (see map in Fig. 5.6.1). The K/T boundary happens to be located in the upper part of the second cycle, about seven meters below the MegaT. The boundary is bracketed by two couples of turbidites, and it is overlain by a 2-m-thick interval of thin-bedded, dark pink platy limestones representing the lowermost Tertiary (Fig. 5.7.6.12).

Among the many good outcrops of the K/T boundary interval known in the Furlo area, the Pietralata section is perhaps the best exposed and most easy to reach. The outcrop is located along a mountain road past the church of Pietralata (see map in Fig. 5.6.1 and panoramic view in Fig. 5.7.6.13a). A characteristic of this K/T boundary that catches the eye at first sight is the intense bioturbation decorating the top, light pink limestone of the Cretaceous (Fig. 5.7.6.13b).

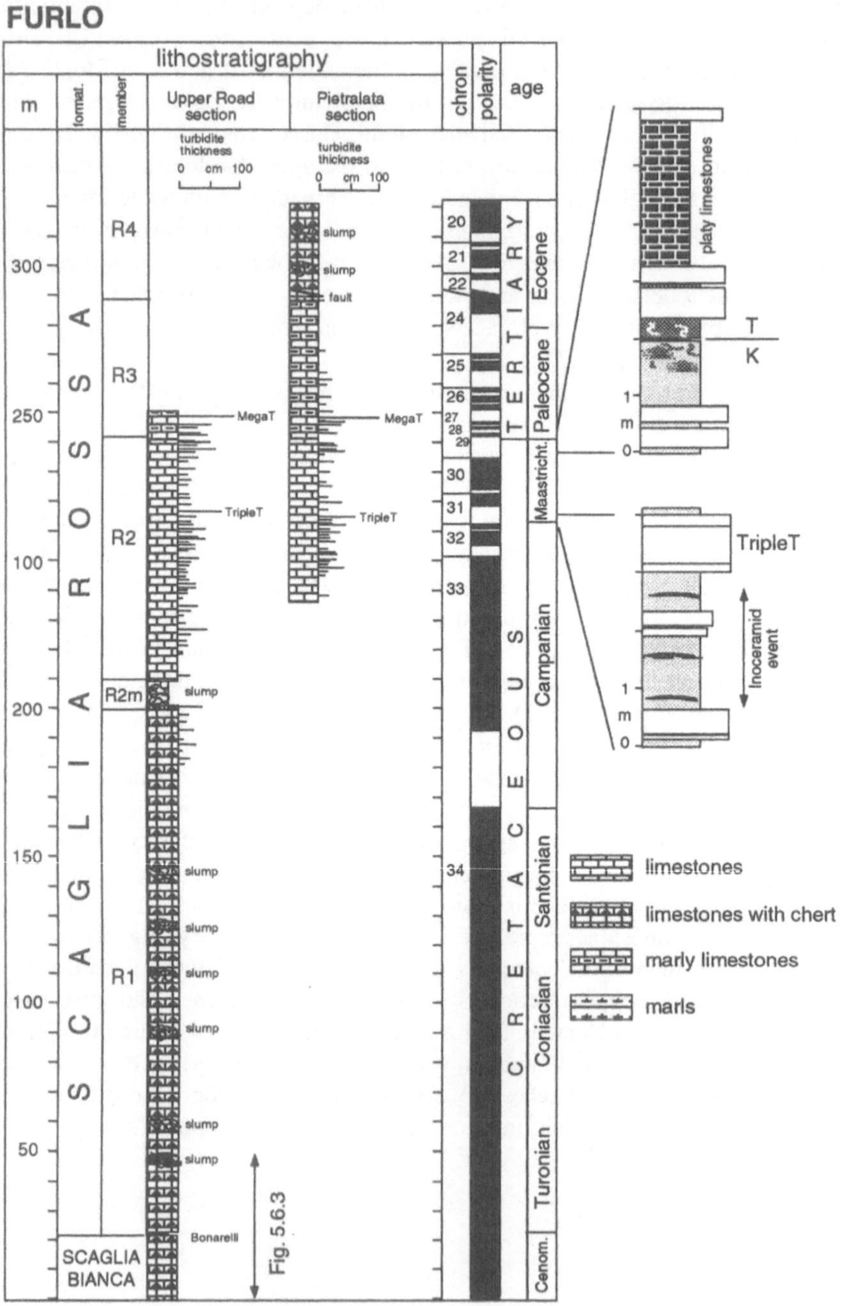

238

Fig. 5.7.6.12. Stratigraphic synthesis of the Scaglia Rossa sequence at Furlo.

Numerous traces of *Planolites* and *Zoophycos*, large and small, are filled with dark pink sediment, and penetrate the top Cretaceous limestone for 20–30 cm below the boundary clay. The microfossils filling the burrows are the tiny globigerinids found in the first Tertiary limestone, but larger Cretaceous microfossils, as well as lowermost Tertiary tests belonging to the upper *G. eugubina* Zone, are also found in the burrows (Fig. 5.7.4.6). As for the actual basal Tertiary, it also appears strongly bioturbated. Reworked tests of Cretaceous foraminifers are much more abundant here than in any other K/T boundary site in the region.

Altered spherules and occasional pieces of boundary clay are also found in the burrows below the boundary, and sparse in the basal portion of the Tertiary limestone. The strong bioturbation and vertical mixing characterizing the Furlo K/T boundary is probably the result of good oxygenation of the sea floor, which was frequently swept by bottom currents in this area of the U–M basin. Winnowing is also the probable reason for a strongly condensed basal Tertiary, which represents a total accumulation rate of 1–2 mm/ka, which is a factor of 3–5 slower than at Gubbio (Alvarez and Lowrie 1994). Despite these particular sedimentological conditions, the 8 mm thick K/T clay layer is still preserved here, although it lacks the typical green sole, and contains an unusually low concentration of Ir (1–2 ppb) compared with other sites in the region. As mentioned in the preceding chapter of this book, winnowing may have removed some of the K/T clay, and even the low-Eh film of stagnating water at the water–sediment interface. However, temporary redox conditions favoring the authigenesis of glaucony must have occurred here as in any other site in the U–M basin, as it is demonstrated by the presence of altered glauconitic impact spherules.

Monte Cònero. The Monte Cònero promontory is the easternmost anticline of the northern Apennine thrust-and-fold belt (Fig. 4.1.1). The U–M carbonate sequence is here exposed from the middle part of the Maiolica Formation all the way up to the Schlier Formation which, in this area, reaches the very top of the Tortonian and is, therefore, laterally heteropic with the flysch formations of the Marnoso–Arenacea and the Laga exposed elsewhere in the U–M Apennines (see Chap. 4).

The sedimentary facies of the Scaglia Rossa at Monte Cònero is different from the equivalent unit of the classic Gubbio sequence. First of all, the Cretaceous portion of this formation exhibits a whitish-yellowish color instead of the typical pink of the U–M facies, and secondly, coccolith-foraminiferal biomicrites are interbedded with white calcarenitic and calciruditic turbidites derived from a carbonate platform, which was located not far to the east of the present Cònero area. Thus, the Scaglia Rossa represents here a proximal facies characterized by lobate and strongly channelized carbonate turbidites and grain flows interfingering, at the toe of a carbonate apron (Colacicchi and Baldanza 1986), with the pelagic muds of the U–M deep-water basin (Montanari et al. 1989).

Fig. 5.7.6.13. a) Panoramic view of the K/T boundary outcrop at Pietralata. b) Close-up of the K/T boundary at Pietralata. Note the numerous, dark colored burrows (P = *Planolites*; Z = *Zoophycos*) in the top Cretaceous limestone.

The K/T boundary is exposed in numerous localities in the Cònero area, but the best sections can be found in the quarries of Fornaci, near the village of Poggio, and at Fonte d'Olio, near the town of Sirolo (Fig. 5.7.6.14). The biostratigraphy of the interval covering the boundary was originally defined by Luterbacher and Premoli Silva (1964), and only recently restudied in several sections at Fornaci and Fonte d'Olio by Casoni (1988) and Mazzanti (1988). A stratigraphic synthesis and correlation among the most representative of these sections is shown in Fig. 5.7.6.15.

The K/T boundary is located about 80 cm below a 2–3 m thick calcirudite marker bed called "Mega T" by Montanari et al. (1989), or "Marchesini Level" by Coccioni et al. (1994), and is marked by a 18 mm greenish clay layer sandwiched between pelagic limestones. Apart from their whitish or creamy color, the texture and microfacies of the top Cretaceous *mayaroensis* limestone, and the basal Tertiary, porcellaneous *eugubina* limestone are identical to those seen in other U–M sections.

These light colors are due to a relatively more reduced state of the paleosea floor compared to the well-oxygenated environments expressed by the typically pink Scaglia Rossa. Higher plankton productivity and upwelling in this proximal environment may have been the main reasons for more reduced conditions on the sea floor in respect to other open sea, deeper areas of the paleobasin.

The spectacular exposures in the Fornaci quarry offer the possibility to follow the K/T boundary along strike for about 500 m. In the west side of the quarry (Fig. 5.7.6.16a), the MegaT and the K/T boundary below it are well visible although it may be dangerous to approach the quarry face due to frequent rock falls. In this part of Monte Cònero, the typical pink color of the Scaglia Rossa reappears in the lower part of the R3 member, a few meters above the MegaT.

The higher level of the central sector of the quarry offers a safer and more approachable exposure of the uppermost part of the Cretaceous. This section can be reached following a trail hidden in the thicket, and departing from the east end of the quarry. Here the calcarenitic turbidites appear strongly channelized, and in some cases they show a convex upper surface indicating a lobate geometry of these detrital bodies. A closer look at these calcareous turbidites reveals a granular texture made of biogenic debris including large benthic foraminifers (orbitoids), rudist fragments, calcareous algae, echinoid spines, pieces of corals, and lithoclasts of pre-existing shallow water limestones. These detritic layers are found below and above the K/T boundary, and may represent excellent material for studying the biofacies evolution of the western Adriatic platform which was completely dismantled after the K/T event. In simpler words, they may function as messengers of an evolving, and eventually killed platform, the incomplete remnant of which is now buried under the Adriatic Sea floor. In fact, large flute casts and inclined laminations indicate a preferential paleocurrent flow from eastern toward western quadrants. Some large flutes with different shapes (i.e., bulb, tongue, curl) are visible at the base of some MegaT boulders piled up at the edge of the quarry rim. The MegaT and the underlying K/T boundary are exposed on a ledge at the very top of this outcrop.

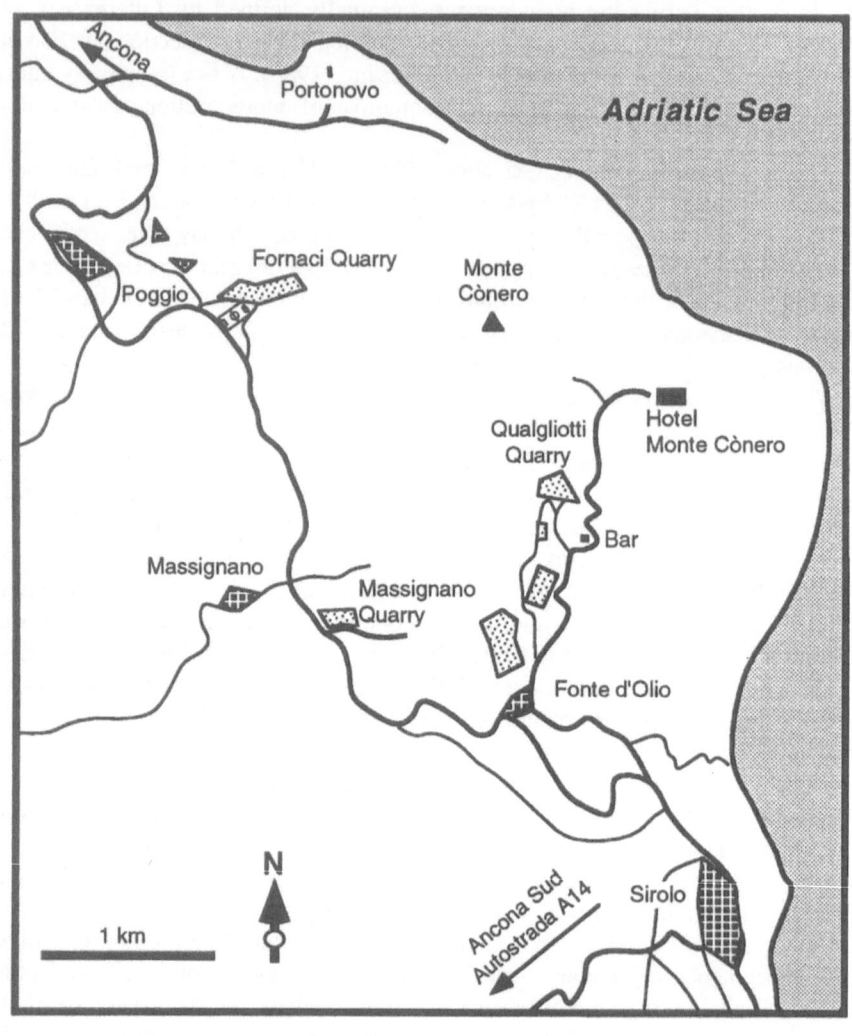

Fig. 5.7.6.14. Location map of the Monte Cònero K/T and E/O sections described in the text.

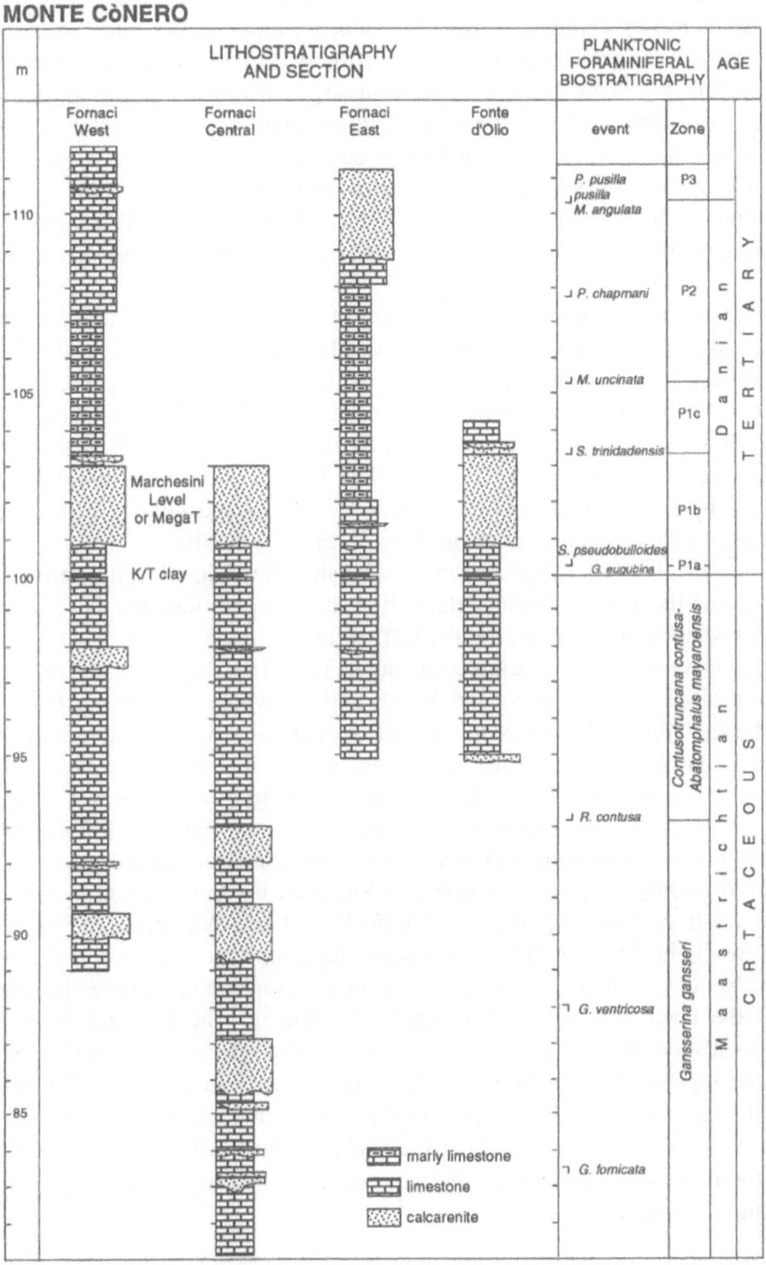

Fig. 5.7.6.15. Stratigraphy of four representative sections in the Cònero area, covering the K/T boundary. Lithostratigraphy after Montanari (1979); Cretaceous biostratigraphy after Mazzanti (1988); Tertiary biostratigraphy after Casoni (1988).

At the base of the trail, now hidden behind a thick grove of young pine trees, there is an interesting exposure of the K/T boundary on the easternmost face of the quarry. The boundary is located up high on the quarry face and can be reached with a ladder (Fig. 5.7.6.16b). The peculiar aspect of this outcrop is that the MegaT is here reduced to a 10 cm thin, discontinuous calcarenitic layer. The lower R3 member of the Scaglia Rossa in this outcrop exhibits its typical dark pink, marly facies. This exposure is topped by another thick calcarenitic turbidite very similar to the MegaT which is not present in the Fornaci West section, a further demonstration of the strongly channelized geometry of these detrital bodies.

The best and most approachable exposure of the K/T boundary in the Cònero area is located in the upper ledge of the so-called Quagliotti quarry in the Fonte d'Olio valley (see location map in Fig. 5.7.6.14 and outcrop panoramic in Fig. 5.7.6.17a). This outcrop can be reached following, for a short distance, a trail departing from a forest road near a resort facility half way up along the Fonte d'Olio–Monte Cònero main road. Here, the K/T boundary exhibits the exact same facies as in the Fornaci West and Central quarries, and is located at practically the same distance (about 80 cm) below the MegaT (Fig. 5.7.6.17b).

However, the Tertiary Scaglia shows a whitish-yellowish color, indicating that in this part of the Cònero basin, relatively reduced conditions on the sea floor lasted for several million years after the K/T event.

The clay layer has a green-ochre color and is, therefore, different from the same (red) layer observed at Gubbio and Furlo, and anywhere else in the Umbria–Marche basin. More reducing conditions of the paleosea floor are also indicated by the fact that goethite spherules are absent in this K/T boundary layer. Only altered microkrystites made of slightly evolved, pale-green glaucony, and K-feldspar are present, along with shocked quartz grains, and well differentiated arenaceous benthonic foraminifers, which are somewhat larger than those found in other K/T boundary layers in the region. Moreover, the boundary clay contains 10 ppb Ir, the largest anomaly measured in the U–M basin (Montanari 1990).

Trace fossils in the whitish top Cretaceous limestone are very thin and scarce compared to other U–M K/T sites. They are mostly concentrated in the uppermost 10 cm of the layer, and filled with *eugubina* sediment. The top surface of this layer also shows evidence of corrosion, as in the most typical K/T boundary sites in the U–M region. The *eugubina* limestone, which has a light ochre color, and a reduced thickness of 22 cm (Mazzanti 1988), is thoroughly homogenized by bioturbation, although discrete trace fossils are seldom visible. The lower part of the layer contains frequent reworked Cretaceous foraminiferal tests and fragments of altered impact spheroids.

Fig. 5.7.6.16. a) Panoramic view of the K/T interval exposed in the Fornaci West quarry face. **b)** Panoramic view of the K/T interval in the Fornaci East quarry face. Note the absence of the MegaT marker above the K/T boundary. Y = yellowish; P = pink; W = white; K = Cretaceous; T = Tertiary; MT = MegaT marker.

Fig. 5.7.6.17. a) Panoramic view of the Quagliotti quarry near Fonte d'Olio. The K/T boundary is located about 80 cm below the 2 m thick MegaT marker. **b)** Close-up of the K/T interval in the Quagliotti quarry outcrop.

Montagna dei Fiori. In the Montagna dei Fiori area, like at Monte Cònero, the R2 and R3 members of the Scaglia Rossa exhibit a relatively proximal facies characterized by frequent and thick calcarenitic turbidites (Crescenti 1969; Montanari et al. 1989). These turbidites are made of calcareous debris derived from the margins of the Abruzzi carbonate platform located not far to the south. As at Monte Cònero, the biomicritic, pelagic limestones interbedded with the turbidites have a whitish color, whereas the typical pink color of the Scaglia Rossa can be seen only in the lower and upper members of this formation. Thin and very fine calcarenitic turbidites are found also in the Scaglia Bianca Formation, and they are usually associated with black tabular chert layers. The Bonarelli Level, at the Cenomanian–Turonian boundary, has a reduced thickness of 60 cm, but its lithology and internal stratigraphy is practically identical to those of the typical Bonarelli marker exposed anywhere else in the U–M Apennines.

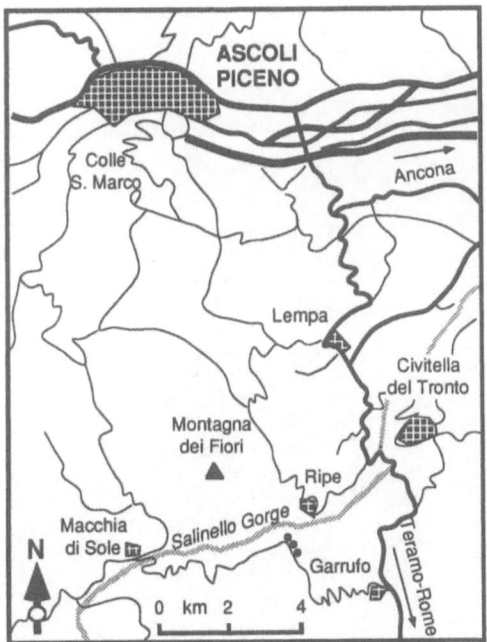

Fig. 5.7.6.18. Location map of the K/T boundary section in the Montagna dei Fiori area.

The K/T boundary is located in a pelagic interval, on a section stretching along the road that leads from the village of Garrufo up to the remote village of Macchia di Sole (see map in Fig. 5.7.6.18). The outcrop in the region of the K/T boundary is strongly tectonized and it cannot be used for a high-resolution stratigraphical study (Fig. 5.7.6.19a, b). Despite strong shearing, this outcrop yields interesting information about the particular sedimentary environment in this part of the paleobasin at the time of the K/T boundary event. First of all, the limestones just above and below the K/T boundary are white indicating low oxygen conditions at the water–sediment interface prior and after the K/T event. The K/T clay is green and practically barren of benthic foraminifers (only a few broken pieces of *Ammodiscus* were found by Montanari 1990). Moreover, K-feldspar and goethite impact spheroids are absent in this clay.

Fig. 5.7.6.19. a) Panoramic view of the Monte dei Fiori section containing the K/T boundary. b) Close-up of the strongly tectonized K/T interval in the same section.

Instead, euhedral grains of pyrite make up nearly 90% of the HCl-insoluble sand-size fraction of the clay (Fig. 5.7.6.20), while the rest is made up by evolved glaucony. These observations indicate that the environment during the deposition of the K/T clay was strongly reducing, to the point of suppressing benthic life, and allowing authigenic growth of iron sulfide.

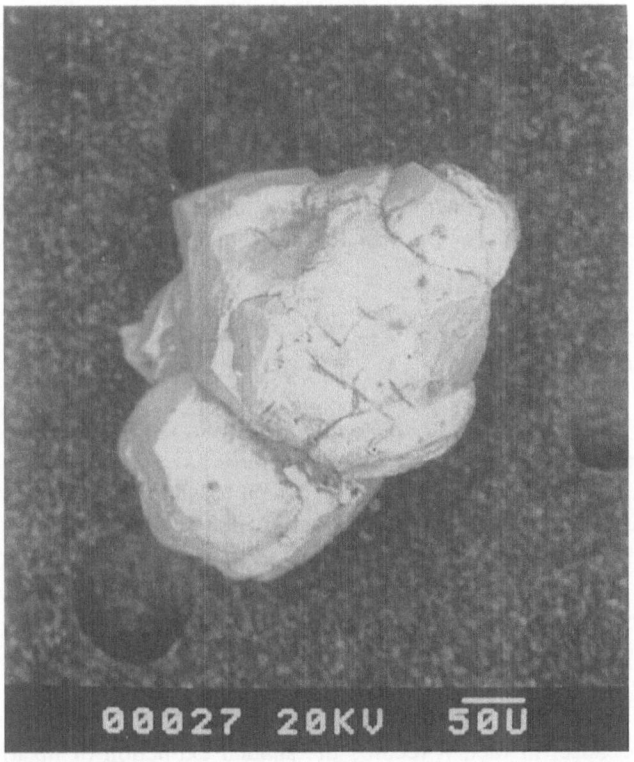

Fig. 5.7.6.20. SEM image of euhedral pyrite from the K/T boundary clay of the Montagna dei Fiori outcrop.

5.8
The Paleogene Impact Record

The diagram of impact energy flux in Fig. 5.1.2 shows two prominent peaks in the Paleogene, the smaller one at the base of the Eocene (i.e., around 52–54 Ma), and the larger in the Late Eocene (~36 Ma). At present there are no known large impact craters, nor prominent peaks of marine extinctions with Oligocene ages, in the Sepkoski (1996) analysis. Therefore, in this chapter we will focus on the U–M record across the Paleocene–Eocene (P/E) boundary and the Late Eocene, which are both well represented in several outcrops of undisturbed, pelagic sediments near the intraformational boundaries between the Scaglia Rossa, Scaglia Variegata, and Scaglia Cinerea formations, respectively.

The most continuous exposure of the Paleogene in the U–M Apennines stretches along road cuts and quarries in the Contessa Valley near Gubbio (see location map in Fig. 5.5.1). An integrated stratigraphic synthesis of this sequence, from the K/T to the Eocene–Oligocene boundaries, is shown in Fig. 5.8.1, along with radioisotopic age tie points, which permit to calibrate the whole sequence by interpolation. This figure reports also the ages and sizes of the largest craters known in this time span. The age uncertainty of the impact events can be directly projected against the stratigraphy of the Contessa sequence, permitting the narrowing down of discrete outcrop intervals that may contain the signatures of these impact events. The discussion and illustration of known information on impact signatures will be given separately for the Paleocene–Eocene, and the terminal Eocene intervals in the following chapters.

5.8.1
The Paleocene–Eocene Boundary

Whereas the integrated stratigraphy across the E/O boundary in the U–M region has been studied in great detail, and impactoclastic signatures have been found in the upper Eocene portion of this interval (see Sect. 5.8.2), to our knowledge no studies have ever focused on impact stratigraphy across the Paleocene–Eocene (P/E) boundary. Also for this interval, the nearly complete and continuous pelagic sequence of Gubbio may offer the rare opportunity to approach the investigation of impact signatures and/or the effects they may have had on the marine biota, which is here represented by an excellent microfossil record.

Although no prominent peak of extinction in marine genera shows up in Sepkoski's (1996) diagram (see Fig. 5.1.1), the P/E boundary has been the focus of recent detailed studies, because, in fact, it records the sudden extinction of up to 50% benthic foraminiferal species in the southern oceans (Thomas 1990; Kennett and Stott 1991), as well as in higher latitude oceans (Douglas and Woodruff 1981; Tjalsma and Lohman 1993; Kaiho 1991), and the Tethys (Ortiz 1994; Speijer 1994). This extinction event seems to have affected only bathyal and abyssal environments, and coincides with a sharp negative shift of the $\delta^{13}C$ record.

The extinction event occurs in a very narrow stratigraphic interval in nannofossil Zone CP8 (i.e., around 55 Ma), which may represent a few tens of thousand years (Thomas 1990), or as little as 3,000 years (Kennett and Stott 1991). This event is apparently not coupled to extinctions or drastic turnovers in the planktonic realm.

It is widely accepted that a global climatic change following a gradual warming of Antarctic surface waters at a time when Antarctica was not yet glaciated, would have decreased the temperature difference between surface and deep waters in all the oceans, significantly slowing down the thermally driven oceanic water circulation. This may have caused dysoxia in deep water environments, which would have affected the benthic community. However, the sudden and drastic benthic extinction at the P/E boundary would have required an equally sudden and short-lived climatic or environmental global event not necessarily consequent to the ongoing, gradual climatic change during this period of time. Eldholm and Thomas (1993) proposed that CO_2 produced by North Atlantic volcanism would have raised surface water temperatures at high latitudes, causing rapid water circulation and environmental changes throughout the world. However, no increase in atmospheric CO_2 could be deduced from the geochemistry of organic matter and carbonate of late Paleocene–early Eocene terrestrial sediments in the Paris Basin analyzed by Sinha and Stott (1994). Therefore, the cause of the sudden event at the P/E boundary remains a mystery, but see Katz et al. (1999) and Norris and Kõhl (1999) for an interesting new proposal.

At Gubbio, the upper Paleocene–Early Eocene interval is exposed along the Contessa Road section (the CR section in location map of Fig. 5.5.1; see also panoramic view in Fig. 5.8.1.1) the magneto- and biostratigraphy of which have been defined by Lowrie et al. (1982) and Monechi and Tierstein (1985). The boundary is found in the upper part of the Scaglia Rossa R3 member (chert-free, marls and marly limestones). The color of these carbonate rocks is dark pink, indicating a well-oxygenated deep-water environment.

High-resolution magnetostratigraphic, micropaleontologic, geochemical, and paleoecologic studies across this boundary in the CR section are now in progress by a team of interdisciplinary researchers (Galeotti et al. 1988), and we are confident that new information will be produced by this team effort, which will certainly contribute to a better understanding of the changes occurring during this interesting episode of Earth's history.

Following the objective of this book, we may ask the question whether the P/E climatic and ecologic events might be related to extraterrestrial causes. The four or five impact craters known through this span of time are too small to be considered effective causes of global change. The largest of them is the Montagnais structure with a diameter of about 45 km, and an age of 50.5 ± 0.8 Ma, which is closely spaced in time with the Kamensk impact crater (25 km diameter; 49.2 ± 0.2 Ma). These impacts are about 5–6 Ma younger than the P/E benthic extinction and the associated $\delta^{13}C$ shift. Moreover, it appears that not even in areas close to the ground zero of the Montagnais crater the microfossil record shows any sign of extinction or evolutionary discontinuity (Aubry et al. 1990).

INTEGRATED STRATIGRAPHY OF THE PALEOCENE-EOCENE PELAGIC SEQUENCE OF THE U-M APENNINES (METER LEVEL REFERRED TO THE CONTESSA CH SECTION)

Fig. 5.8.1.1. Panoramic view of the Contessa Road (CR) section containing the Paleocene–Eocene boundary, which is located about 5 m below the first chert layer of the Scaglia Rossa R4 member, near the base of the marly interval in the middle of this outcrop.

As a working hypothesis, it may be considered that the nearly coeval Kamensk and Montagnais impacts are part of a comet shower, such as the one which has been inferred in the Late Eocene (Farley et al. 1988; see next chapter), and possibly in the lower-mid Maastrichtian (see Sect. 5.7.1). In fact, Farley (1995) reported a significant increase of extraterrestrial ^{3}He through the lower Eocene in a Giant Piston Core in the central North Pacific. However, a hypothetical comet shower in the lower Eocene would be much younger than the benthic extinction at the P/E boundary and, therefore, can not be considered as a cause for this biotic crisis. Nevertheless, the particularly favorable stratigraphic qualities of the Contessa CR section constitute certainly a strong invitation to approach high-resolution investigations on the impact record of this critical interval of geologic time.

← **Fig. 5.8.1.** Integrated stratigraphic synthesis of the Paleocene–Eocene Contessa Highway (CH) section, and locations of major impact events. The age of the impact glass from Beloc and Mimbral is from Swisher et al. (1991). Radioisotopic age tie points of Agost (biotite-rich volcanic ash), Greenland (basaltic spherules), and DSDP 516F (biotite-rich volcanic ash) are from Montanari and Swisher (1994). The ages of biotite-rich volcanic ashes from Massignano and Contessa are from Montanari et al. (1988b). Bio- and magnetostratigraphy, and simplified lithostratigraphy, are from Lowrie et al. (1982) and Monechi and Thierstein (1985).

5.8.2
The Terminal Eocene Events

In the years immediately following the publication by Alvarez et al. (1980) of the hypothesis that the impact of a giant extraterrestrial object was the cause of the K/T boundary mass extinction, an obvious question crossed the mind of those researchers who found such a hypothesis at least interesting: are there other mass extinction events throughout Earth's history that could be related to impacts?

It was with this question in mind that Alvarez et al. (1982) undertook a trace element geochemical investigation across two interesting stratigraphical boundaries, namely the Permian–Triassic (P/TR), and the Eocene–Oligocene (E/O). The first records the most severe Phanerozoic mass extinction, whereas near the latter, in some Caribbean and Gulf of Mexico deep sea cores, glassy microspherules were known to exist, and they were already recognized as impactoclastic material (e.g., Glass et al. 1973; Maurrasse and Glass 1976; Glass and Zwart 1979). Moreover, with the first compilation and analysis of a data base and the plotting of marine families extinction rates through the past 250 Ma, Raup and Sepkoski (1982) showed that, in addition to the K/T boundary, there were some other ten statistically significant extinction peaks through this span of Earth history, and one of them was, in fact, at or near the E/O boundary.

While the P/TR boundary yielded no Ir anomaly, the samples from the microspherule layer near the E/O boundary contained up to 0.4 ppb Ir. This was one order of magnitude less than the K/T boundary, but undoubtedly a strong indication that an anomalous accretion of extraterrestrial matter occurred at that time (Alvarez et al. 1982). These coincidences (Ir, microspherules, and mass extinction) were rendered even more intriguing by the fact that the largest impact structure in the past 250 Ma record known on Earth in the early 1980s was the 100 km diameter Popigai crater, in central Arctic Siberia (see compilation by Grieve 1982). Early radioisotopic dates gave an imprecise age of 39 ± 9 Ma for the Popigai structure, covering the long stratigraphic interval between the Middle Eocene to the Early Oligocene.

The whole picture of the terminal Eocene events in the early 1980s was somewhat out of focus compared to the controversial, but undoubtedly exciting, emerging story of the K/T boundary. That the Late Eocene was a time of major change in Earth's history was felt by many researchers (see Pomerol and Premoli Silva 1986). During the Eocene, the separation of Antarctica from the southern continents led to the formation of the circum-Antarctic current and the gradual glaciation of the continent. As the Antarctic polar ice cap spread until reaching the ocean, cold water masses formed along the continent's coasts. Being denser than warmer surface waters, these cold water masses started to sink moving away from the isolated continent for long distances toward North, eventually changing water circulation and temperature in the Earth's oceans. With time, the glaciation and a renewed oceanic circulation affected climate. Climatic cooling began to affect productivity of abyssal organisms as well as the plankton, and eventually altered the equilibrium of the global ecosystem.

The global cooling model, so strongly supported by geologic, geochemical, and tectonic evidence, was a classical example of gradualism, which could stand by itself, and explain the observed global changes without having to invoke a catastrophic extraterrestrial event. Moreover, there was neither any discrete boundary clay nor a dramatic extinction of dinosaurs (or of anything else) across the E/O boundary that could really attract the attention of the earth science community toward a possible extraterrestrial cause (or co-cause) for these events. And yet, impact evidence near the boundary were already available since the early 1970s. This was enough to encourage a few researchers to investigate further on these impact signatures.

One of the most prohibitive factors against a prompt development of research on the possible cause-and-effect relationship between impacts and mass extinction near the E/O boundary was, since the very beginning, a somewhat confusing stratigraphic and geochronologic calibration of this time interval. In fact, ages ranging between 36 and 34 Ma for the North American tektites (see Sect. 2.4.5.1), to which the Caribbean and Gulf of Mexico microtektites are related (Chao 1973; Glass et al. 1973), were determined by fission track and K/Ar datings by a number of researchers (e.g., Zähringer 1963; Fleischer and Price 1964; Garlick et al. 1971; Storzer et al. 1973). On the basis of biomarkers, the microtektites in Caribbean and Gulf of Mexico cores were assigned to the upper Eocene NP18 (nannofossil) Zone or to the P15 (foraminiferal) Zone in some cores, and to the NP19 Zone or P16 Zone in others. Further accurate $^{40}Ar/^{39}Ar$ dating of microtektites and tektite fragments from a land-based section in Barbados yielded an age of 35.4 ± 0.6 (Glass et al. 1986b). The confusion was that the chronostratigraphic time scales most used in the early 1980s were assigning an age around 37 Ma to the E/O boundary (e.g., Berggren et al. 1978, 1985) and, therefore, they were in contrast with an age of 34.7–36 Ma for the microtektites, which were apparently contained in upper Eocene sediments. In other words, on the basis of geochronology the impact that generated the tektites and microtektites occurred in the lower Oligocene, but microfossils associated with them indicated an upper Eocene age.

Such stratigraphic and geochronologic uncertainties did not help in searching for the needle in the haystack. Precise and accurate means for world wide time correlation are the *sine qua non* for assessing the uniqueness of a sudden event such as an impact, and locate it within the stratigraphic–paleontologic record. Moreover, the time correlation between marine sediments and an impact crater, upon which any kind of inference on a possible cause-and-effect relationship can be drawn, has to be based on numerical geochronology, and, therefore, on a reliable radioisotopic calibration of the time scale. The problem lies in the fact that marine sediments are rarely dated by radioisotopic means, just as impact rocks are rarely dated by biostratigraphic means.

The U–M sequence, once again, offered the unique opportunity to untangle the apparent geochronologic paradox of the E/O boundary. In fact, Lowrie et al. (1982) reported the presence of several biotite-rich layers interbedded with the pelagic marly limestones of the Scaglia Variegata–Scaglia Cinerea sequence in the Contessa Quarry (CQ) section. These discrete layers represented the fallout of material erupted from distant volcanoes and, thus, they could be used to date directly, by radioisotopic means, the sediments covering the biostratigraphically and magnetostratigraphically calibrated Eocene–Oligocene interval (Lowrie et al. 1982).

Early K/Ar dates from two biotite layers bracketing the E/O boundary at Contessa yielded an interpolated age for the boundary younger than 35 Ma (Montanari 1983). Further K/Ar, Rb/Sr, and $^{40}Ar/^{39}Ar$ dating on other biotite layers from the sections of Contessa, the Monte Cagnéro (near Piobbico), and the Massignano (at Monte Cònero near Ancona; Montanari et al. 1985, 1988b; Odin et al. 1988; Deino et al. 1988), corroborated by detailed geochemical and mineralogical analyses of the biotite used for these datings (Montanari et al. 1988b; Montanari 1988b; Odin et al. 1991), finally lead to a precise and accurate interpolated radioisotopic age for the E/O boundary of 33.7 ± 0.4 Ma. This new age determination lowered the dates in use in the 1980s by as much as 3 Ma, and undoubtedly placed the North American tektite strewn field in the upper Eocene.

Whereas a great effort was made by a number of researchers to resolve the integrated stratigraphy of the E/O boundary in the Apennines (e.g., Premoli Silva et al. 1988), investigations focused on impact spherules and Ir anomalies in deep-sea cores revealed the presence of several (at least two, and possibly four) impactoclastic layers in upper Eocene sediments throughout the world. First of all, it soon became clear that in Caribbean and Gulf of Mexico cores there were not one, but two closely spaced microspherule layers (see Sect. 3.3): the older one contained clinopyroxene-bearing microkrystites, and the younger one only the crystal-free microtektites (Glass et al. 1985). Iridium was found associated with the microkrystite layer in several sections in the Caribbean–Gulf of Mexico area, as well as in the Pacific and Indian oceans, whereas the microtektite strewn field, with little or no Ir, was seemingly confined within the Caribbean, Gulf of Mexico, and northwestern Atlantic areas (Keller et al. 1983, 1987). Shocked quartz and coesite were found to be associated with this Ir-free, microtektite layer (Glass 1987; Bohor et al. 1988; Glass and Wu 1993).

In addition to the closely spaced spherule layers (about 20 ka apart) in the *Globorotalia cerroazulensis* Zone (i.e., the P16 Zone), Keller et al. (1983b, 1987) reported an older microkrystite layer, also associated with Ir, in the upper part of the *Globigerinateka semiinvoluta* Zone (i.e., the P15 Zone) in equatorial Pacific and Indian Ocean sites. This finding was questioned by Glass and Burns (1987), who argued that the biostratigraphic determination by Keller and co-workers was faulty due to reworking of critical index fossils.

Hazel (1989) used the graphic correlation technique on available biostratigraphic data to determine the stratigraphic position of iridium anomalies and microspherule layers in sections from the North Atlantic, Barbados, Venezuela Basin, Gulf of Mexico, central and western equatorial Pacific, and eastern equatorial Indian Ocean. He proposed at least six microspherule layers in the upper Eocene, three of which were restricted to the eastern American region, and the other three in the equatorial Indo–Pacific region. According to Hazel (1989), all of the Late Eocene microspherule layers studied occur in the *G. semiinvoluta* Zone.

In contrast, Glass (1990) argued that petrographic and compositional data indicate two, or at most four upper Eocene microspherule layers: two closely spaced layers with microkrystites and microtektites in the *G. cerroazulensis* Zone, and possibly two similar layers in the *G. semiinvoluta* Zone (as represented in DSDP Site 612). This conclusion (only two layers) was reinforced by Glass and Koeberl (1999) in their study of ODP 689B core samples.

Late Eocene sediments in Core 21-5 of DSDP 612, in the north Atlantic off New Jersey, contain two spherule-bearing layers separated by ~5 cm. The lower layer is represented mainly by microkrystites, although it contains also tektite fragments, whereas the upper layer contains only microtektites and tektite fragments. These microspherule layers were attributed to the *G. semiinvoluta* Zone by Miller et al. (1991), rather than the *G. cerroazulensis* Zone, in which microtektites and tektite fragments are found at Barbados (Saunders et al. 1984). However, the chemistry of tektite glass in DSDP 612 is almost identical to that of the Barbados section (D'Hondt et al. 1987; Koeberl 1988; Koeberl and Glass 1988; Stecher et al. 1989; Glass 1990). Moreover, a $^{40}Ar/^{39}Ar$ plateau age of 35.5 ± 0.3 Ma for the tektitic glass in DSDP 612 (Obradovich et al. 1989) is comparable to a $^{40}Ar/^{39}Ar$ age of 35.4 ± 0.6 Ma obtained from tektite fragments and microtektites from the Bath Cliff section in Barbados (Glass et al. 1986b). However, the analytical uncertainty of these dates does not permit to discern whether or not they are the same impactoclastic material or rather different impactoclastic events closely spaced in time.

Miller et al. (1991) advanced the hypothesis that the Barbados and DSDP 612 tektites represent different impact events separated by ~0.5–1.0 Ma, in the upper *G. semiinvoluta* Zone, and the middle of *G. cerroazulensis* Zone. As an alternative hypothesis, Miller et al. (1991) proposed that tektite glass in the two sites was produced by the same impact event within the *G. cerroazulensis* Zone, and that very high sedimentation rates at DSDP 612, and diachroneity of biomarkers defining zonal boundaries in the Late Eocene, would make them appear as two different events.

With this background of confusing information on how many impacts, and of which age, occurred in the Late Eocene, Montanari et al. (1993) undertook a systematic search for Ir in 5 cm spaced samples from the Massignano section, near Ancona, which was at the time the best exposure of the Eocene–Oligocene transition known in the U–M region. A prominent Ir anomaly was found in a level dated at 35.7 ± 0.4 Ma, and an other peak in a sample higher in the section. This encouraged further high-resolution impact stratigraphic investigations of this section, the exciting results of which are reported below.

5.8.3
The Global Stratotype Section and Point for the E/O Boundary at Massignano

With a meeting of the IUGS Subcommission on Paleogene Stratigraphy at Monte Cònero (Ancona, October 1–3, 1987), a long period of prolific interdisciplinary stratigraphic research began in the abandoned quarry of Massignano, carried out by a number of international research teams. The quarry is located along the provincial road of the Monte Cònero Park, about 4 km north of the resort town of Sirolo (Fig. 5.7.6.14), right at the terminal of the Ancona–Massignano city bus line No. 93. The quarry face exposes a 23 m thick, continuous, and complete sequence of pelagic marly limestones and calcareous marls, rich in well-preserved benthic and planktonic microfossils, spanning the upper Eocene and the lowermost Oligocene, and interbedded with several biotite-rich volcano-sedimentary layers. This was regarded as an ideal situation for the application of an integrated stratigraphic approach aimed at the precise and accurate calibration of the litho-, bio-, magneto, and chemostratstratigraphic records with direct radioisotopic datings.

An IUGS Subcommission on Paleogene Stratigraphy special publication (Premoli Silva et al. 1988), containing several technical, interdisciplinary papers first discussed at the Monte Cònero meeting in 1987, was presented at the 28th International Geological Congress in Washington, along with formal proposals for promoting the Massignano quarry as the Global Stratotype Section and Point (GSSP) for the E/O boundary (Premoli Silva and Montanari 1989; Montanari and Odin 1989). The Massignano GSSP was formally established by the IUGS–SPS at the 1993 IGC in Kyoto (Premoli Silva and Jenkins 1993).

A number of studies, further refining the mineralogic and lithostratigraphic (Mattias et al. 1992), biostratigraphic and paleontologic (Gonzalvo and Molina 1992; Molina et al. 1993), magneto-stratigraphic (Lowrie and Lanci 1994), chemostratigraphic (Montanari et al. 1993; Vonhof et al. 1998), and geochronologic (Odin and Montanari 1989; Oberli and Meier 1991; Odin et al. 1991) characteristics of this section were published independently in various international journals in the years after the Monte Cònero meeting. In particular, recent high-resolution stratigraphic studies lead to the discovery, at 5.6 m in the GSSP section, of an impactoclastic layer, possibly of worldwide occurrence, containing an iridium anomaly (Montanari et al. 1993), shocked quartz (Clymer et al. 1996; Langenhorst 1996), extraterrestrial Ni-rich spinel and altered microkrystites (Pierrard et al. 1998), and a prominent anomaly of ^{3}He (Farley et al. 1998), which was interpreted by Farley and co-workers as evidence of a comet shower that lasted some 2.5 million years.

These new discoveries encouraged further detailed studies on the paleontologic and paleobiologic records of the Massignano GSSP aimed at the verification of the effects that the inferred cosmic events (impacts and comet shower) may have had on the terminal Eocene marine biota (Brinkhuis and Coccioni 1995; Vonhof et al. 1995, 1998; Gardin et al. 1999).

In 1993, A. Montanari commissioned a 40 m deep, continuous core, the so-called MASSICORE, which was drilled in the immediate vicinity of the GSSP (Montanari et al. 1994), covering the stratigraphic interval of the exposed section, as well as some additional 15 m of the lower Oligocene. The MASSICORE is permanently stored at the Osservatorio Geologico di Coldigioco and it is available, upon request, for sampling and cooperative research. A detailed magne-tostratigraphic study of the MASSICORE was carried out by Lanci et al. (1996), and studies on the foraminiferal, nannofossil, dinocyst, and stable isotope records are currently in progress under the leadership of R. Coccioni of the University of Urbino.

Finally, on May 9, 1997, the Cònero Regional Park inaugurated the Massig-nano quarry as an accessible "didactic-scientific site". Eugene and Carolyn Shoe-maker participated in the inaugural ceremony as guests of honor. The reclamation work of the GSSP included consolidation and cleaning of the quarry face, land-scaping of the adjacent area, and the instalment of two geological signs, with color pictures and explanations, in Italian and English, accessible to the general public. A 1999 updated integrated stratigraphic model of the Massignano GSSP is shown in Fig. 5.8.3.1.

In summary, the Massignano GSSP constitutes an excellent playground for in-vestigating on the effects that extraterrestrial events had on the marine, pelagic environment. Studies with this focus are currently in progress and, therefore, there is no definitive picture of the biologic changes, turnovers, or extinctions through the terminal Eocene in this section. For sure, we can say that nothing as severe as the K/T catastrophe occurred at this time. The terminal Eocene has been always regarded has a time of significant, but not drastic, change, a decline spread through a span of several millions of years, and seen by many researchers as a sequence of accelerated, step-wise extinctions (e.g., Keller 1986). Therefore, in the next chapter of this book, we will concentrate on the description of the impact stratigraphy at Massignano, leaving the more complex, yet fundamentally impor-tant, paleobiologic aspects of the terminal Eocene events to the several researchers currently working on them.

GLOBAL STRATOTYPE SECTION AND POINT FOR THE EOCENE-OLIGOCENE BOUNDARY: THE GSSP OF MASSIGNANO [1]

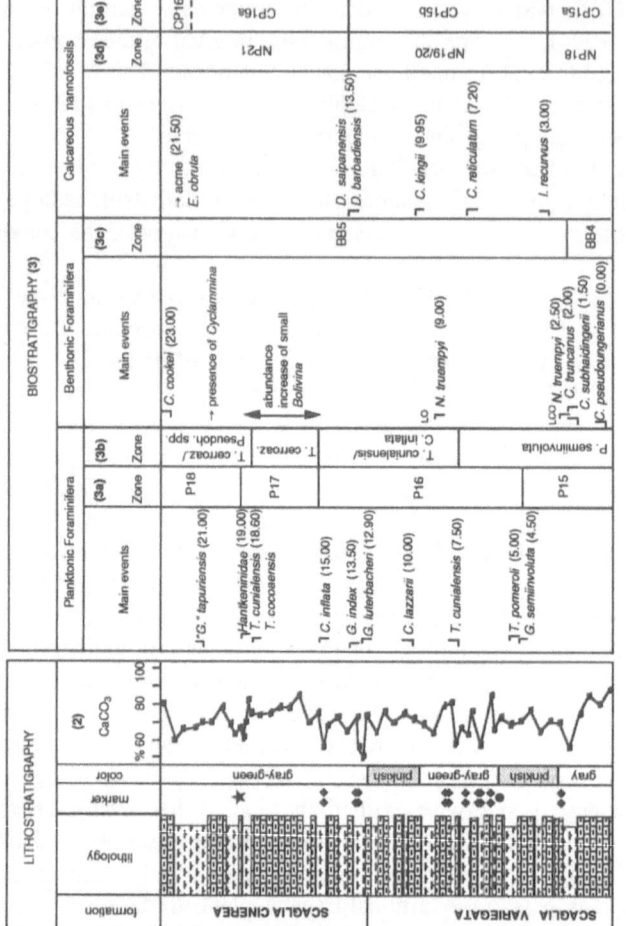

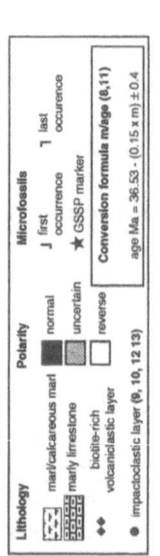

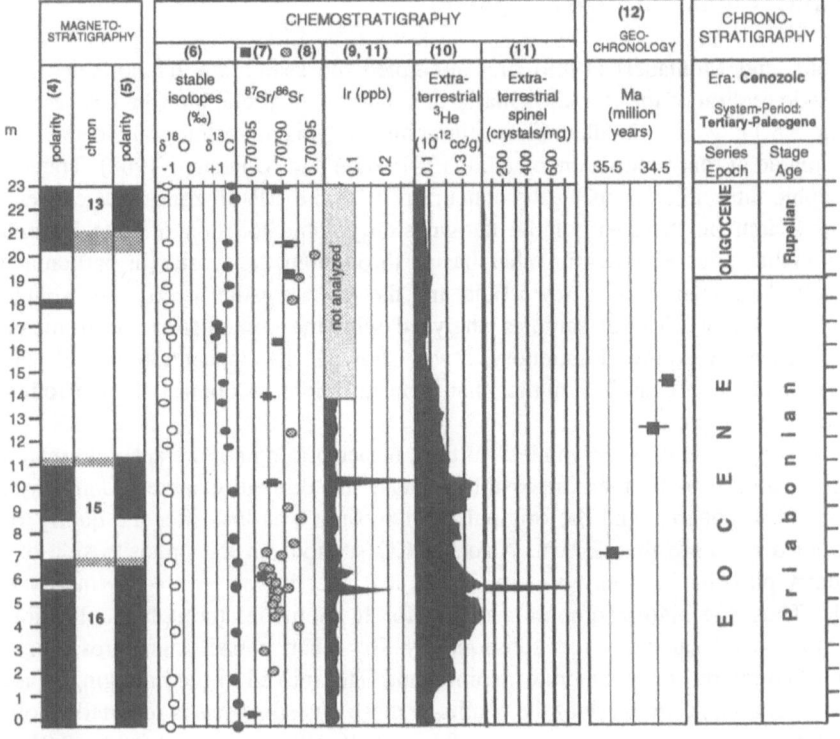

REFERENCES

(1) Premoli Silva and Jenkins 1993
(2) Mattias et al. 1992
(3) Coccioni et al. 1988
(3a) Blow 1969; Berggren et al. 1985
(3b) Berggren et al. 1995
(3c) Berggren and Miller 1989
(3d) Martini 1971
(3e) Okada and Bukry 1980
(4) Bice and Montanari 1988

(5) Lowrie and Lanci 1994
(6) Odin et al.1988
(7) Montanari et al. 1991
(8) Vonhof et al. 1998
(9) Montanari et al. 1993
(10) Farley et al. 1998
(11) Pierrard et al. 1998
(12) Montanari et al. 1985; Odin et al. 1988; Odin et al. 1991
(13) Clymer et al. 1996; Langenhorst 1996

Fig. 5.8.3.1. Integrated stratigraphic model of the GSSP for the Eocene–Oligocene boundary at Massignano (this page and facing page).

5.8.4
Impact Signatures at Massignano

Asaro and Montanari (1988) first attempted the search for Ir anomalies in the classic section of the Barbetti Quarry at Contessa, near Gubbio (the CB section in Fig. 5.5.1), where detailed biostratigraphic and magnetostratigraphic data were available at that time (Monaco et al. 1987; Bice and Montanari 1988). The stratigraphic interval comprising the concurrent P16 and CP15b zones was chosen for this search on the basis of the known stratigraphic location of Late Eocene Ir anomalies and spherules in other basins throughout the world (in particular the North American tektite strewn field and the microkrystite layers; see Montanari 1990). Out of 20 pilot samples analyzed with the coincidence spectrometer at Lawrence Berkeley Laboratory, two showed Ir concentrations of about 0.07 ± 0.004 ppb, slightly higher than concentrations of about 0.04 ppb found in the other samples.

Testing these two "anomalies" further, or performing a more accurate search in this section, was by then impossible because the Barbetti Concrete Company expanded the quarry, and the original CB exposure was lost after the quarry front receded by more than 200 m. Also the CQ section, on the opposite side of the quarry pit (Fig. 5.8.4.1), was temporarily lost after it was covered by quarry rubble. Thus, the high-resolution scanning for Ir anomalies through a collection of 5 cm spaced samples, and subsequently for other impact signatures, such as shocked quartz, microspherules, spinel, and ^{3}He, moved to the emerging Massignano section. A summary of the results of these investigations is reported below.

Iridium. Two prominent peaks of Ir of 0.19 ± 0.02 ppb, and 0.33 ± 0.01 ppb, were found at meter levels 5.61 and 10.25, respectively, in the Massignano section by Montanari et al. (1993). The lower peak, located in the upper part of Chron 16n2, and in the lower part of foraminiferal Zone P16, mid-lower part of nanno-fossil Zone CP15b (see Fig. 5.8.3.1), covered a 25 cm interval and comprised 5 samples with Ir concentrations higher than 0.08 ppb. On the other hand, the peak at 10.25 m (mid C15n, mid P16, and upper CP15b) was defined by a single sample. In addition to these, the Ir profile at Massignano showed another minor peak of about 0.1 ppb, defined by two samples at around 6.2 m in the section.

Ratios among other trace elements, including Fe, Cs, Sb, and Se, were used by Montanari and co-workers to distinguish the characteristic background chemical profile of these pelagic carbonates from the several biotite-rich volcanic ashes interbedded with them, and to test whether the three Ir anomalies had a volcanic rather than extraterrestrial origin. The result was that the Ir peaks were true anomalies, in the sense that they were not derived from volcanic material, nor were they produced by diagenetic processes and/or precipitation from sea water.

Fig. 5.8.4.1. a) Re-excavated lower part of the Contessa Quarry (CQ) section; **b)** Close-up of the quarry face outcrop containing the Late Eocene impactoclastic Ir, and shocked quartz-bearing layer discovered by Clymer (1996) right at the top of a red band (RB) interval.

However, the peak at 10.25 m, represented by a single sample, remained somewhat suspect. The unusually high abundances of Ir, Se, and Sb, and, possibly, Ag in this sample could have been the result of sulfide precipitation in the sediment, but there is not enough iridium in the ocean water to accommodate 0.3 ppb Ir, and the samples just above and below 10.25 m have a normal (i.e., not leached) background around 0.04 ppb Ir.

Further samples, covering the interval between 5 and 6 m, were re-analyzed by F. Asaro in 1994, and once again in 1995 (Clymer 1996). In this last scan, one sample at exactly 5.6 m yielded about 0.4 ppb Ir, a factor of two higher than in previous analyses.

In 1995, the Barbetti Co. quarry re-excavated the lower part of the CQ section at Contessa (Fig. 5.8.4.1). This allowed Clymer (1996) to collect a suite of samples through a 2 m thick stratigraphic interval in which, in a marly layer located at the top of a pink colored band, he found shocked quartz (see below). This level was correlated with the Ir peak at meter level 5.6 in the Massignano section on the basis of the color change and available magnetostratigraphic data. The INA analyses performed by F. Asaro at LBL (in Clymer 1996, unpublished dissertation) revealed, in a sample from this shocked-quartz level, a relatively high Ir concentration of 1.06 ± 0.04 ppb, a factor of 3 to 5 higher than in any of the previous analyses at Massignano.

Iridium was measured independently at Massignano by Pierrard et al. (1998) through the same stratigraphic interval analyzed already three times at LBL. A maximum concentration of 0.28 ± 0.02 was detected at 5.6 m, thus confirming the results of Montanari et al. (1993) and Clymer (1996). No further analyses to confirm the relatively high concentration of 1 ppb Ir were ever performed, to our knowledge, in the Contessa CQ section, nor across meter level 10.25 at Massignano, where Montanari et al. (1993) found an anomalous concentration of 0.3 ppb Ir.

Shocked Quartz. Shocked quartz was first reported by Clymer et al. (1995) in the Ir-rich layer at meter level 5.6 of the Massignano section, and in the correlative layer in the CQ section at Contessa (Clymer 1996). The marly limestones throughout the upper Eocene at Massignano, normally contain 2% of sand-size mineral grains, 80% of which are quartz. Only 4% of the quartz faction, typically in the size range between 50 and 150 μm, display shock features (Fig. 5.8.4.2), and of these ~5% show one single set of planar deformational features (PDFs), 28% contain two, and 67% have three or more sets (Clymer et al. 1996).

Other samples representing genuine pelagic carbonates and biotite-rich volcaniclastic layers were also investigated by Clymer and co-workers in this section, but no shocked quartz grains were found.

An independent crystallographic study by Langenhorst (1996) on the Massignano shocked quartz, using the transmission electron microscope (TEM), confirmed the shock metamorphic origin of the PDFs, which, according to this author, formed at shock pressures between 10 and 25 GPa, in a non-porous, quartz-rich, near-surface rock. This last inference was derived by the fact that high-pressure silica phases, such as stishovite and coesite, which are known to form in predominantly porous impact target rocks (Grieve et al. 1996), are absent in the Massignano shocked quartz.

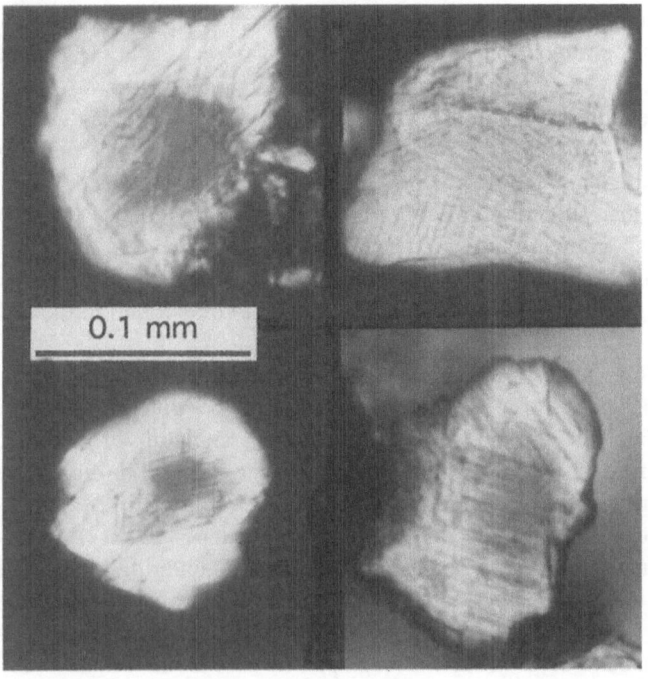

Fig. 5.8.4.2. Shocked quartz grains from the Late Eocene impactoclastic layer at 5.6 m in the Massignano section. Scale bar = 0.1 mm.

Ni-rich Spinel and Microspherules. After an Ir anomaly and shocked quartz were pinpointed at 5.6 m in the Massignano section, a systematic search for impact spinel and microspherules was undertaken by Pierrard et al. (1998) through the whole 23 m thick section, with particular attention for the interval across the Ir- and shocked quartz-bearing layer. At meter level 5.6 Pierrard and co-workers found up to 800 Ni-rich spinel microcrystals per milligram of bulk rock, and no spinel was found in any other level analyzed in this section.

Spinel microcrystals have sizes in the range of 10 μm, a common dendritic habit often characterized by crossed arms (Fig. 5.8.4.3a, b), and can be seen with a powerful magnifying lens on a hand sample looking like very fine pepper dusting sparse throughout the gray, marly biomicrite. While sampling, Pierrard and co-workers noticed that in the layer at 5.6 m there were also round, flattened objects, mostly of a black color, fewer with a green color, and with diameters in the range of 0.2–0.4 mm. They were scattered throughout a 3–5 cm interval, and in a few instances, they were concentrated in *Planolites* burrows (Fig. 5.8.4.3c). After a closer microscopic examination, they turned out to be flattened microkrystites (Pierrard et al. 1998).

These spherules, which eluded for years the many researchers who worked at the Massignano GSSP, are extremely delicate, flattened to disks with 1:20 axial ratios (Fig. 5.8.4.3d), and they cannot survive routine sediment washing procedures such as the ones micropaleontologists usually apply with sieves and cloths under running water to separate and clean microfossils. Nor they can possibly survive rougher washing and acid treatments, such as those normally used to concentrate quartz and other silicate grains. The black, strongly magnetic spherules are made of clustered spinel microcrystals (Fig. 5.8.4.3e), whereas the green spherules are made of a smectitic clay which, in thin section, exhibits internal fibral structures very similar to those of the K-feldspar spheroids found in the K/T clay (Fig. 5.8.4.3f).

Extraterrestrial ^{3}He. A conspicuous increase in extraterrestrial ^{3}He was first detected by Farley (1995) in the upper Eocene portion of the Giant Piston Core from the North Pacific. This anomaly was interpreted as a signature of increased influx of interplanetary dust particles (IDPs) during a span of a few million years which covered the critical time of multiple impacts. This helium isotope is extremely rare in terrestrial crustal rocks, but relatively abundant in IDPs. However, particles larger than a few tens of micrometers would heat up entering the Earth's atmosphere to the point of releasing their helium, which, being very light, would be lost into the outer space. Thus, ^{3}He is supposedly derived from very small particles, and a significant increase through a long span of time in the order of millions of years is suggestive of a comet shower, which not only would increase the possibility of multiple impacts (Hut et al. 1987), but implies a significant increase in the accretion on Earth of interplanetary dust .

Poor age control in the carbonate-free, abyssal clay of the North Pacific GPC precluded precise correlation with the known Late Eocene impact record. Therefore, to better constrain the timing of this Late Eocene ^{3}He event, Farley et al. (1998) performed a high-resolution scanning of the isotope content through the Massignano GSSP and the MASSICORE as well. An abrupt increase of the ^{3}He content was detected starting at meter level 2, and reached a maximum, with a concentration a factor of about 8 higher than background, at meter level 5.6, exactly at the impactoclastic layer containing the iridium anomaly, shocked quartz, and impact spinels. A second ^{3}He peak a factor of six higher than background was detected by Farley and co-workers at meter level 10.3, corresponding to the second, still unconfirmed, Ir peak measured by Montanari et al. (1993). The broad ^{3}He anomaly gradually decreased to background levels at 16 m.

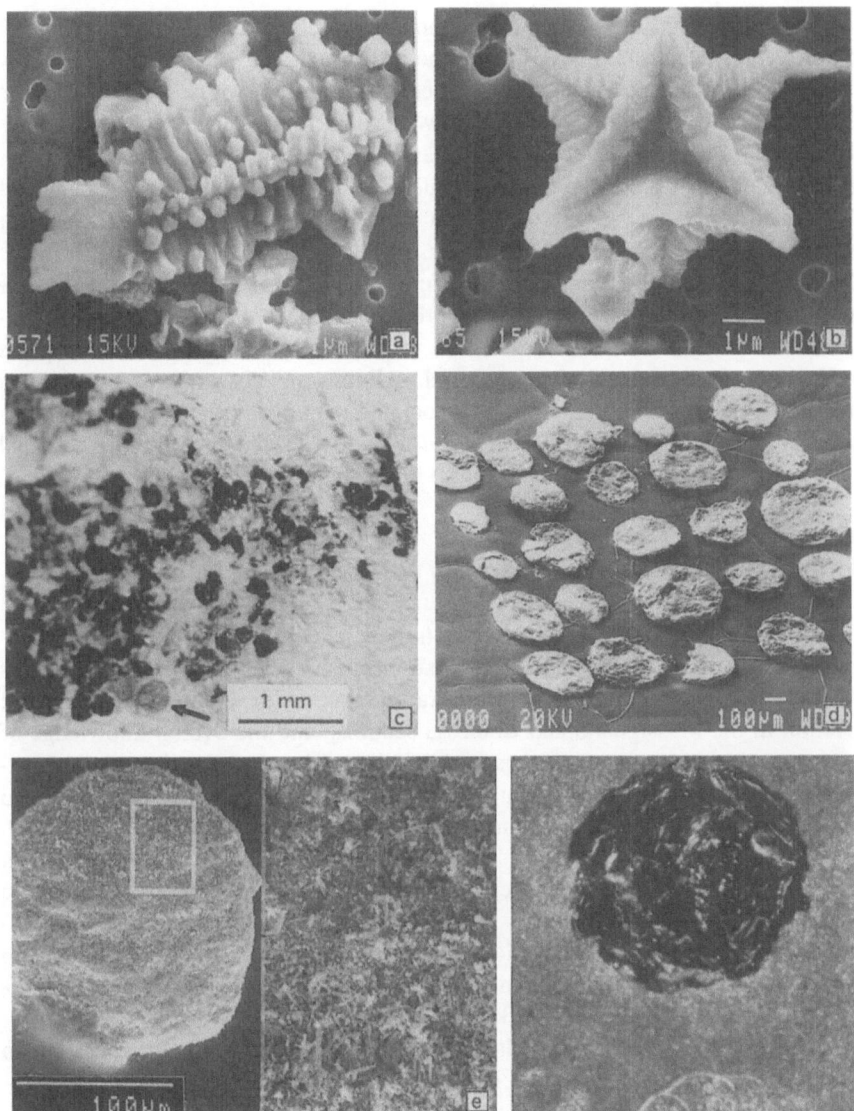

Fig. 5.8.4.3. Ni-rich spinel and microkrystites from the upper Eocene impactoclastic layer at Massignano. The most common crystal habit of Ni-rich impact spinel is dendrite, which is typical of rapid growth under nonequilibrium conditions. This habit (a; SEM image) is obtained by quenching in air of meteoritic material heated above liquidus (>1560 °C). On the other hand, skeletal and octahedral (b; SEM image) habits indicate crystallization from melt heated below liquidus. Microkrystites are often found in *Planolites* burrows (c; close-up photograph). Most of them are black, but some are made of a green (smectitic) clay (see arrow). Strongly flattened black micro-krystites (d; SEM image) are made of clustered microcrystals of spinel (e; SEM image), whereas the altered, smectitic microkrystites show, in thin section (f; crossed polarizers), fibrous structures similar to the altered K-feldspar microkrystites found in the K/T boundary.

In summary, the Late Eocene ^{3}He anomaly, which covers a span of about 2.2 million years, from about 36.4 Ma to about 34.2 Ma, was interpreted by Farley et al. (1988) as strong evidence for a comet shower during which multiple impacts may have occurred. At present, we know of some four to five craters with ages covering this span of time (see Fig. 5.1.1 and Table 1.9.1). Two of these, the Popigai and the Chesapeake Bay impact structures, have diameters in the range of 100 km and fairly precise ages consistent with that of the impactoclastic layer of Massignano. The other impact structures, the Mistastin (28 km; 38 ± 4 Ma), the Logoisk (17 km; 40 ± 5 Ma), and the Wanapitei (7.5 km; 37 ± 2 Ma) craters, are much smaller, and they would not have produced impactoclastic signatures recognizable in distant sections throughout the world. However, these smaller impacts may be part of the comet shower proposed by Farley and co-workers. A prolonged time of extraterrestrial bombardment, associated with a significant influx of interplanetary dust particles, could have been a main causal factor for the terminal Eocene global environmental changes and biological crises.

5.8.5
High-resolution Impact stratigraphy at Massignano

The Late Eocene record in the U–M constitutes today an ideal situation for investigating the possible cause-and-effect relationship between extraterrestrial events and bio-environmental changes. The stratigraphic evidence for extraterrestrial accretion in the Massignano and Contessa sections, and the overall knowledge of the impact record (craters and impactoclastic layers in coeval sediments throughout the world) undoubtedly indicate that the Late Eocene witnessed a prolonged time of crisis, perhaps a comet shower, such as the one predicted by Hut et al. (1987) and seemingly confirmed by the ^{3}He record revealed by Farley et al. (1998). However, no giant impact crater the size of the Chicxulub is known with a Late Eocene age, and no boundary layer like the K/T can be pinpointed and correlated with a sharp mass extinction. The possible cause-and-effect relationship between impact and bioenvironmental changes has to be searched through a stratigraphic interval several meters thick.

However, the discrete impactoclastic layer at meter level 5.6 in the Massignano type section is good evidence that invites a closer look at what may be the sedimentologic and biologic response to a "non-lethal" impact event.

In terms of physical stratigraphy, this impactoclastic layer is located at the very top of a pinkish band, which, in the Scaglia Variegata Formation of the U–M Apennine, can be considered as a precise and consistent regional lithostratigraphic marker (Fig. 5.8.5.1; see also outcrop images in Figs. 5.8.4.1 and 5.8.5.2). This interval is contained in the upper part of magnetic Chron 16n, although several magnetostratigraphic studies by different researchers in the CQ section at Contessa (Lowrie et al. 1982), the Massignano section (Bice and Montanari 1988; Lowrie and Lanci 1993), and through the MASSICORE (Lowrie et al. 1996), have yielded slightly different results (see Fig. 5.8.5.1). This is probably due to interlaboratory differences in demagnetization and analytical techniques, and ways of interpreting paleomagnetic data.

As for biostratigraphy, the impactoclastic layer is located in the upper part of planktonic foraminiferal *P. semiinvoluta* Zone (i.e., lower P16 Zone), and in the middle–lower part of nannofossil zones NP19/20 and CP15b. Therefore, it is not located at a biozonal boundary, across which a faunal or floral change could be readily recognized with a magnifying lens directly on hand samples (as for the K/T) or even by looking at thin sections and smear slides with a standard microscope.

As for the numerical age of this event, it is important to note that these marine U–M sections are, to our knowledge, the only ones in the world that have been calibrated with direct radioisotopic dating of volcaniclastic material (biotite, zircon, and monazite) interbedded with pelagic carbonates (e.g., Montanari et al. 1988b; Odin et al. 1991).

An interpolated age of 35.7 ± 0.4 Ma for the impactoclastic layer is derived from all the radioisotopic dating performed in a few of these biotite-rich volcaniclastic layers at Contessa and Massignano. Additional radioisotopic dating may provide, in the future, a more precise and accurate age for this impact event. Nevertheless, an age of 35.7 ± 0.4 Ma is indistinguishable, within analytical uncertainty, from the ages of the North American tektites and microtektites (Glass et al. 1986b; Obradovich et al. 1989), the inferred age for the Chesapeake Bay impact structure (Poag and Aubry 1995; Koeberl et al. 1996c), and the radioisotopic age of impact melt rocks from the Popigai crater (Bottomley et al. 1997) (see Fig. 5.8.5.1).

There is no direct evidence that indicates whether or not the impactoclastic layer at Massignano is related to the Popigai or rather to the Chesapeake Bay impact events. However, the fact that iridium, in Caribbean and Gulf of Mexico sections, is found in the microkrystite layer and not in the closely spaced microtektite layer, and that the latter is attributed to the Chesapeake Bay impact, suggest that the Massignano Ir-rich impactoclastic layer may be related to Popigai. Moreover, the absence of high-pressure silica phases in the shocked quartz of Massignano, which, on the other hand, are present in the Chasepeake Bay-related microtektite layer of DSDP 612 (Bohor et al. 1988), lead Langenhorst (1996) to suggest that Massignano shocked quartz was derived from the non-porous, crystalline target rock of the Popigai crater (see also Sect. 3.3 for further discussion).

In any case, standing in front of the Massignano section, and looking at the interval between meter marks 5 and 6 (Fig. 5.8.5.2a), one would hardly imagine that some 35.5 million years ago two enormous extraterrestrial objects the size of Mont Blanc collided with the Earth in remote localities 5–6,000 km from here, excavating craters with diameters in the range of 100 km (i.e., roughly the distance between Massignano and Gubbio), each liberating energies equivalent to some 10 million megatons of TNT. From a distance, the visitor can only see a faint change in the color of these marly limestones, from pink to gray.

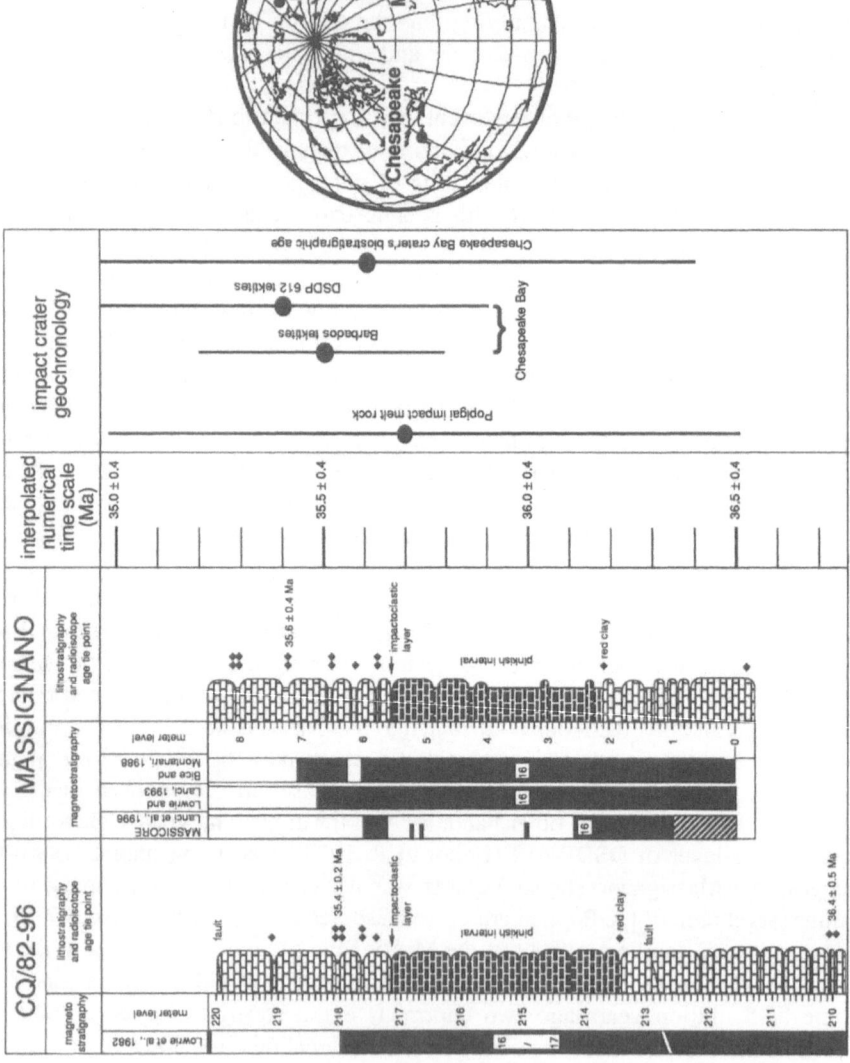

Fig. 5.8.5.1. Integrated litho- and magnetostratigraphic correlation between the Contessa CQ and Massignano sections across the late Eocene impactoclastic layer, with chronostratigraphic and present-day geographic locations of the Popigai and Chesapeake Bay impact structures. Diamonds indicate biotite-rich volcaniclastites.

Getting closer to the outcrop, one can see a clear break in sedimentation at 5.8 m, which is marked by a softer, more clay-rich layer. This is a biotite-rich volcaniclastic bed resting on top of a 50 cm thick marly limestone with a faint pinkish color. The pinkish limestone looks fairly homogeneous except for a 10 cm thick, hardly noticeable grayish band at 5.6 m (Fig. 5.8.5.2b).

This band is the impactoclastic layer. On hand samples, it is possible to notice scattered mineralized spherules. In some rare cases, these spherules are concentrated in *Planolites* and *Zoophycos* burrows (Fig. 5.8.4.3c). However, many of the trace fossils through the whole marly limestone layer at the top the pinkish interval, including the thin impactoclastic horizon, contain biotite flakes that were derived from the volcaniclastic layer at 5.8 m. This gives an idea of the degree of vertical mixing in these sediments caused by intense bioturbation activity. Worms and other detritus feeding organisms that were living at or above the biotite level at 5.8 m penetrated the subbottom sediment for as much as half a meter, passing through the impactoclastic layer. On the other hand, the few burrows that contain spherules were probably dug by organisms that witnessed the impact event: they were pasturing on the impactoclastic layer immediately after its deposition.

A polished rock slab of the top pinkish marly limestone reveals that the color change from pink to gray occurs right at the base of the impactoclastic horizon (Fig. 5.8.5.3a). Moreover, the intricate network of fossil traces seems to be interrupted between 5.61 and 5.63 m (Fig. 5.8.5.3b) by a relatively undisturbed band of sediment. These observations suggest that there has been a very brief interruption or decrease in bioturbation activity during the deposition of the impactoclastic layer which coincided also with a brief episode of relative reduction of the sea floor sediment. Normal oxidizing conditions were restored at around 5.67 m, as it is indicated by a return to a pinkish color. The arrival on the sea floor of volcaniclastic material at 5.83 m marks the definitive end of these slightly oxidizing conditions, and from here up to 9.5 m these marly limestones exhibit a gray color.

All in all, the situation around the impactoclastic layer of Massignano, characterized by a decrease of bioturbation activity during a short episode of relatively low-Eh conditions, is somewhat similar to the K/T boundary with the difference that in the latter these events are much more marked and severe (see Sect. 5.7.4). Given this, the question remains on whether there has been a significant response of the plankton and/or benthos to the distant impact(s) represented by the impactoclastic layer. Preliminary results from a high-resolution, quantitative analysis across the Massignano section, using a multivariate statistical approach on planktonic Foraminifera and calcareous nannofossils, indicate that during a time of general, gradual cooling, an environmental break took place right after the impactoclastic layer was deposited, leading to accelerated pulses of temperature drops, which may represent the relatively long-term effects of this and other terminal Eocene impact events (Gardin et al. 1999).

272

Fig. 5.8.5.2. a) Panoramic view of the Late Eocene interval at Massignano. The frame refers to the close-up picture in **b)**, which covers the interval between 5 and 6 m. The impactoclastic layer is found at a faint color change, from pink below, to gray above, at 5.61 m. The soft, receding layer at 5.8 m is a biotite-rich volcaniclastite.

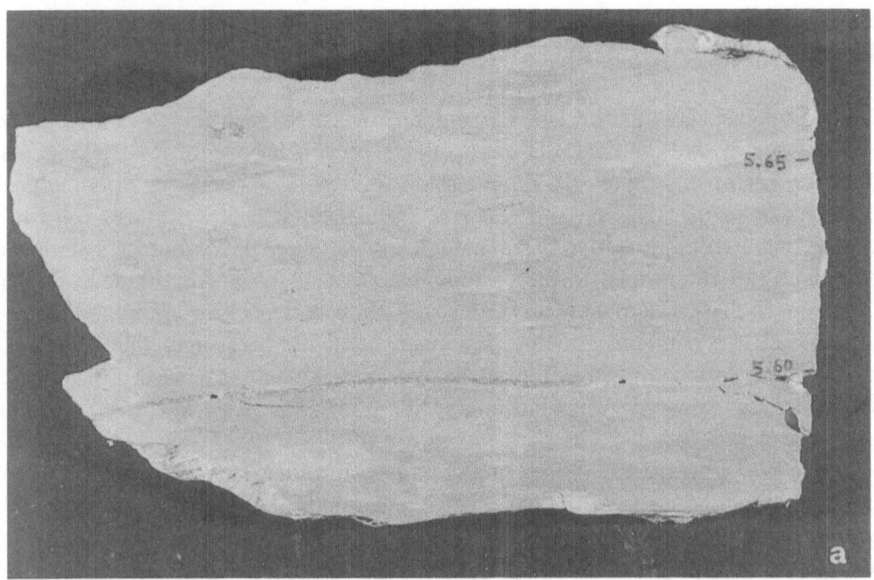

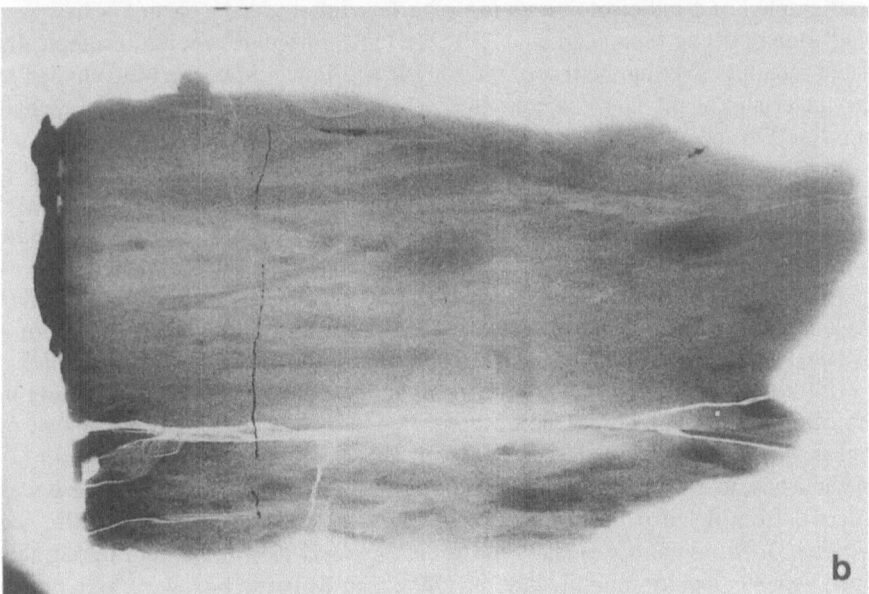

Fig. 5.8.5.3. a) Polished slab of the marly limestone layer containing the Late Eocene impacto-clastic layer at Massignano (5.61 m); **b)** X–ray contact print of the same slab (courtesy of the medical staff of the Civil Hospital in Cingoli), showing the complex network of ichnofossil burrows, which are apparently less abundant in the interval containing the impactoclastic debris.

5.9
The U–M Miocene Record

During the Miocene, as the U–M Apennine fold-and-thrust belt started to grow and migrate toward the East, the onset of flysch deposition in NW–SE elongated foredeep basins interrupted the quiet carbonate, deep-water sedimentation, which begun back in the Triassic (see Chap. 4). While siliciclastic turbidites were invading this deforming Adrian basin, pelagic and hemipelagic carbonate sediments continued to be deposited in the easternmost edge of it, outside the deep flysch foredeep. In these external areas, open sea, carbonate deposition continued undisturbed until the beginning of the Messinian. After the Messinian salinity crisis, which in the eastern U–M basin was characterized by the deposition of evaporitic sequences, syn-orogenic siliciclastic deposition took over even in the most external, far eastern areas.

Therefore, a nearly continuous and complete, turbidite-free Miocene sequence is found along the cliffs of the Cònero Riviera, south of Ancona (Figs. 5.9.1 and 5.9.2). The integrated stratigraphy of this sequence, from the Langhian up to the Messinian, including calcareous plankton biostratigraphy, O, C, and Sr isotope stratigraphy, and radioisotopic dating of a few interbedded volcanic ashes, was studied in detail by Montanari et al. (1997a). Other turbidite-free, radioisotopically dated sections covering the rest of the Middle and Lower Miocene were studied in the hinterland of the U–M region, and were documented in detail by Montanari et al. (1997b and c), Coccioni et al. (1997), and Deino et al. (1977).

In summary, in the Alpine–Himalayan orogenic domain, where the Miocene is usually represented by poorly dated, syn-orogenic and/or post-orogenic flysch and molasse deposits, the pelagic sequence of the U–M Apennine constitutes a rare, fortunate case. It represents that kind of ideal stratigraphic situation where a given geochronologic interval can be confidently and accurately localized in a discrete sequence of undisturbed layers (Fig. 5.9.3). In our specific case, the ejecta of a distant, well dated impact event can be searched in an interval a few meters thick.

Unfortunately (for us) only a few, medium size craters are known at present in the Miocene impact record. One of these, Haughton in Canada, has a diameter of 24 km and a $^{40}Ar/^{39}Ar$ age of 23.3 ± 1.0 Ma. This age range corresponds, in our U–M sequence, to a hiatus at the very top of the Scaglia Cinerea Formation, as it was inferred from detailed biostratigraphic analysis by Montanari et al. (1997b; see Fig. 5.9.3). No biotic crises show up in the Sepkoski (1996) diagram during this time, which coincides roughly with the Oligocene–Miocene boundary, and, in any case, the Houghton impact crater is too small and distant to deliver an impactoclastic signature to the U–M basin, even if this span of geologic time is here represented by continuous deposition.

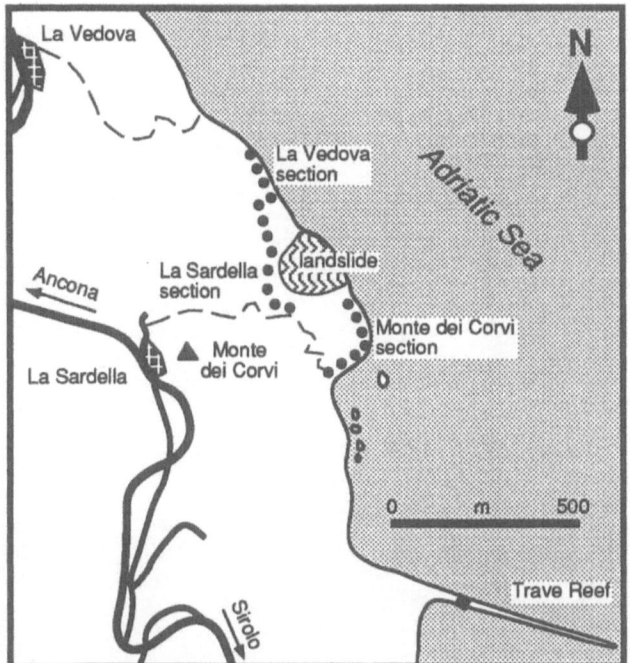

Fig. 5.9.1. Location map of the stratigraphic sections exposed along the cliffs of Monte dei Corvi, near Ancona, which cover the interval from the mid Langhian to the lower Messinian.

On the other hand, the Ries crater in southwestern Germany, which also has a diameter of about 24 km, is located only 600 km north of the Cònero Riviera, where a complete sequence of coeval marine, marly limestones are exposed in the La Vedova section (Montanari et al. 1997a). This 40 m thick section is located at the base of the Monte dei Corvi cliff and contains a volcaniclastic biotite-rich level, which, on the basis of calcareous plankton stratigraphy, was correlated with a similar volcaniclastic marker in the L'Annunziata section near Apiro (Montanari et al. 1997c). At L'Annunziata, this level yielded a preliminary laser fusion $^{40}Ar/^{39}Ar$ age of about 15 Ma from single sanidine crystals (Montanari et al. 1988a), which is practically identical to the age of the Ries crater. Therefore, the La Vedova section may represent a promising area for investigating the regional effects that a medium size impact may have had on a proximal marine environment. To our knowledge, no studies focused on this interesting problem were undertaken to date in this section (although we have recently started an effort to remedy this situation).

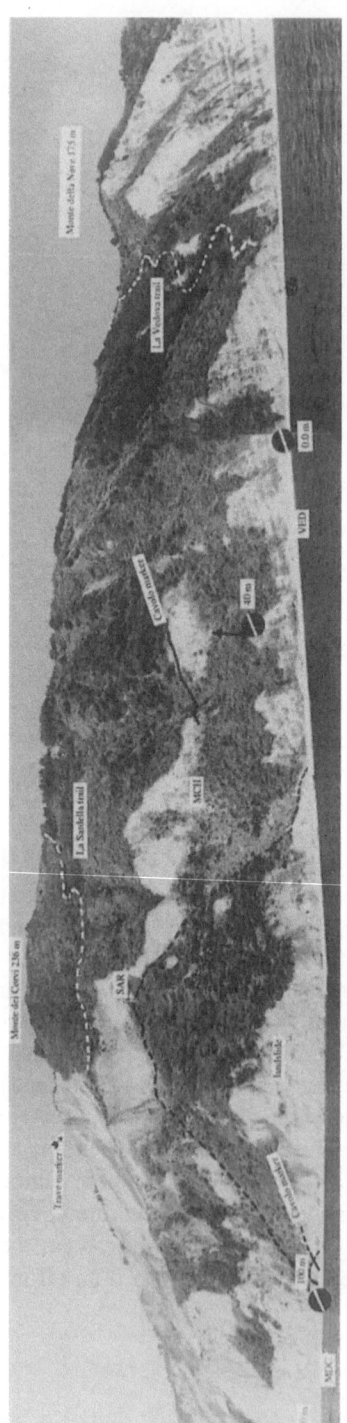

Fig. 5.9.2. Panoramic view of the Monte dei Corvi cliffs and location of the sections studied by Montanari et al. (1997a).

In addition to these two well dated medium size impact events, the Miocene record features two other small, distant, and poorly dated impact craters, Karla in Russia (12 km diameter; 10 ± 10 Ma), and Bigach in Kazakhstan (7 km diameter; 6 ± 3 Ma). However, these events are so inconspicuous that they could not have produced an impactoclastic signature in our U–M basin, nor they can be considered responsible for an environmental crisis recognizable at a global scale.

Nevertheless, a small Ir anomaly was detected in coeval oceanic sediments in two distant sites (i.e., ODP 689B in the Weddell Sea, and ODP 588B in the Tasman Sea) by Asaro et al. (1988). This iridium signature, which in ODP 588B has a concentration of 0.15 ± 0.01 ppb, was attributed by Asaro an co-workers to a large impact event, which would have occurred near the Middle–Late Miocene boundary, between 11.7 and 10 Ma. This stratigraphic interval corresponds also to a minor, actually the smallest extinction peak in Sepkoski's (1996) diagram. In the Monte dei Corvi section, the Middle–Late Miocene boundary, which corresponds to the Serravallian–Tortonian interstage boundary, has been precisely dated at 11.0 ± 0.2 Ma (Odin et al. 1997). Once again, the U–M sedimentary sequence offers a remarkably complete, well exposed, and accurately dated stratigraphic section, where these events could be studied in great detail.

278

MIOCENE STRATIGRAPHY OF THE U-M APENNINES

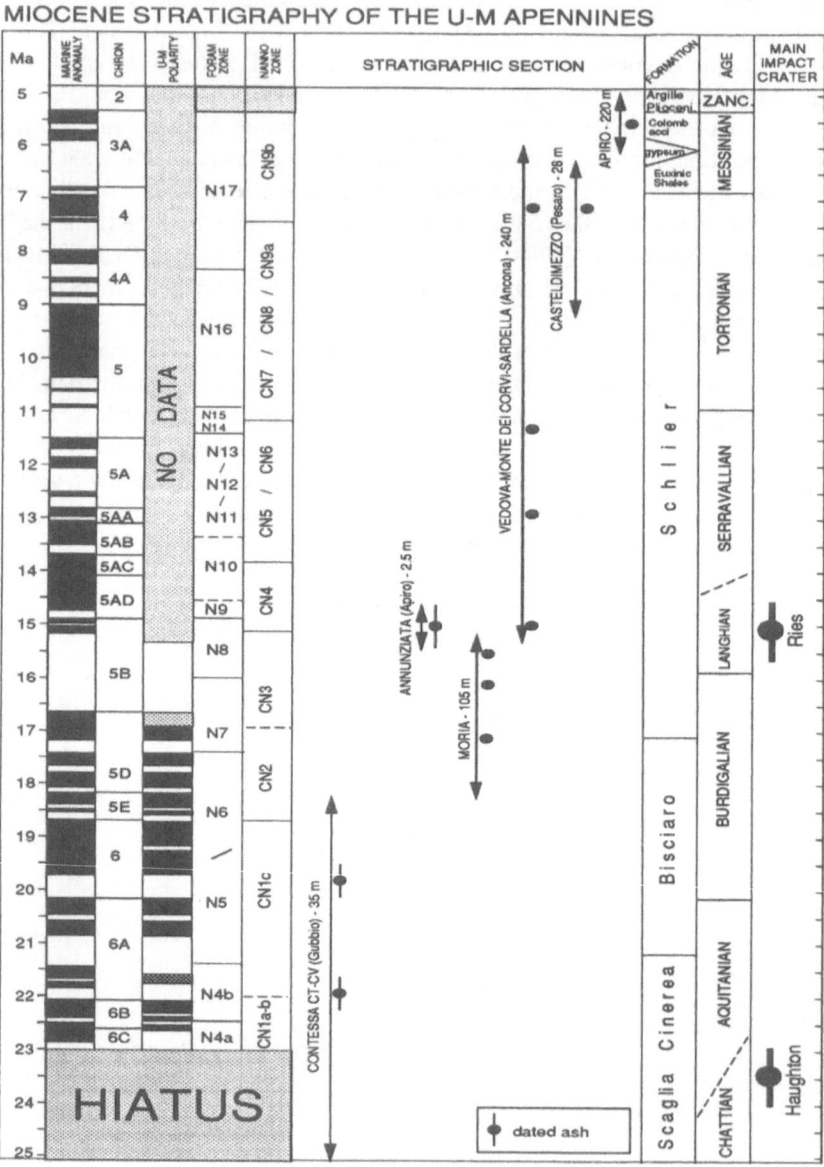

Fig. 5.9.3. Integrated stratigraphy of the Miocene Epoch as represented in exposed stratigraphic sections of the U–M Apennines. The stratigraphic position of the two main impact craters of this time interval, Haughton and Ries, can be derived from the precise and accurate radioisotopic age calibration of this sedimentary sequence (Montanari et al. 1977a, b, c; Deino et al. 1977).

6 Documentation and Laboratory Techniques

In this section we want to provide a very short overview of some of the simple field and sample preparation methods that are used to take and isolate samples for impact-related studies (including techniques that have been employed on samples described in Chap. 5), and review a few of the methods used in the study of shocked minerals and meteoritic components in impactites. It is beyond the scope of this treatment to cover standard petrographic and geochemical methods in any detail, or to discuss any paleontological and geophysical methods.

6.1
Documentation and Sampling in the Field

The work of data acquisition in any geological research project starts in the field. However, in many cases laboratory analyses are performed by technicians or researchers who did not collect the samples personally, and may not be familiar with the field and the general geology of the collected samples. Therefore, correct sampling and exhaustive documentation of the field work becomes as important as the precision and accuracy in routine laboratory analysis in the complex process of data production.

Unlike an error in laboratory analysis, which can be (in theory) easily found and the analysis re-performed correctly, an error in the field documentation may not be as evident and, if found or suspected, it will require to go back on the outcrop, which may be difficult, expensive, or even impossible. Needless to say, an error in the field work is inevitably transferred into the lab leading to incorrect results and misleading interpretations.

Whereas laboratory analyses are performed following precise routine procedures in a controlled and unchanging environment, field work (including outcrop observation and sample collection) may occur in such different environmental conditions (i.e., topography, weather, vegetation, etc.) and about so many different geological subjects (i.e., different rock types, outcrop exposure, accessibility, specific goals), that makes it difficult to establish a standard procedure of what to do and how to do things. Nevertheless, there are a few guidelines that can be taken as useful suggestions for correct documentation and sampling in the field.

Once an outcrop is chosen as the place for investigating the possible presence of a meteoritic component and/or impact ejecta, either because it contains the paleontologic record of a biotic crisis or because it covers the time span of a major impact event, the first thing to do is to locate it as accurately and precisely as possible on a map. If detailed topographic or geologic maps are not available, a standard road map can be used and coupled with a detailed description in the field note book of the location and appearance of the exposure. If available, a satellite global positioning (GPS) receiver can be used to obtain, directly in the field, the geographical coordinates of the locality. A sketch of the outcrop, with highlighting of reference items such as a tree or a prominent layer or a road sign, as well as a scale, is something always useful to do. Even better would be to take an instantaneous (i.e., Polaroid) picture of the outcrop to be pasted in the notebook. High-quality color pictures, digital or movie camera pictures, should be taken before, during, and after having finished the work on the outcrop.

The next step would be to measure the section, and to gather information about the lithostratigraphy. As it often occurs in well-studied areas, such as the U–M Apennines, lithostratigraphic columns representing known sections are already available in literature, and some exposed sections may be provided with permanent markers (metal plaques or spikes driven into the rock, or signs painted on the outcrop). These signs, along with paleomagnetic drill holes, should be highlighted on the outcrop sketch or instant photo, and eventually in the log sheet. In those cases where permanent meter markers are present, a lithologic log should maintain the existing meter system. The best way of doing this is to use pre-made log sheets, in which lithologic information can be taken layer by layer. The information should be taken respecting a stratigraphic order on the log sheet, i.e., older layers below and younger above.

If the section is not provided with permanent meter markers, then it is better to measure it before making a lithologic log. Measuring a section layer by layer while doing the log may introduce a cumulative thickness error. The lowermost exposed layer is usually taken as marker for the base of the section. However, it is always convenient not to call it "meter level 0" but rather "meter level 100.00". In this way, if later on an extension of the section is discovered stratigraphically below the one that is being measured, it may be numbered in continuity and with positive numbers.

Before starting the acquisition of observations in the log sheet, the outcrop should be cleaned using a broom or a sturdy brush, and then marked every meter using a Jacob staff or a meter stick, trying the best to measure the actual stratigraphic thickness, that is the distance normal to bedding planes. In between meter marks, 10 cm subdivisions can be drawn on the outcrop with a permanent ink marker or a pencil. Of course, the distance between two meter markers on the surface of the outcrop is always greater than the real stratigraphic thickness, except in the unique case where the surface of the outcrop is perpendicular to the bedding plane, as well as the line between the two consecutive meter markers. So, in order to have accurate marking every 10 cm it is sufficient to divide by ten the actual distance on the outcrop face between two meter markers.

The essential lithologic characteristics to be reported on the log sheet are the rock type (limestone, marl, clay, chert, sandstone, etc.), the rock color, the type of contact between consecutive layers (i.e., gradational vs. sharp), the presence and relative abundance of macrofossils (including trace fossils), and general notes regarding the state of preservation of the rock (weathering, tectonization), and the presence of sedimentary structures (soft-sediment slumping, lamination). Faults should be reported specifying whether or not the offset can be measured. The log sheet should also report the date, the name of the locality, and the name(s) of the operator(s). The section can be designated with an abbreviated code representing the name of the locality, and the year. For instance, a section logged in the Contessa valley in 1999 can be abbreviated with a code like CON/99. This code, followed by a meter level, will unequivocally designate a sample, which can be recognized immediately at any time in the future. In short, a sample coded CON/99–113.52 indicates that it was collected in 1999 at Contessa at 13.52 m above the base of the measured section.

Once the log is completed, sampling can proceed quickly and accurately. For relatively hard rocks such as limestone, hammer and chisel are the needed tools. Gloves and goggles are recommended for protection. Parts of the rock samples showing alteration or contamination should be chiselled out, and only the inner, fresh part collected. Sample size and spacing depend on which kind of analysis will have to be performed, and on the stratigraphic resolution one wants to obtain.

Packing is best done in sturdy plastic bags. Zip lock-type bags, such as the ones used for preserving frozen foods, are the most popular and serve the purpose perfectly. The sample code should be written with a permanent ink marker both on the hand rock sample and on the plastic bag label. If the sample is larger than a few centimeters, a tick with the exact meter level, and an arrow indicating the stratigraphic up direction, should be written directly on it.

For systematic and efficient sampling of limestones destined for geochemical analyses, which require very small amounts of material (i.e., stable isotopes, trace element analyses, etc.), we have often and successfully used a power drill with a standard 8 or 10 mm masonry drill bit. The best tool we used so far is a cordless 18 volt hammer drill, but a standard electric drill with a generator and a long extension cord has been successfully used in easily accessible sections (i.e., road cuts, quarry faces). This sampling technique requires also a small plastic funnel, and brushes to clean funnel and drill bit after each sampling. By placing the funnel directly against the exposed limestone layer and holding a plastic bag or a vial under it with one hand, a hole can be drilled right above it into the limestone. The fine limestone powder produced by the hole drilling will fall into the funnel and directly into the sample bag. The sampling operation will take just a few seconds per sample. A 2 cm deep hole with a 10 mm drill bit would produce about 4 grams of powdered sample. Beside the rapidity of the sampling, this simple technique produces already powdered, clean samples that will cut down tedious and long hours of preparation in the lab usually involving crushing and grounding whole rock samples with mortar and pestle.

More care has to be taken in collecting soft rock samples, such as marl or clay. These rocks tend to be easily contaminated either by surficial outcrop contaminants (soil, plant roots, particles from industrial or engine pollution, debris from overlying layers; see Montanari 1986), or by the actual sampling procedure. The soft layer should be excavated some 5 to 10 cm before being collected, and only the larger, clean pieces taken. For thin clay layer sandwiched between hard limestones, like in the case of the K/T boundary, the best way is to remove the overlying limestone layer using hammer and chisel (or even a crowbar), and then collecting the clay with a spatula. Whole slabs of the clay layer can be preserved by wrapping them in aluminium foil, and carefully packing them into a plastic bag, which help preserve moisture and prevent crumbling. The clay layer preserved in this way can be carefully unwrapped in the lab and taken apart with tweezers for high-resolution (millimetric) study and sampling.

Plastic tools are recommended in order to prevent any possible contamination of soft samples, especially for PGEs, although standard tool steel (hammer, chisel, spatula, tweezers, masonry drill bits) has been proven safe from contamination. On the other hand, jewellery such as gold or platinum rings may be a source for PGEs contamination (see Alvarez 1997, p. 67, note 4), and they should be removed while handling the samples in the field and in the lab as well.

6.2
Sample Preparation

6.2.1
Bulk Samples

We only consider the necessary steps for preparing samples for standard techniques and those that involve platinum group elements. As a general rule, the less sample preparation is required, the better. Each preparation and treatment step increases the chance of contamination or loss. In geochemical analyses one has to always enter a compromise between available sample mass and what constitutes a representative sample. In the study of impactoclastic layers, this problem is even more severe, because of the low abundance of the impact-derived debris within a large amount of local matrix. It is necessary to consider if the goal of the analyses is to search for an extraterrestrial component in the first place, or to study in detail a known impactoclastic layer. In the first case, a large number of samples need to be scanned for impact-characteristic signatures, whereas the second case usually requires breaking down a large sample from a known location into its components. If proximal ejecta, or crater rocks, are studied, care has to be taken to obtain a representative set of target rock and impactite lithologies for chemical analyses and search for shock effects. In some cases, the characteristics of rocks from a particular crater structure are to be compared with the corresponding characteristics of distal ejecta to determine if there is a connection between the two.

Taking all these considerations into account leads to a variety of sample preparation requirements, but they all have some points in common. The most important requirement is to avoid contamination. The search for siderophile element anomalies commonly involves very low elemental abundances, and cross-contamination from other samples, from crushers and mills, and during chemical treatment, can introduce severe problems that may not occur during standard geochemical work. In general it is advantageous to avoid using steel jaw crushers and any mills involving metal parts (e.g., those that use tungsten carbide or similar components and alloys, which are very common in standard swing mills). Coarse crushing can be done with the samples wrapped in thick sheets of plastic foil, followed by crushing in jaw crushers with ceramic jaws (e.g., alumina). Powdering of the samples to the required particle size for mineral separation (sieving) or bulk analysis is best done in alumina ceramic, agate, or boron carbide mills (of various designs). Boron carbide is brittle and very expensive, but it is useful for very hard materials, and our own experience has shown that it seems to be the mill material with the least amount of abrasion, which, therefore, leads to the least amount of sample contamination. Nevertheless, it is always a good idea to run blanks for each crusher and mill (using, e.g., pure quartzite), and to insert a cleaning step in between each sample (for example, by grinding clean, commercially available, sea sand). No equipment (for example, in central sample processing facilities) that has been used in the preparation of mineralized samples should be employed for the crushing and grinding of impact-derived material, as it is very easy to introduce contamination at the $\leq 10^{-9}$ g/g (sub-ppb) levels of interest here.

6.2.2
Isolation of Grains and Inclusions

Sand-size grains of shocked quartz and impact spherules (microtektites and microkrystites) are the distal impactoclastic components recognizable with a low-magnification optical microscope. Following the experience gained from working on the K/T boundary and the Late Eocene impactoclastic horizons in the U–M Apennines, we give herein a brief account of simple laboratory techniques that can be used to concentrate and isolate shocked quartz grains and microspherules from the types of sedimentary rocks encountered in this region (i.e., limestone, marl, and clay). We note that some of the procedures mentioned below involve dangerous acids and substances, and no responsibility can be taken for improper use of these substances. Following these procedures is entirely at the risk of the reader.

Whereas spherules can sometimes be seen in a hand sample with a hand lens, shock quartz and other impact-metamorphosed mineral grains are usually very scarce and of small size in distal ejecta, and they require concentration from relatively large bulk rock samples (i.e., 100 to 1000 g) in order to be seen and studied with a petrographic microscope. If the sample to be investigated is a limestone or an indurated marl (like the Late Eocene impactoclastic layer at Massignano), it has to be crushed first with mortar and pestle down to a grain size of few millimeters to 1 cm. A solution of concentrated hydrochloric acid can be used to dissolve the carbonate into a large plastic bucket. Half a liter of cheap, industrial HCl (the kind that is available in hardware or drug stores for domestic purposes) is more than sufficient to digest entirely a 100 g carbonate sample. The reaction of HCl with the carbonate is fairly violent and it is best done outdoors or in a fume hood.

Safety glasses should be used. Frequent stirring helps the reaction and the carbonate fraction of the sample can be dissolved in less than an hour.

The muddy insoluble residue left in the bucket is first rinsed with abundant tap water three or four times, decanting the clay and silt that remain in suspension, and allowing sand grains (i.e., >63 μm) do settle down. The rest of the residue has to be wet sieved under running water in order to eliminate all the remaining silt and clay. This is best done with a polyester cloth with 63 μm openings. The cloth is placed into a plastic funnel into which the wet residue is pored. The cloth is then folded and twist wrapped around the residue like a candy bar and, holding it tight with one hand, it is brush-washed under running tap water using the other hand (the same way one would wash a pair of dirty socks) until all the silt and clay is rinsed out of the mesh. The >63 μm washed residue left in the cloth is then rinsed out into a ceramic bowl or a sturdy plastic can, and dried at low temperature (i.e., less than 100 °C) under a heat lamp or in an oven. Rinsing the wet residue with alcohol or acetone speeds up the drying.

At this point, further concentration of quartz would require a Franz isomagnetic separator. If this machine is not available, then quartz grains will have to be searched among many other grains of silicate minerals, fragments of arenaceous benthic foraminifera, and pieces of radiolarian skeletons or other HCl-insoluble particles, such as fragments of impact spherules. Knowing that in distal impacto-clastic layers, such as those of the K/T boundary and Late Eocene in the U–M Apennines, shocked quartz grains have an average size of around 100 μm, the residue can be dry-sieved with stacked 80 and 150 μm sieves. Quartz grains are recognizable under a reflecting binocular microscope, but they may be confused with clear sanidine crystals or other clear feldspars. They can be hand picked individually using a moist, very fine paint brush, and placed in a cardboard tray of the kind used for collecting microfossils. Later on, the grains are picked individually with the pin of a U-stage, and observed under a petrographic microscope to assess whether or not they contain planar features (see Sect. 6.3.5).

A quicker way to check a sample for possible shocked quartz is to smear a small amount of sieved residue in transparent immersion oil onto a glass slide (the kind used for thin sections). Cheap Vaseline oil or clove oil would serve the purpose perfectly. Clove oil has the advantage over Vaseline of giving immersed sanidine grains a pale yellow tinge, and to quartz a light blue tinge, thus making them immediately recognizable with plain, transmitted light. Uncovered oil smear slides allow for moving the grains with a fine pin-point, and even to separate and recover them for further spindle- or U-stage study. Permanent and covered smear slides can be made using UV cement, the kind that hardens in a few seconds under ultraviolet light and is commonly used in car body shops to repair small cracks or holes in windshields. UV cement is routinely used by micropaleontologists for preparing nannofossil smear slides.

The insoluble residue can be prepared also for a standard thin section using a simple grain mount technique. The residue is incorporated into fluid epoxy (the kind used for standard thin sections) and smeared over a support made of a cut block of marble or micritic limestone. Ready-to-use two component epoxy in small quantities is today available at low cost in any hardware store, and serves perfectly our purpose. The block with the smeared epoxy residue mixture is then placed under a low temperature heat lamp for a quick curing. When hardened, the top surface of the grain mount is polished over spinning metal laps using water and carborundum grits in decreasing grades (from 320 to a polish finish of 1000 grade), until cut sections of the mounted grains are exposed on the surface of the polished grain mount. Alternatively, grinding and polishing can be performed with a paste of grit and water over thick glass plates, or even, without the grit paste, over wet polishing sand paper of the kind used in car body shops. Finally, the polished mount surface is glued over a thin section glass slide using UV cement or epoxy, sliced with a diamond saw, and then ground down to a standard 30 μm thickness petrographic thin section.

If the purpose is to separate particles other than resistant quartz grains, such as altered microkrystites or delicate microfossils, the wet sieving procedure described above has to be performed very gently, especially when washing the residue in the polyester cloth under running water. A long immersion of ground marlstone in hydrogen peroxide would produce a certain amount of loose mud without damaging carbonate and other more or less delicate grains. In this case, also calcareous microfossils and fish teeth will be preserved, and can be concentrated after gentle washing with the polyester cloth or a standard sieve.

Marls and marly limestones (up to 75 wt.% $CaCO_3$) can be disaggregated using the following technique. A sample made of small chunks of rock (the size of walnuts) is heated to a temperature of about 300–400 °C (a cast iron frying pan over a kitchen stove or a Bunsen burner would do just fine). The hot samples are then thrown into a bucket filled with kerosene (i.e., diesel fuel) using a pair of long tongs (safety glasses should be worn). The hot sample will first sizzle for a second and then it will cool down quickly (best is to do this operation outdoors and in an open space). Once the sample is soaked with kerosene, it is transferred into a bucket of hot water, and in a short time it will disaggregate to a loose mud. Calcareous and silicate particles can then be wet sieved or washed using a polyester cloth as described above.

For clay-rich samples, such as the K/T boundary clay, grinding with mortar and pestle, digestion in HCl, or soaking in kerosene is not necessary. The clay can be disaggregated quickly and gently in a beaker of water or even better in a 30 vol.% solution of hydrogen peroxide. If laboratory grade H_2O_2 is not available, standard 10–15 vol.% hydrogen peroxide of the kind sold in a pharmacy for medical use is good enough to help break down the clay sample. In the case of the K/T clay, the wet-sieving procedure using a polyester cloth has to be performed very gently in order to prevent fragile particles, such as altered impact spherules made of glaucony, to break down. Cleansing in an ultrasonic bath is not advised because glaucony spheroids, as well as some arenaceous benthic foraminifera, would disintegrate. Spherules and microfossils can be hand-picked using a moist, very fine paint brush under a reflected-light binocular microscope as described above. An aliquot of the bulk washed residue can be digested in HCl, which will produce an insoluble residue containing impact spherules, silicate grains (including shocked quartz), and arenaceous benthic foraminifera. The various particles of these residues can then be treated as described above for making grain-mount thin sections or smear slides, or prepared for SEM or TEM study.

The spinel-bearing and smectite spherules in the Late Eocene impactoclastic layer are very fragile and extremely flattened, and cannot be separated from the enclosing hardened calcareous marl using the wet-sieving, polyester cloth washing, H_2O_2 disaggregation, or HCl-dissolution techniques described above; they would disintegrate inevitably. The best way to separate and isolate them is to pluck them with a needle or the edge of a razor blade (or a cutter) directly from a raw hand sample under a reflected-light binocular microscope. For making thin petrographic sections of these spherules, a small, dry slice of the soft marly limestone should first be impregnated with fluid epoxy in a vacuum, and then polished. In absence of a impregnation vacuum set-up, a thin coat of fluid epoxy, or even an impregnation with fluid used for masonry coating on the surface of the sample exhibiting such flattened spherules may be sufficient to keep them together while gently polishing them on a glass slab smeared with fine grit paste, or on a sand paper sheet of the kind used for metal polishing. If the sample to be cut or polished contains water soluble components, it is better to use oil for lubrication rather than water.

6.3
Laboratory Techniques

It is beyond the scope of this book to discuss all techniques that have been, or can be, used in the recognition and study of impact-derived rocks and distal ejecta. However, we felt that it was appropriate to give at least brief summaries of some of the more important methods for the determination of a meteoritic component and the identification of shocked quartz, as these determinations are of crucial importance to confirm an impact origin of the investigated rocks.

6.3.1
Standard Bulk and Microprobe Analysis Methods

After the preparation of bulk samples for analysis, almost any method for major, trace, and isotopic analysis available to modern geochemistry can be used. This type of analysis will be most often used on rocks taken directly from a (suspected) impact structure, such as suevites, impact melt rocks, and target rocks (for comparison). The first step, which is unrelated to actual analytical methods, though, is to study any rock that should be chemically analyzed by standard petrographic methods, i.e., by doing a classical thin section description with the petrographic microscope in transmitted light. This study is necessary to define the rock type(s) and to determine if, and how much, melt clasts (or matrix) there might be present, and/or to study the clast population in a target rock or breccia, which is necessary for interpreting chemical data (for an example, see French and Nielsen 1990).

Bulk analyses are usually of little importance for the study of the components in impactoclastic layers, as in these cases we are only looking for a few grains (of, e.g., quartz or spinel) in a foreign matrix, but they can also be important for the search of siderophile elements (e.g., Ir) in a stratigraphic profile.

Once the rocks to be studied have been prepared for analysis (usually as powders; see Sect. 6.2.1), a variety of methods for major and trace element analysis can be employed. The most common ones are X–ray fluorescence (XRF) spectrometry (with sample powder pressed as pills, or with glass disks made by fusion of sample powder and a flux material), inductively coupled plasma atomic emission spectrometry (ICP–AES), inductively coupled plasma mass spectrometry (ICP–MS), and instrumental neutron activation analysis (INAA). All of these methods allow the simultaneous (rarely sequential) determination of a large number of major and trace elements. Requirements for sample mass vary somewhat, depending on the method (XRF tends to need several grams of material, whereas a few tens to hundreds of mg are sufficient for the latter three methods). An introduction to the basics of these techniques can be found in, e.g., Gill (1997), and more details on the methodology are provided in the excellent textbook of Potts (1995).

Neutron activation analysis has the advantage that sample preparation is minimal: sample chips or fragments (or whole grains or spherules, such as microtektites) can be used as they are, or powdered samples are analyzed without any pressing, melting, or dissolution (each of which could introduce contamination). Small sample sizes can cause problems if samples are not homogeneous (which they rarely are). Analysis of single grains or fragments are representative of this one particular component of a mixture (which may be desired), but not of the mixture as a whole. Samples analyzed by INAA can, after completion of the counting procedures, be mounted for other analyses, such as electron microscopy, electron microprobe, or ion microprobe. Details on the use of INAA for small sample analysis are given by Koeberl (1993c).

Electron microbeam techniques (e.g., scanning electron microscopy [SEM] in combination with energy dispersive X–ray [EDX] analysis, or electron probe microanalysis [EPMA]) are used to determine the mineral composition of a mixed sample, usually from a petrographic thin section that has previously been studied by optical microscopy), or the properties of single grains or spherules. Regarding the study of single mineral grains and spherules of possible impact origin, it cannot be emphasized enough that simple (SEM) photographs, or even semiquantitative EDX spectra of such samples, have no diagnostic meaning whatsoever. Unfortunately the literature (especially the unrefereed literature, such as abstracts) is full with claims of "new types" of spherules or grains, all of which have supposedly formed by some impact process or are derived from some (always "new" or "usual") extraterrestrial source, with all these claims based on the interpretation of photographs and/or crude surface analyses.

Detailed studies of such claims have revealed a host of "down-to-earth" explanations (e.g., Montanari 1986; Byerly et al. 1990), ranging from industrial contamination, natural concretions, opaque minerals, remnants of abrasion or erosion, or even insect eggs. Most of these mis-identifications could have been avoided if proper analytical procedures had taken precedent over enthusiasm or wishful thinking. After surface photography (where desired), all such samples have to be mounted, sectioned, and polished for mineral phase identification and quantitative analyses. Surface analyses may be useless and meaningless. Once sectioned, mineral grains and spherules reveal their internal structure (mineral assemblages, zonation, etc.) and can be analyzed by EDX or EPMA methods (note that in some cases the resolution of EDX systems is not good enough, and before very unusual compositions are reported it is prudent to check them with more sophisticated equipment, such as wavelength-dispersive EPMA). Back-scattered electron (BSE) images, taken with an SEM instrument, are also very useful for such investigations. The misidentification of meteor ablation spherules, which are common on Earth in a variety of sedimentary environments, as impact-derived spherules can easily be avoided by detailed analyses (BSE images of sectioned samples and EDX analyses), as most ablation spherules have a characteristic composition (often dominated by olivine and magnetite, but they may also be glassy or metallic in composition; see, e.g., Blanchard et al. 1980; Brownlee 1981).

With any analytical method, two points are of specific importance: a) the method used in a study must be properly documented and referenced, and b) international standard reference materials must be used for quality control. Details that should be reported for any analytical procedure used are given in Table 6.3.1.1.

Table 6.3.1.1. Information required for proper documentation of analytical techniques.

- Sample preparation procedure
     (give details of the types of mills)
- Original sample quantity
- Sample mass used in analysis
- Instrumentation (equipment used)
- Analytical conditions
- (for example, in EPMA: acceleration voltage, sample current, beam diameter, counting time)
- Data reduction methods
- Type of standards
- Precision (reproducibility) of method
- Accuracy of method
- Detection limit

When reporting analytical data (e.g., in a table or a figure), some items require attention. First, units have to be indicated for all results (taking into account that "percent" alone can be ambiguous, as there are wt.%, atom.%, vol.%, etc.). Second, it is important to print only a realistic number of significant digits. To give all data to two decimal places (simply because of the convenience of spreadsheet output) is annoying and betrays a precision of the data that is not supported by the analytical method. For example, to report a chromium content of, say,

751.58 ppm, when the method has a precision of 3 rel.% (i.e., about 23 ppm!), is misleading and clutters the data table with useless numbers. To report EPMA data of, for example, 0.0212 wt.% MnO, when the detection limit of the method is on the order of 0.01 to 0.02 wt.%, is equally unrealistic. A final note refers to the unsettling practice of reporting "zero" values in data tables (which is another artefact of the uncritical use of spreadsheets or direct instrument output). Zero values do not exist in nature and cannot be measured; there is always one atom or another of a certain element in a finite sample mass. Thus, such values should be reported either as "below detection limit" (in such cases it would be desirable that the actual detection limit is reported in the table, as a "less or equal than" value) or a note should be made that this element was "not analyzed" in this particular sample. All these comments refer to good geochemical analytical practice, but should especially be observed when dealing with the low abundances (of, e.g., the siderophile elements) in impact-derived rocks.

6.3.2
Iridium Determinations

The determination of the PGE iridium is the most commonly used scanning technique in the search for chemical impact markers. This element gained notoriety especially in the early days of the K/T boundary dispute (see Sects. 1.1, 1.8, and 3.2). Although this point has been made before, it should be emphasized that the reason that often only the Ir content is measured is because this element can be determined with greater sensitivity and more ease than any of the other PGEs. Thus, Ir acts as a marker for the other PGEs, which, if an Ir anomaly is found, may then also be determined in a limited subset of samples (given the much greater analytical efforts for complete PGE analyses).

There is a variety of techniques for measuring Ir contents, with the most commonly used ones based on neutron activation. The reason is the very high neutron capture cross section of the isotope ^{191}Ir, resulting in the radioactive ^{192}Ir isotope, which decays (half life: 74.2 days) by combined ß-γ decay, with several gammalines in the region of the gamma-spectrum in which detectors are most sensitive (e.g., major lines at 296, 308, 316, and 468 keV). Detection limits of routine INAA procedures vary according to the neutron flux during irradiation, duration of irradiation, counting duration, detector sensitivity, and sample composition (the latter influencing the background of the spectrum), but are commonly on the order of 0.5 to 2 ppb Ir (see, e.g., Koeberl 1993a, c). Radiochemical methods (e.g., Palme et al. 1978, 1979; French et al. 1989) can be used to measure lower abundances, but are tedious. Detection limits can be significantly improved by using γ-γ coincidence spectrometry, as described by, e.g., Meyer (1987), Rocchia et al. (1990), and Koeberl and Huber (2000). The technique makes use of the coincident decay of ^{192}Ir by counting a radioactive sample simultaneously with two detectors and plotting the resulting three-dimensional coincidence spectrum (Fig. 6.3.2.1).

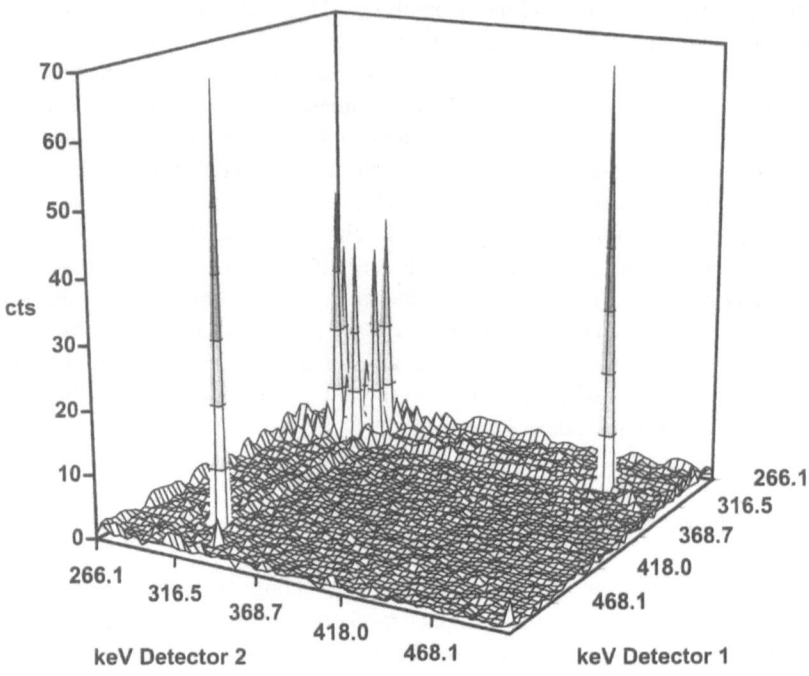

Fig. 6.3.2.1. Part of a gamma-gamma multiparameter coincidence spectrum, showing the spectral region from about 260 to 500 keV, with the coincidence peaks at 316 x 468 and 468 x 316 keV.

In such a spectrum, the coincidence space is represented by a plane and only signals that are recorded within a short coincidence window (usually a few hundred nanoseconds) appear as peaks that grow out of the plane. In the case of ^{192}Ir, the two most intense lines are at 316 and 468 keV, and each of the photons emitted (at 180° of each other) in the coincident decay can cause a coincidence signal in both detectors, resulting in prominent peaks at 316 (detector 1) x 468 (detector 2) and 468 (detector 1) x 316 (detector 2) keV. The volume of the peaks is a measure of the intensity. This method drastically reduces the background that limits detection in a single spectrum. Alvarez et al. (1980) employed a similar technique, but with additional background reduction using a large guard detector, resulting in relatively short measuring times.

The coincidence method allows the analysis of a fair number of samples with detection limits that are on the order of 5 to 30 ppt (10^{-12} g/g or pg/g) Ir, again depending on the irradiation, instrumentation, counting, and sample parameters (e.g., Rocchia et al. 1990; Huber et al. 1999, 2000; Koeberl and Huber 2000). Recently, ICP–MS techniques have been used for Ir analyses, but require larger samples, chemical sample preparation (dissolution, separation), and reach the detection limits obtained by coincidence counting only with great analytical effort.

Isotope dilution techniques (with the measurements made by thermal ionization mass spectrometry or ICP–MS) are very precise and yield very low detection limits, but the great experimental effort makes them impractical to use for reconnaissance purposes. Also, after completion of the counting for INAA the samples can be re-used for other analyses.

6.3.3
Platinum Group Element Determinations

Most methods for PGE analysis used in geochemistry have been developed for the study of, or reconnaissance for, commercially important ore deposits; more recently, some methods were developed for the study of mantle rocks. The abundance levels of interest in impactite studies are several orders of magnitude lower than those in ore deposit studies. Chondritic meteorites have average abundances of (for example) about 700 ppb Ir, 1800 ppb Pt, 700 ppb Pd, and 900 ppb Ru (see Anders and Grevesse [1989] for cosmic abundance values). Typical mantle xenoliths have abundances on the order of 2 ppb Ir, 5 ppb Pt, 2 ppb Pd, and 4 ppb Ru. The upper continental crust, into which impacts of meteoritic projectiles occur, has abundances that are about a factor of 100 below those of mantle rocks. As discussed in Chap. 1.8, impactites contain very small meteoritic contributions (usually"well below 1 wt.%). Assuming (for example) that 0.2% by weight of a chondritic projectile (the most common meteorite type) is mixed in with upper crustal rocks yields (for Ir) $700 \times 0.002 = 1.4$ ppb in the resulting impactite (and we note, as mentioned in the previous section, that Ir is the easiest of the PGEs to determine). Abundances in impactoclastic layers are commonly much lower (see discussion in Chap. 5). Quantitative PGE determinations at sub-ppb abundance levels are analytically very challenging.

Earlier studies (e.g., Morgan et al. 1975, 1979; Morgan 1978; Palme et al. 1978, 1979, 1981; Wolf 1980) used radiochemical methods, where the PGEs were separated from the sample matrix, and in some cases from each other, after neutron irradiation (i.e., radiochemical neutron activation analysis, RNAA). This procedure involved (for example) dissolution of the radioactive sample by alkali fusion or acid digestion, followed by ion exchange, liquid extraction, and/or precipitation steps that may be repeated several times for better separation efficiency. Counting and data reduction were then done by standard INAA techniques. Such procedures are not much used anymore due to the reluctance of handling radioactive samples in liquid form, and because of the radioactive decay of short-lived isotopes during the time-consuming separation steps, which lead to a loss of signal.

More recently, the development of reliable ICP–MS instruments, both with quadrupole and high-resolution MS, has provided the geochemist with another tool for low level PGE work. The sample digestion and separation procedures are still similar to those used before in RNAA work. One of the most common methods involves a Ni sulfide fire assay with Te co-precipitation, followed by ICP–MS measurements. A typical recipe can be summarized as follows (Koeberl et al.

2000a, modified after Jackson et al. 1990): thoroughly mix 6 g of Na-carbonate, 12 g borax, 0.7 g sulfur, 2.10 g Ni carbonate, and 3.5 g of silica with 10 g of sample powder; transfer into a fire clay (or porcelain) crucible; fire for 90 minutes at 1060°C; remove, let cool, and break open the crucibles; separate the Ni sulphide buttons, break, and transfer fragments to a 500 ml beaker containing 120 ml of concentrated HCl; dissolve on a hotplate; co-precipitate PGEs from solution with Te using $SnCl_2$ as a reductant; filter insoluble noble metal-bearing residue through a 0.45 µm nitrocellulose filter paper and wash; place filters inside a 100 ml screw-top Teflon vial; add 4.0 ml of concentrated HCl and 3 ml of concentrated HNO_3; seal and place overnight at 40°C in a waterbath; cool (ice), open, and dilute with water to 100 g; spike solutions with internal standard monitors for instrumental drift; measure on ICP–MS instrument. Detection limits obtained with this method by Koeberl et al. (2000a) are (in ppb) 0.02 Ir, 0.05 Ru, 0.02 Rh, 0.14 Pt, 0.10 Pd, and 0.02 Au. A somewhat similar procedure was recently developed by Hassler et al. (1999), which allows not only the determination of the PGEs, but also of the Os isotopic ratio in the same sample (see next section). Their method uses PGE isotope spikes, which are introduced at the dissolution step, providing the advantages of isotope dilution methods (e.g., elimination of yield determination).

Several problems with all these methods result from sample size and detection limits. First, due their low abundances the PGEs are often inhomogeneously distributed in rocks, which, together with sensitivity limits of the measurements, require a large enough sample for a representative analysis. This causes problems with complete dissolution of such large samples, and, because of the larger amount of chemicals required in the following steps, leads directly to the second problem: potentially interfering blank levels of the PGEs in the reagents used. For example, great care has to be employed to obtain NiS that has low enough levels of the PGEs so that blank values do not interfere with sample values. Procedural blank values of standard NiS/Te-coprecipitation methods can reach 0.1 ppb. Reduction in sample weight to 5 g and digestion of the sample powder in a modified Carius tube (Shirey and Walker 1995; see also next section) led to improved procedural blanks of (for example) <3 pg/g for Ir and <6 pg/g for Ru and Pd (Rehkämper et al. 1998). However, all such determinations are at the current limits of analytical geochemistry and require a dedicated laboratory and great analytical experience. No PGE measurements at the sub-ppb level are routine.

6.3.4
Osmium Isotope Measurements

The basics of the method are discussed in Sect. 1.8.3. In principle, it is necessary to determine the ratio of the radiogenic Os isotope, ^{187}Os, compared to a non-radiogenic Os isotope. Osmium has seven naturally occurring isotopes, which are all stable. The isotopes and their abundances (in parentheses; given in rel.%) are: ^{184}Os (0.024), ^{186}Os (1.600), ^{187}Os (1.510), ^{188}Os (13.286), ^{189}Os (16.252), ^{190}Os (26.369), and ^{192}Os (40.958) (Faure 1986).

The amount of ^{187}Os increases with time as a result of the decay of ^{187}Re. This decay can be described by normalizing to an Os isotope not affected by radioactive decay:

$$^{187}\text{Os}/^{188}\text{Os} = (^{187}\text{Os}/^{188}\text{Os})_i + (^{187}\text{Re}/^{188}\text{Os})(e^{\lambda t} - 1)$$

where ^{187}Os/^{188}Os and ^{187}Re/^{188}Os are the measured ratios of these isotopes, $(^{187}\text{Os}/^{188}\text{Os})_i$ is the initial isotopic ratio at the time when the system became closed for Re and Os, λ is the decay constant for ^{187}Re ($\ln 2/t_{1/2}$, with $t_{1/2}$ being the half-life), and t is the time elapsed since system closure for Re and Os. Normalization to ^{188}Os is preferred over ^{186}Os (which was used earlier, e.g., Allègre and Luck 1980; Luck and Turekian 1983), because ^{186}Os also has a (rare) radiogenic source, and because ^{188}Os has higher abundances and is the isotope actually measured by most workers. With this equation it is possible to calculate an "age" if the initial isotopic ratio of Os is known. Alternatively, it is possible to calculate the initial Os isotopic ratio from a suite of samples with the same age, in analogy to the Rb–Sr method (Faure 1986; Shirey 1991). The use of this isotope system to determine an isochron requires a very high precision of the analytical measurements. In the 1980s, the available techniques did not allow such precise (or sensitive) measurements.

The situation changed in the early 1990s with the development of the negative thermal ionization mass spectrometry (NTIMS) technique for Os isotope measurements (Völkening et al. 1991; Creaser et al. 1991). This method allows the determination of abundances and isotopic ratios of Os and Re at the low abundance levels found in most terrestrial rocks. However, two crucial points are the complete dissolution of the samples and the equilibration of the sample with the isotope spike. The addition of a spike of known isotopic composition is necessary to determine the concentration of Os (and Re) by isotope dilution (Faure 1986; Potts 1995).

There is a variety of methods for the determination of the Os content and isotopic composition, depending on the goals of the analyses. For example, it is possible to determine only the ^{187}Os/^{188}Os ratio, or the isotope ratio plus the Os content, or both plus the Re content. There are only two naturally occurring Re isotopes, ^{187}Re and ^{185}Re, and ^{187}Re is used (normalized to ^{188}Os) in a standard Os isotope diagram (see Fig. 1.8.3.1). It should be noted, though, that these diagrams were designed for isochron work, where samples of the same age fall on a straight line. In impact studies, this diagram is used in a different way. Here, different components (e.g., several different target rocks and impact melt rocks or breccias) plot in widely dispersed locations in the Re–Os diagram. Target rocks commonly have high ^{187}Os/^{188}Os and ^{187}Re/^{188}Os ratios, whereas meteorites plot in a specific location in the lower left of the diagram, at very low values for these ratios. The target rock and meteorite values define the endmembers in a mixing relation, with impact breccias and melt rocks commonly plotting in between the endmember values (if all endmember compositions were analyzed). This type of application places different constraints on the analytical precision compared to mantle geochemistry (e.g., isochron work; cf. Shirey 1991). In the latter case, differences in the isotopic ratios by fractions of a percent can have important implications, whereas in impact studies differences more likely are in the order-of-magnitude range (e.g., meteorites have a $^{187}Os/^{188}Os$ ratio around 1, whereas ratios for target rocks can vary between 5 and 100; cf. Esser and Turekian 1993; Koeberl and Shirey 1997). It may not be necessary to determine the Re content, as the Os content and isotopic ratio alone yield the desired answer.

Most modern analyses involve a high temperature acid digestion step for dissolution of the rocks and isotope equilibration between the spike and the sample Re and Os, followed by separation of the Os in some sort of distillation, and isolation of Re by anion exchange chromatography. However, it has usually been difficult to obtain complete digestion of mineral constituents and isotope equilibration between spike and sample Re and Os. Only recently have these problems been solved by the application of the so-called Carius tube digestion method, which involves sealing sample powder with mixed concentrated acids into a thick walled quartz or borosilicate glass tube (Shirey and Walker 1995). It is clearly important that the samples do not come into contact with metal and are ground in a contamination-free mill (e.g., agate or alumina). The low abundances of Os in terrestrial samples require the use of fairly large samples, about 2–10 g of sample powder. Samples are spiked with enriched ^{190}Os and ^{185}Re (Shirey and Walker 1995). For the NTIMS measurements, Re and Os fractions are loaded (separately) on filaments (commonly Pt) that have to be cleaned to remove any Re background, and are measured as negative ReO_4^- and OsO_3^- ions, respectively. Measurements can also be done by ICP–MS (see previous section). After correction for oxygen isotopic composition and fractionation, and normalization of the Os isotopic ratios ($^{192}Os/^{188}Os = 3.0826$), abundances of Os and Re, and the $^{187}Os/^{188}Os$ ratio, can be calculated.

A different method was recently described by Hassler et al. (1999). These authors isolate the PGEs in more or less the same way as for "standard " ICP–MS analyses (see previous section), by adding a multi PGE isotopic spike and then using a Ni-sulfide fire assay preconcentration procedure. However, after dissolution of the metal beads in a Teflon container, volatile OsO_4 is purged (with Ar) directly in an ICP–MS instrument (without a nebulizer), allowing direct, rapid (measurement time is a few minutes), and very sensitive measurement of Os abundance and isotopic composition. The remaining solution can then be used for "normal" ICP–MS measurement of the abundances of the PGEs. This method has several advantages over the standard NTIMS Os isotope measurements by being much faster and very sensitive, and also allowing the determination of the PGE abundances in the same sample. A disadvantage is that the analytical precision is not as good as in NTIMS analyses, which is, however, of little concern in impact studies. The rapid technical progress in this field will likely allow even more sensitive and faster analyses in the near future.

6.3.5
Methods for Shocked Quartz Measurements

As described in Sect. 1.6.4, planar microdeformations in rock forming minerals are probably the features that are most characteristic of shock pressures related to an impact event and, thus, provide diagnostic evidence for an impact origin. The mineral most commonly studied for shock features is quartz (e.g., Stöffler and Langenhorst 1994; Martinez and Agrinier 1998), because it is abundant in many crustal rocks, is stable over long periods of geological time, is an optically simple (uniaxial) mineral, and displays a wide range of planar features whose development is correlated with shock pressure. Other rock forming minerals also show microdeformations, but either they are difficult to quantitatively measure, such as in feldspar (because it is a biaxial mineral), or the minerals are relatively rare in crustal rocks, as with olivine (see, e.g., Kieffer et al. 1976b; Bischoff and Stöffler 1992). It should be noted that, under some conditions, shocked minerals can weather faster than their unshocked equivalents (Boslough and Cygan 1988; Boslough 1991).

The characteristic appearance of planar microdeformations in quartz, such as PDFs (see Sect. 1.6.4), is evident under the optical and electron microscope and can reveal, to the experienced worker, confirming evidence of an impact origin. Care has to be taken, however, in the proper identification of PDFs, and possible confusion with tectonic deformation lamellae (Böhm lamellae) has to be excluded (see Sect. 1.6.4 for criteria). Basically, tectonic lamellae are relatively thick, widely spaced, and curved bands, whereas PDFs are narrow, closely and regularly spaced, straight features that extend though the whole quartz grain and often occur in more than one set. PDFs can be studied either in thin sections of rocks or in single grains isolated from a matrix.

Besides (qualitative) identification of these features in the petrographic microscope, it is possible to use the electron microscope to distinguish between tectonic features and PDFs. Gratz et al. (1996) described a method in which quartz grains are etched and then studied in the SEM. Etching procedures include the following options: a) rinsing of grains for a few minutes in a 30 vol.% HF solution (followed by neutralization of the samples, e.g., with Na_2CO_3, to avoid etching of microscope parts by HF vapor); b) HF vapor etching of a thin section or grain mount (e.g., on top of an open HF bottle); or c) hot alkaline solutions (e.g., 0.001–0.01 M KOH at about 200°C for 4 hours in a sealed Teflon container). As PDFs are filled with amorphous quartz, they are preferentially etched and appear as narrow and straight lines with so-called internal "pillaring", which is visible at high magnifications (Gratz et al. 1996). Tectonic deformation lamellae are either not etched or show broad arrays of curvy dislocation loops.

However, once PDFs are indicated by optical or SEM work, more definitive (quantitative) work should follow. Ideally, this would involve a combination of optical and transmission electron microscope (TEM) studies, but the tedious sample preparation techniques and expensive and specialized instrumentation for TEM work precludes the wider use of this method (for examples, see Müller 1969;

Kieffer et al. 1976a; Gratz et al. 1988, 1992a; Goltrant et al. 1991; Leroux et al. 1994; Cordier and Gratz 1995). Optical (and related) parameters for the investigating shock metamorphic effects in quartz include the following: a) decrease of refractive index, density, and birefringence with increasing shock pressure (measurements to determine refractive indices, using the spindle stage, are described by Bloss [1981] and Medenbach [1985]); b) mosaicism (either by optical measurements [e.g., Dachille et al. 1968] or X–ray diffraction [e.g., Hörz and Quaide 1971]); c) infrared and Raman spectroscopy (e.g., Velde and Boyer 1985); and d) NMR and related structural methods (e.g., Cygan et al. 1990; Boslough et al. 1995). For details on these parameters, see Chap. 1.6 and the excellent reviews by Stöffler and Langenhorst (1994) and Langenhorst and Deutsch (1998). Cathodoluminescence analyses may be of potential interest (e.g., Owen and Anders 1988; Seyedolali et al. 1997), but not enough data are presently available.

For reasons of space and simplicity, we discuss only one very important method here: the measurement of the crystallographic orientations of PDFs using a universal stage. The goal of these measurements is to measure the angle between the c-axis and the pole of the plane of the PDFs, as well as the azimuthal angle, in a quartz grain. A stereographic projection of the rational crystallographic planes in α-quartz (Fig. 6.3.5.1) illustrates this point. The c-axis is at the center of the projection, and the outer limit of the diagram marks a polar angle of 90°. The azimuthal angle is measured between the three a-axes (a1, a2, a3); for example, the $\{10\bar{1}3\}$ (or ω) plane has an azimuthal angle of 30°, whereas the $\{11\bar{2}1\}$ (or s) plane has an azimuthal angle of 60°.

The measurements can be done using a so-called universal stage (U-stage). These microscope attachments came into use in the early decades of the 20th century for petrofabric work and for compositional measurements on plagioclase. Two different types exist: one with four, and one with five axes. The four-axis model (e.g., Reinhard 1931) consists of an inner vertical axis I–V, an outer vertical axis O–V, and two horizontal axes [the inner north-south (N–S) and the outer east-west (O–E–W) axis], whose point of intersection lies in I–V. The order of the axes from the inside out is: I–V, N–S, O–V, and O–E–W (followed by the vertical microscope stage). All axes have graduated circles that allow the precise reading of the amount of rotation. The sample is placed on a glass plate at the center of the stage, with glass hemispheres of known refractive index (for quartz: 1.554) above and below the glass plate. The fifth axis in the five-axis stage (e.g., Emmons 1943) is useful for the measurement of biaxial minerals; it is not needed if uniaxial minerals (such as quartz) are measured. Fig. 6.3.5.2 shows a four-axis U-stage.

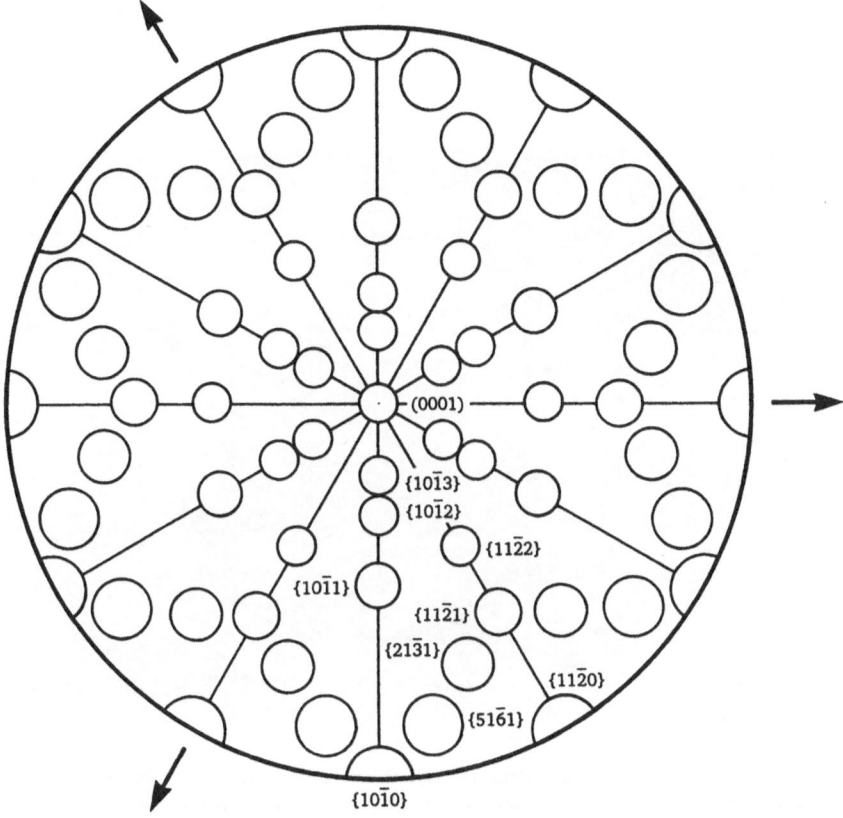

Fig. 6.3.5.1. Standard stereographic projection (lower hemisphere) of rational crystallographic planes in α-quartz, which is used to index crystallographic planes of PDFs based on universal-stage measurements. The arrows indicate the three a-axes of quartz, and the c-axis (the (0001) plane) is in the center of the projection. Also indicated are the low Miller indices in a part of the diagram (other indices can be derived from crystal symmetry). The circles are about 5° in diameter and indicate the accuracy of the U-stage measurements (see, e.g., Engelhardt and Bertsch 1969; diagram after Koeberl 1997a).

Fig. 6.3.5.2. Four-axis universal stage (Leitz), showing the glass hemispheres at the center (which contain the sample thin section), followed by the four axes; from inside out: inner vertical, inner horizontal (N–S), outer vertical, and outer horizontal (O–E–W). The latter is operated by the large knob on the right side of the U-stage.

A detailed lab manual of the operation of a universal stage is beyond the scope of this treatment. However, in short, the measurement of PDF orientations involves the following steps: a) set-up of U-stage (place special objectives into microscope and center; place thin section into stage and assemble stage; place U-stage on microscope stage and fix with screws; center U-stage; adjust height of plate); b) measure orientation of quartz c-axis (set all horizontal axes to zero; rotate grain on I–V to extinction; test orientation by rotating on N–S; rotate O–E–W by 15–30° – grain should leave extinction; rotate on N–S to extinction; rotate on O–E–W back to zero and note readings on I–V and N–S); and c) measure orientations of PDFs (reset axes to zero; rotate I–V until set of PDFs is parallel to crosshairs, then rotate N–S until planes appear as sharp and narrow as possible; note readings on axes). The measurements have then to be plotted using a standard stereonet (Wulff net), assuming the lower hemisphere of the stereoplot. In principle, the optical axis (c-axis) is rotated into the center of projection and the locations of the poles of PDFs are plotted; this is practically done either by a computer program, or manually with a transparent overlay on the stereonet and the angle between the c-axis point and the pole point is measured once the two are rotated so both are on a N–S great circle (meridian). This angle can then be compared to the rational crystallographic planes in quartz (as shown in Fig. 6.3.5.1).

The precision of such measurements has been estimated from practical work at about ±4° (B.M. French, personal communication, 1994). For practical plotting purposes, the "bin" size, with which the angles are plotted in a histogram, with the measured angle on the x-axis, should be about 4 to 6°. The measured angles that fall within 5° of the theoretical polar angle of the plane are considered valid and can be indexed. Unequivocal assignment of the specific Miller indices is impossible if only one plane per grain is observed, as some of the crystallographic planes occur (as a result of the symmetry class of α-quartz) in positive and negative or right and left forms with three symmetrically equivalent planes. A distinction is only possible if at least one plane of each form is present. If the unequivocal assignment of planes to Miller indices is not possible, general forms are identified as the low index planes and written in curly (or wavy) brackets. Only the basal orientation (0001) can be determined unequivocally (Carter 1965). In a histogram, only the polar angle information is plotted (on the x-axis). There is a variety of histogram types, depending on how the data are plotted on the x-axis and what is plotted on the y-axis. Different workers have used somewhat different diagrams in the past. Some workers plot all data, giving "number of observations" (i.e., number of measured planar microstructures) on the y-axis. Others present frequency in percent on the y-axis, again using all data (i.e., absolute frequency). Others again plot frequency in percent, using only the total number of indexed planes for normalization. Definitions are given by Engelhardt and Bertsch (1969) and Stöffler and Langenhorst (1996). In most cases, plots are presented as bar diagrams, with the width of the bars equivalent to the bin size mentioned above (compare Fig. 1.6.4.3).

More recently, Grieve and Therriault (1995) and Grieve et al. (1996) proposed to use fence diagrams, in which all observations within the error of the measurement are assigned to the corresponding crystallographic orientation and plotted as a single line with a height corresponding to the relative frequency (i.e., without plotting unindexed planes or the variation in measurement). Fig. 6.3.5.3 shows the results of this procedure. It is obvious from the literature that different workers have different plotting preferences, but no matter which method is used, the details on how a specific diagram was constructed (i.e., how indexing was done and what type of frequency is plotted) must be given. Two or more PDF planes per grain allow, in principle, specific indexing. Results from this procedure can be shown in single grain, multiple plane stereoplots, which are useful only if three or more sets of PDFs are found per grain.

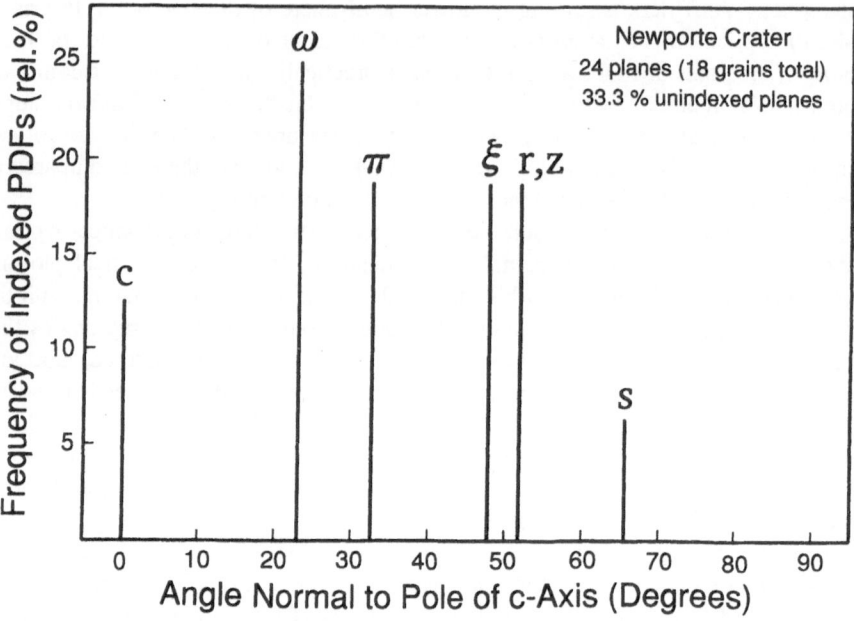

Fig. 6.3.5.3. Crystallographic orientation of PDFs in quartz from the Newporte (North Dakota) impact structure, shown as a histogram giving the frequency of indexed PDFs versus angle between c-axis and poles of PDFs, without plotting unindexed planes (see Grieve and Therriault 1995 and Grieve et al. 1996); the shock-characteristic orientations (0001), {1013}, {1012}, {1122}, {1011}, {0111}, and {1121} (c, ω, π, ξ, r, z, and s, respectively), are dominating (diagram after Koeberl and Reimold 1995a).

In summary, the study of shocked quartz is not as simple and straightforward as showing a photograph of a quartz grain with some kind of (sub?)planar features. If optical microscopy shows the clear and unambiguous presence of multiple sets of straight, closely spaced planar features, there is a good chance that these represent PDFs. SEM studies of etched samples might amplify such a conclusion. Confirmation, and shock barometry studies, however, require the determination of crystallographic orientations by using U-stage and optical microscopy work, or by TEM investigations, which also allows verification that the planes are filled with an amorphous phase. The study of PDFs in other minerals follows equivalent procedures, but can be more complicated depending on the crystal structure of the minerals, and be impeded by the fact that much less is known about the development and shock pressure dependence of microstructures in these minerals compared to quartz.

References

Abate B, Koeberl C, Kruger FJ, Underwood JR (1999) BP and Oasis impact structures, Libya, and their relation to Libyan Desert Glass. In: Dressler BO, Sharpton VL (eds) Large Meteorite Impacts and Planetary Evolution II. Geological Society of America, Special Paper 339, pp 177–192

Abbate E, Sagri M (1982) Le unità torbiditiche cretacee dell'Appennino Settentrionale ed i margini continentali della tetide. Mem Soc Geol It 25: 115–126

Aggrey K, Tonzola C, Schnabel C, Herzog GF, Wasson JT (1998) Beryllium-10 in Muong Nong-type tektites. Meteoritics and Planetary Science 33: A8–A9

Albertao GA, Martins PP Jr (1996) A possible tsunami deposit at the Cretaceous-Tertiary boundary in Pernambuco, northeastern Brazil. Sed Geol 104: 189–201

Albertao GA, Kotsoukos EAM, Regali MPS, Attrep M Jr, Martins PP Jr (1994) The Cretaceous-Tertiary boundary in southern low-latitude regions: Preliminary study in Pernambuco, northeastern Brazil. Terra Nova 6: 366–375

Allegre CJ, Luck JM (1980) Osmium isotopes as petrogenetic and geological tracers. Earth Planet Sci Lett 48: 148–154

Alexopoulos JS, Grieve RAF, Robertson PB (1988) Microscopic lamellar deformation features in quartz: Discriminative characteristics of shock-generated varieties. Geology 16: 796–799

Alt D, Sears JM, Hyndman DW (1988) Terrestrial maria: the origins of large basalt plateaus, hotspot tracks and spreading ridges. J Geol 96: 647–662

Alvarez LW, Alvarez W, Asaro F, Michel HV (1980) Extraterrestrial cause for the Cretaceous-Tertiary extinction. Science 208: 1095–1108

Alvarez W (1989a) Evolution of the Monte Nerone seamount in the Umbria-Marche Apennines 1. Jurassic-Tertiary stratigraphy. Boll Soc Geol It 108: 3–22

Alvarez W (1989b) Evolution of the Monte Nerone seamount in the Umbria-Marche Apennines 2. Tectonic control of the seamount-margin transition. Boll Soc Geol It 108: 23–39

Alvarez W (1997) T. rex and the Crater of Doom. Princeton University Press, Princeton, 185 pp

Alvarez W, Lowrie W (1984) Magnetic stratigraphy applied to synsedimentary slumps, turbidites, and basin analysis: The Scaglia Limestone at Furlo (Italy). Geol Soc Am Bull 95: 324–336

Alvarez W, Muller RA (1984) Evidence from crater ages for periodic impact on the Earth. Nature 308: 718–720

Alvarez W, Arthur M, Fischer A, Lowrie W, Napoleone G, Premoli Silva I, Roggenthen W (1977) Upper Cretaceous-Paleocene magnetic stratigraphy at Gubbio, Italy. V: Type section for the Late Cretaceous-Paleocene geomagnetic reversal time scale. Geol Soc Am Bull 88: 367–389

Alvarez W, Asaro F, Michel HV, Alvarez LW (1982) Geochemical anomalies near the Eocene-Oligocene and Permian-Triassic boundaries. In: Silver LT, Schultz PH (eds) Geological Implications of Impacts of Large Asteroids and Comets on the Earth. Geological Society of America, Special Paper 190, pp 517–528

Alvarez W, Alvarez LW, Asaro F, Michel HV (1984) The end of the Cretaceous: Sharp boundary or gradual transition? Science 233: 1183–1186

Alvarez W, Colacicchi R, Montanari A (1985) Synsedimentary slides and bedding formation in Apennines pelagic limestones. J Sed Petrol 55: 720–734

Alvarez W, Asaro F, Montanari M (1990) Iridium profile for 10 million years across the Cretaceous-Tertiary boundary at Gubbio (Italy). Science 250: 1700–1702

Alvarez W, Claeys P, Kieffer SW (1995) Emplacement of Cretaceous-Tertiary boundary shocked quartz from Chicxulub crater. Science 269: 930–935

Amelin Y, Lee D-C, Halliday AN, Pidgeon RT (1999) Nature of the Earth's earliest crust from hafnium isotopes in single detrital zircons. Nature 399: 252–255

Amorosi A, Coccioni R, Tateo F (1994) The volcaniclastic bodies in the lower Miocene Bisciaro Formation (Umbria-Marche Apennines, Italy. In: Coccioni R, Montanari A, Odin GS (eds) Miocene Stratigraphy of Italy and Adjacent Regions. Giorn Geol 56: 33–46

Anders E, Grevesse N (1989) Abundances of the elements: Meteoritic and solar. Geochim Cosmochim Acta 53: 197–214

Anderson RR, Witzke BJ, Hartung JB (1996) Impact materials recovered by research core drilling in the Manson impact structure, Iowa. In: Ryder G, Fastovsky D, Gartner S (eds) New Developments Regarding the KT Event and Other Catastrophes in Earth History. Geological Society of America, Special Paper 307, pp 527–540

Appel PWU (1979) Cosmic grains in an iron-formation from the Early Precambrian Isua supracrustal belt, West Greenland. J Geol 87: 573–578

Arinobu T, Ishiwatari R, Kaiho K, Lamolda MA (1999) Spike of pyrosynthetic polycyclic aromatic hydrocarbons associated with an abrupt decrease in $\delta^{13}C$ of a terrestrial biomarker at the Cretaceous-Tertiary boundary at Caravaca, Spain. Geology 27: 723–726

Arndt N, Chauvel C (1991) Crust of the Hadean Earth. Bull geol Soc Denmark 39: 145–151

Arnold G, Anbar A, Mojzsis SJ (1998) Iridium and platinum in early Archean metasediments: Implications for sedimentation rate and extraterrestrial flux. GSA Annual Meeting, Abstracts with Programs 30(7): A82–A83

Arthur MA, Fischer AG (1977) Upper Cretaceous-Paleocene magnetic stratigraphy at Gubbio, Italy, I. Lithostratigraphy and sedimentology. Geol Soc Am Bull 88: 367–371

Arthur M, Premoli Silva I (1982) Development of widespread organic carbon-rich strata in Mediterranean Tethys. In: Schlanger SO, Cita MB (eds) Nature and Origin of Cretaceous Carbon-rich Facies. Academic Press, New York, pp 9–54

Asaro F, Montanari A (1988) Small Late Eocene iridium anomalies in the Contessa Highway III section. In: Premoli Silva I, Coccioni R, Montanari A (eds) The Eocene-Oligocene Boundary in the Marche-Umbria Basin (Italy), IUGS Special Publication, F.lli Aniballi Publishers, Ancona, pp 187–188

Asaro F, Alvarez LW, Alvarez W, Michel HV (1982) Geochemical anomalies near the Eocene/Oligocene and Permian/Triassic boundaries. In: Silver LT, Schultz PH (eds) Geological Implications of Impacts of Large Asteroids and Comets on the Earth. Geological Society of America, Special Paper 190, pp 517–528

Asaro F, Alvarez W, Michel HV, Alvarez LW, Anders MH, Montanari A, Kennett JP (1988) Possible world-wide Middle Miocene iridium anomaly and its relationship to periodicity of impacts and extinctions. In: Global Catastrophes in Earth History: an interdisciplinary conference on impacts, volcanism, and mass mortality (Snowbird, Utah). Lunar and Planetary Institute Contribution 673, pp 6–7

Attrep M, Orth CJ, Quintana LR, Shoemaker CS, Shoemaker EM, Taylor SR (1991a) Chemical fractionation of siderophile elements in impactites from Australian meteorite craters. Lunar Planet Sci 22: 39–40

Attrep M Jr, Orth CJ, Quintana LR (1991b) The Permian-Triassic of the Gartnerkofel-1 core (Carnic Alps, Austria): Geochemistry of common and trace elements II - INAA and RNAA. In: Holser WT, Schönlaub HP (eds) The Permian-Triassic boundary in the Carnic Alps of Austria (Gartnerkofel Region). Abhandlungen der Geologischen Bundesanstalt 45, Wien, pp 123-137

Aubry MP, Gradstein FM, Jansa LF (1990) The late Early Eocene Montagnais meteorite: no impact on biotic diversity. Micropaleontology 36: 164-172

Badjukov DD, Lobitzer H, Nazarov MA (1987) Quartz grains with planar features in the Triassic-Jurassic boundary sedimens from northern calcareous Alps, Austria. Lunar Planet Sci 18: 38

Baines KH, Pope KO, Ocampo AC, Ivanov BA (1996) Long-term environmental effects of the Chicxulub impact. Lunar Planet Sci 28: 55-56

Baksi AK (1990) Search for periodicity in global events in the geologic record: Quo Vadimus? Geology 18: 983-986

Baldanza A, Colacicchi R, Parisi G (1982) Controllo tettonico sulla deposizione dei livelli detritici della Scaglia cretacico-terziaria (Umbria orientale). Rend Soc Geol It 5: 11-14

Baldwin RB (1974) Was there a "Terminal Lunar Cataclysm" 3.9-4.0 x 10^9 years ago? Icarus 23: 157-166

Barnes VE (1961) Tektites. Scientific American 205(5): 58-65

Barnes VE (1963a) Tektite Strewn-Fields. In: O'Keefe JA (ed) Tektites. University of Chicago Press, Chicago, pp 25-50

Barnes VE (1963b) Detrital mineral grains in tektites. Science 142: 1651-1652

Barnes VE (1989) Origin of tektites. Texas Journal of Science 41: 5-33

Barnes VE (1990) Tektite research 1936-1990. Meteoritics 25: 149-159

Barnes VE, Pitakpaivan K (1962) Origin of indochinite tektites. Proceedings of the National Academy of Sciences of the U.S. 48: 947-955

Barnes VE, Underwood JR Jr (1976) New investigations of the strewn field of Libyan Desert Glass and its petrography. Earth Planet Sci Lett 30: 117-122

Barnouin-Jha OS, Schultz PH (1998) Lobateness of impact ejecta deposits from atmospheric interactions. J Geophys Res 103: 25739-25756

Barrat JA, Jahn BM, Amosse J, Rocchia R, Keller F, Poupeau G, Diemer E (1997) Geochemistry and origin of Libyan Desert Glasses. Geochim Cosmochim Acta 61: 1953-1959

Barrera E (1994) Global environmental changes preceding the Cretaceous-Tertiary boundary: Early-late Maastrichtian transition. Geology 22: 877-880

Barrera E, Savin SM (1999) Evolution of late Campanian-Maastrichtian marine climates and oceans. In: Barrera E, Johnson CL (eds) Evolution of the Cretaceous Ocean-Climate System. Geological Society of America, Special Paper 332, pp 245-281

Bartolini A, Nocchi M, Baldanza A, Parisi G (1992) Benthic life during the early Toarcian anoxic event in the southwestern Tethyan Umbria-Marche basin, Central Italy. Studies in Benthic Foraminifera, Benthos '90, Sendai Japan. Tokai Univ Press, pp 323-338

Beaudoin B, M'Ban EP, Montanari A, Pinault M (1996) Stratigraphie haute resolution (<20 ka) dans le Cenomanien du bassin de Marches-Ombrie (Italie): Comptes Rendus Acad Sci (Paris) 323: 689-696

Becker L, Bada JL, Winans RE, Hunt JE, Bunch TE, French BM (1994) Fullerenes in the 1.85-billion-year-old Sudbury impact structure. Science 265: 642-645

Becker L, Poreda RJ, Bada JL (1996) Extraterrestrial helium trapped in fullerenes in the Sudbury impact structure. Science 272: 249-252

Becker L, Poreda RJ, Bunch TE (1999) Fullerenes, helium and impact events on the Earth throughout geologic time: Implications for changes in the biostratigraphic record. GSA Annual Meeting, Abstracts with Programs 31(7): A63

Bentley CR (1979) No giant meteorite crater in Wilkes Land, Antarctica. J Geophys Res 84: 5681–5682

Beran A, Koeberl C (1997) Water in tektites and impact glasses by FTIR spectrometry. Meteoritics and Planetary Science 32: 211–216

Berggren WA, Miller KG (1989) Paleogene tropical planktonic foraminiferal biostratigraphy and magnetobiochronology. Micropaleontology 34: 362–380

Berggren WA, McKenna MC, Hardenbol J, Obradovich JD (1978) Revised Paleogene polarity time scale. J Geol 86: 67–81

Berggren WA, Kent DV, Flynn JJ, Van Couvering JA (1985) Cenozoic geochronology. Geol Soc Am Bull 96: 1419–1427

Berggren WA, Kent DV, Swisher C III (1995) A revised Cenozoic geochronology and chronostratigraphy. In: Berggren WA, Kent DV, Hardenbol J (eds) Geochronology, Time Scales and Global Stratigraphic Correlations: A Unified Temporal Framework for an Historical Geology. SEPM Spec Publ 54, pp 129–212

Bhandari N, Shukla PN, Ghevariya ZG, Sundaram SM (1995) Impact did not trigger Deccan volcanism: Evidence from Ajar K/T boundary intertrappean sediments. Geophys Res Lett 22: 433–436

Bhandari N, Shukla PN, Ghevariya ZG, Sundaram SM (1996) K/T boundary layer in Deccan intertrappeans at Anjar, Kutch. In: Ryder G, Fastovsky D, Gartner S (eds) New Developments Regarding the KT Event and Other Catastrophes in Earth History. Geological Society of America, Special Paper 307, pp 417–424

Bice DM, Montanari A (1988) Magnetic stratigraphy of the Massignano section across the Eocene-Oligocene boundary. In: Premoli Silva I, Coccioni R, Montanari A (eds) The Eocene-Oligocene Boundary in the Marche-Umbria Basin (Italy). IUGS Special Publication, F.lli Aniballi Publishers, Ancona, pp 111–117

Bice DM, Stewart K (1985) Ancient erosional grooves on exhumed bypass margins of carbonate platforms: Examples from the Apennines. Geology 14: 565–568

Bice DM, Newton CR, McCauley S, Reiners PW, McRoberts CA (1992) Shocked quartz at the Triassic-Jurassic boundary in Italy. Science 255: 443–446

Bigazzi G, De Michele V (1996) New fission-track age determinations on impact glasses. Meteoritics and Planetary Science 31: 234–236

Bischoff A, Stöffler D (1992) Shock metamorphism as a fundamental process in the evolution of planetary bodies: Information from meteorites. Eur J Mineral 4: 707–755

Blanchard MB, Brownlee DE, Bunch TE, Hodge PW, Kyte FT (1980) Meteoroid ablation spheres from deep-sea sediments. Earth Planet Sci Lett 46: 178–190

Bloss FD (1981) The Spindle Stage - Principles and Practice. Cambridge University Press, Cambridge (UK), 340 pp

Blow WH (1969) Late Middle Eocene to Recent planktonic foraminiferal biostratigraphy. Proc 1st Int Conf Plankt Microfossils, Geneva, 1, pp 199–442

Blum JD, Chamberlain CP (1992) Oxygen isotope constraints on the origin of impact glasses from the Cretaceous-Tertiary boundary. Science 257: 1104–1107.

Blum JD, Papanastassiou DA, Koeberl C, Wasserburg GJ (1992) Nd and Sr isotopic study of Australasian tektites: New constraints on the provenance and age of target materials. Geochim Cosmochim Acta 56: 483–492

Blum JD, Chamberlain CP, Hingston MP, Koeberl C, Marin LE, Schuraytz BC, Sharpton VL (1993) Isotopic comparison of K-T boundary impact glass with melt rock from the Chicxulub and Manson impact structures. Nature 364: 325–327

Bohor BF (1990) Shocked quartz and more; Impact signatures in Cretaceous/Tertiary boundary clays. In: Sharpton VL, Ward PD (eds) Global Catastrophes in Earth History. Geological Society of America, Special Paper 247, pp 335–347

Bohor BF (1996) A sediment gravity flow hypothesis for siliclastic units at the K/T boundary, northeastern Mexico. In: Ryder G, Fastovsky D, Gartner S (eds) New Developments Regarding the KT Event and Other Catastrophes in Earth History. Geological Society of America, Special Paper 307, pp 183–195

Bohor BF, Glass BP (1995) Origin and diagenesis of the K/T impact spherules - From Haiti to Wyoming and beyond. Meteoritics 30: 182–198

Bohor BF, Izett G (1986) Worldwide size distribution of shocked quartz at the K/T boundary: Evidence for a North American impact site. Lunar Planet Sci 17: 68–69

Bohor BF, Koeberl C (1996) Are microtektites really micro-tektites? Meteoritics and Planetary Science 31: A17

Bohor BF, Foord EE, Modreski PJ, Triplehorn DM (1984) Mineralogical evidence for an impact event at the Cretaceous/Tertiary boundary. Science 224: 867–869

Bohor BF, Foord EE, Ganapathy R (1986) Magnesioferrite from the Cretaceous-Tertiary boundary, Caravaca, Spain. Earth Planet Sci Lett 81: 57–66

Bohor BF, Modreski PJ, Foord EE (1987a) Shocked quartz in the Cretaceous/Tertiary boundary clays: Evidence for global distribution. Science 236: 705–708

Bohor BF, Triplehorn DM, Nichols DJ, Millard HT (1987b) Dinosaurs, spherules, and the "magic" layer: A new K/T boundary site in Wyoming. Geology 15: 896–899

Bohor BF, Betterton WJ, Foord EE (1988) Coesite, glass, and shocked quartz and feldspar at DSDP Site 612: Evidence for nearby impact in the late Eocene. Lunar Planet Sci 19: 114–115

Bohor BF, Betterton WJ, Krogh TE (1993) Impact-shocked zircons: discovery of shock-induced textures reflecting increasing degrees of shock metamorphism. Earth Planet Sci Lett 119: 419–424

Boiko AK, Vailler AA, Vishnyak MM (1985) On the age of the Boltysh depression (in Russian). Geol Zh 45: 86–90

Bollinger K (1993) ^{40}Ar-^{39}Ar Datierung von Tektites. Diploma Thesis, Max-Planck-Institut, Heidelberg, and Ruprecht-Karls-Universität Heidelberg, 84 pp

Bonarelli G (1891) Il territorio di Gubbio. Notizie geologiche. Tipografia Economica, Roma, pp 1–38

Boon JD, Albritton CC (1938) Established and supposed examples of meteoritic craters and structures. Field and Laboratory 6: 44–56

Boslough MB (1991) Shock modification and chemistry and planetary geologic processes. Annual Reviews of Earth and Planetary Science 19: 101–130

Boslough MB, Asay JR (1993) Basic principles of shock compression. In: Asay JR, Shahinpoor M (eds) High-pressure Shock Compression of Solids. Springer Verlag, Berlin, pp 7–42

Boslough MB, Cygan RT (1988) Shock-enhanced dissolution of silicate minerals and chemical weathering on planetary surfaces. Proc Lunar Planet Sci Conf 18th, pp 443–453

Boslough MB, Cygan RT, Izett GA (1995) NMR spectroscopy of quartz from the K/T boundary: Shock-induced peak broadening, dense glass, and coesite. Lunar Planet Sci 26: 149–150

Boslough MB, Chael EP, Trucano TG, Crawford DA, Campbell DL (1996) Axial focusing of impact energy in the Earth's interior: A possible link to flood basalts and hotspots. In: Ryder G, Fastovsky D, Gartner S (eds) New Developments Regarding the KT Event and Other Catastrophes in Earth History. Geological Society of America, Special Paper 307, pp 541–550

Bottomley RJ, York D (1988) Age measurement of the submarine Montagnais impact crater. Geophys Res Lett 15: 1409–1412

Bottomley RJ, York D, Grieve RAF (1990) ^{40}Argon-^{39}Argon dating of impact craters: Proc Lunar Planet Sci Conf 20th, pp 421–431

Bottomley RJ, Grieve RAF, York D, Masaitis V (1997) The age of the Popigai impact event and its relations to events at the Eocene/Oligocene boundary. Nature 388: 365–368

Bourgeois J, Hansen TA, Wilberg PL, Kauffman EG (1988) A tsunami deposit at the Cretaceous-Tertiary boundary in Texas. Science 241: 567–570

Bowring SA, Erwin DH, Jin YG, Martin MW, Davidek K, Wang W (1998) U/Pb zircon geochronology and tempo of the end-Permian mass extinction. Science 280: 1039–1045

Bralower TJ, Monechi S, Thierstein HR (1989) Calcareous nannofossil zonation of the Jurassic-Cretaceous boundary interval and correlation with the geomagnetic polarity time scale. Marine Micropaleontology 14: 153–325

Brenan RL, Peterson BL, Smith HJ (1975) The origin of Red Wing Creek structure, McKenzie County, North Dakota. Wyoming Geological Association Earth Science Bulletin 8: 11–41

Brett R (1992) The Cretaceous-Tertiary extinction: A lethal mechanism involving anhydrite target rocks. Geochim Cosmochim Acta 56: 3603–3606

Brinkhuis H, Coccioni R (1995) Is there a relation between dinocysts changes and iridium after all? Preliminary results from the Massignano section. In: Montanari A, Coccioni R (eds) The effects of impacts in the evolution of the atmosphere and biosphere with regard to short- and long-term changes. ESF, Ancona, pp 40

Brownlee DE (1981) Extraterrestrial components. In: Emiliani C (ed) The Sea, vol. 7. J. Wiley & Sons, New York, pp 733–762

Buchanan PC, Reimold WU (1998) Field and laboratory studies of the Rooiberg Group, Bushveld Complex, South Africa: No evidence for an impact origin. Earth Planet Sci Lett 155: 149–165

Buchanan PC, Koeberl C, Reid AM (1998) Impact into unconsolidated, water-rich sediments at the Marquez Dome, Texas. Meteoritics and Planetary Science 33: 1053–1064

Buchanan PC, Koeberl C, Reimold WU (1999) Petrogenesis of the Dullstroom Formation, Bushveld Magmatic Province, South Africa. Contrib Mineral Petrol 137: 133–146

Bucher W (1963) Cryptoexplosion structures caused from without or from within the Earth? ("Astroblemes" or "Geoblemes"?). Amer J Sci 261: 597–649

Buchwald VF (1975) Handbook of Iron Meteorites. University of California Press, Berkeley, 1418 pp

Buick R (1987) Comment on "Early Archean silicate spherules of probable impact origin, South Africa and Western Australia". Geology 15: 180–181

Bunch TE, Dence MR, Cohen AJ (1967) Natural terrestrial maskelynite. American Mineralogist 52: 244–253

Bunopas S, Wasson JT, Vella P, Fontaine H, Hada S, Burrett C, Suphajunya T, Khositanont S (1999) Catastrophic loess, mass mortality and forest fires suggest that a Pleistocene cometary impact in Thailand caused the Australasian tektite field. Journal of the Geological Society of Thailand 1: 1–17

Byerly GR, Lowe DR (1994) Spinel from Archean impact spherules. Geochim Cosmochim Acta 58: 3469–3486

Byerly GR, Hazel JE, McCabe C (1990) Discrediting the late Eocene microspherule layer at Cynthia, Mississippi. Meteoritics 25: 89–92

Camargo-Zanoguera A, Suárez-Reynoso G (1994) Evidencia sismica del crater de impacto de Chicxulub. Boletín de la Asociacion Mexicana de Geofisicos de Exploracion 34: 1–28

Cameron AGW, Benz W (1991) The origin of the Moon and the single impact hypothesis IV. Icarus 92: 204–216

Canudo JI (1997) El Kef blind test I result. Marine Micropaleontology 29: 73–76

Canuti P, Fazzuoli M, Ficcarelli G, Venturi F (1983) Occurrence of Liassic faunas at Waaney (Uanei) Province of Bay, South-western Somalia. Riv It Pal Strat 89: 31–46

Carlisle DB, Braman DR (1991) Nanometre-size diamonds in the Cretaceous/Tertiary boundary clay of Alberta. Nature 352: 708–709

Carrigy MA (1968) Evidence of shock metamorphism in rocks from the Steen River structure, Alberta. In: French BM, Short NM (eds) Shock Metamorphism of Natural Materials. Mono Book Corp, Baltimore, pp 367–378

Carstens H (1975) Thermal history of impact melt rocks in the Fennoscandian Shield. Contrib Mineral Petrol 50: 145–155

Carter NL (1965) Basal quartz deformation lamellae: A criterion for the recognition of impactites. Am J Sci 263: 786–806

Casoni A (1988) Micropaleontologia, microbiostratigrafia (foraminiferi) e litostratigrafia della Scaglia paleocenica del Monte Cònero (Ancona). Laurea Thesis, Univ Urbino, 62 pp

Cassidy WA, Glass BP, Heezen BC (1969) Physical and chemical properties of Australasian microtektites. J Geophys Res 74: 1008–1025

Castellarin A, Colacicchi R, Praturlon A (1978) Geodinamica della linea Ancona-Anzio. Mem Soc Geol Ital 19: 741–757

Cecca S, Cresta S, Pallini G, Santantonio M (1990) Il Giurassico del Monte Nerone (Appennino marchigiano, Italia centrale): Biostratigrafia, litostratigrafia, ed evoluzione paleogeografica. Mem Descr Carta Geol d'Italia 40: 51–126

Centamore E, Chiocchini M, Deiana G, Micarelli A, Pieruccini U (1971) Contributo alla conoscenza del Giurassico nell'Appennino umbro-marchigiano. Studi Geologici Camerti 1: 1–89

Centamore E, Chiocchini U, Cipriani N, Deiana G, Micarelli A (1978) Analisi dell'evoluzione tettonico-sedimentaria dei "bacini minori" torbiditici del Miocene Medio-Superiore nell'Appennino Umbro-Marchigiano e Laziale Abruzzese: risultati degli studi in corso. Mem Soc Geol It 18: 135–170

Chai CF, Zhou Y, Moa X, Ma S, Ma J, Kong P, He J (1992) Geochemical constraints on the Permo-Triassic boundary event in South China. In: Sweet WC, Yang Z, Dickins JM, Yin HF (eds) Permo-Triassic events in the Eastern Tethys. Cambridge University Press, Cambridge, pp 158–168

Chalmers RO, Henderson EP, Mason B (1976) Occurrence, distribution, and age of Australian tektites. Smithsonian Contributions to the Earth Sciences 17: 1–46

Chan LS, Montanari A, Alvarez W (1985) Magnetic stratigraphy of the Scaglia Rossa: Implications for syndepositional tectonics of the Umbria-Marches basin, Italy. Riv It Paleont Strat 91: 219–258

Channell JET, D'Argenio B, Horvath F (1979) Adria, the African promontory: Mesozoic Mediterranean paleogeography. Earth-Sci Rev 15: 213–292

Chao ECT (1963) The petrographic and chemical characteristics of tektites. In: O'Keefe JA (ed) Tektites. University of Chicago Press, Chicago, pp 51–94

Chao ECT, Fahey JJ, Littler J (1961) Coesite from Wabar Crater, near Al Hadida, Arabia. Science 133: 882–883

Chao ECT, Fahey JJ, Littler J, Milton DJ (1962) Stishovite, SiO_2, a very high pressure new mineral from Meteor Crater, Arizona. J Geophys Res 67: 419–421

Chapman CR, Morrison D (1994) Impacts on the earth by asteroids and comets: Assessing the hazard. Nature 367: 33–40

Chapman DR, Scheiber LC (1969) Chemical investigation of Australasian tektites. J Geophys Res 74: 6737–6776

Chauris H, LeRousseau J (1996) La disparition des Inocérames au Maastrichtien et les variations environmentales autour the la limite Crétacé-Tertiaire. Option Sciences de la Terre et Environment, Ecole des Mines de Paris (unpublished), 210 pp

Chauris H, LeRousseau J, Beaudoin B, Propson S, Montanari A (1998) Inoceramid extinction in the Gubbio basin (northeastern Apennines of Italy) and relations with environmental changes during the mid-Maastrichtian. Palaeogeography Palaeoclimatology Palaeoecology 139: 177–193

Chaussidon M, Koeberl C (1995) Boron content and isotopic composition of tektites and impact glasses: Constraints on source regions. Geochim Cosmochim Acta 59: 613–624

Chaussidon M, Sigurdsson H, Metrich N (1994) Sulfur and boron isotope study of high-Ca impact glasses from the KT boundary: Constraints on source rocks. In: Ryder G, Fastovsky D, Gartner S (eds) New Developments Regarding the KT Event and Other Catastrophes in Earth History. Geological Society of America, Special Paper 307, pp 253–262

Chen G, Tyburczy JA, Ahrens TJ (1994) Shock-induced devolatilization of calcium sulfate and implications for K-T extinctions. Earth Planet Sci Lett 128: 615–628

Chijiwa T, Arai T, Sugai T, Shinohara H, Kumazawa M, Takano M, Kawakami S-i (1999) Fullerenes found in the Permo-Triassic mass extinction period. Geophys Res Lett 26: 767–770

Chyba CF, Thomas PJ, Zahnle KJ (1993) The 1908 Tunguska explosion: atmospheric disruption of a stony asteroid. Nature 361: 40–44

Ciarapica G, Cirilli S, Passeri L, Trincianti E, and Zanetti L (1987) 'Anidriti di Burano' et 'Formation du Monte Cetona' (nouvelle Formation), biostratigraphie de deux series-types du Trias superieur dans l'Apennin septentrional. Rev Paleobio 6: 341–409

Cintala MJ, Grieve RAF (1998) Scaling impact melting and crater dimensions: Implications for the lunar cratering record. Meteoritics and Planetary Science 33: 889–912

Cirilli S, Bucefalo Palliani R, Pontini MR (1994) Palynostratigraphy and palynofacies of the Late Triassic R. contorta facies in the Northern Apennines: II) The Monte Cetona Formation. Rev Paleobio 13: 319–339

Claeys P, Casier JG (1994) Microtektite-like impact glass associated with the Frasnian-Famennia boundary mass extinction. Earth Planet Sci Lett 122: 303–315

Claeys P, Casier JG, Margolis SV (1992) Microtektites and mass extinctions: Evidence for a late Devonian asteroid impact. Science 257: 1102–1104

Claeys P, Kyte FT, Herbosch A, Casier JG (1996) Geochemistry of the Frasnian-Famennian boundary in Belgium: Mass extinction, anoxic oceans and microtektite layer, but not much iridium? In: Ryder G, Fastovsky D, Gartner S (eds) New Developments Regarding the KT Event and Other Catastrophes in Earth History. Geological Society of America, Special Paper 307, pp 491–504

Clayton PA, Spencer LJ (1934) Silica glasses from the Libyan Desert. Mineralogical Magazine 23: 501–508

Clymer AK (1996) Discovery and trace element geochemistry of Late Eocene shocked quartz: Insights into the Late Eocene impacts. Master's Thesis, Univ California, Berkeley, 85 pp

Clymer AK, Bice DM, Montanari A (1995) Shocked quartz in the late Eocene: Bolide impact evidence from Massignano, Italy. In: Montanari A, Coccioni R (eds) The effects of impacts in the evolution of the atmosphere and biosphere with regard to short- and long-term changes. ESF, Ancona, pp 60–61

Clymer AK, Bice DM, Montanari A (1996) Shocked quartz from the late Eocene: Impact evidence from Massignano, Italy. Geology 24: 483–486

Coccioni R, Galeotti S (1994) K-T boundary extinction: geologically instantaneous or gradual event? Evidence from deep sea benthic foraminifera. Geology 22: 779–782

Coccioni R, Galeotti S (1998) What happened to small benthic foraminifers at the Cretaceous/Tertiary boundary? Bull Soc Géol France 169: 271–279

Coccioni R, Savelli D (1983) Osservazioni stratigrafiche sul limite K-T della sezione di Coldorso (Serra S. Abbondio, Appennino umbro-marchigiano. Acta Nat Aten Parm, Parma 19: 199–212

Coccioni R, Monaco P, Monechi S, Nocchi M, Parisi G (1988) Biostratigraphy of the Eocene/Oligocene boundary at Massignano (Ancona, Italy). In: Premoli Silva I, Coccioni R, Montanari A (eds) The Eocene-Oligocene Boundary in the Marche-Umbria Basin, IUGS Special Publication, F.lli Aniballi Publishers, Ancona, pp 50–80

Coccioni R, Franchi R, Nesci O, Perilli N, Wezel FC, Battistini F (1990) Stratigrafia, micropaleontologia, e mineralogia delle Marne a Fucoidi (Aptiano inferiore-Albiano superiore) delle sezioni di Poggio le Guaine e del Fiume Bosso (Appennino umbro-marchigiano). In: Pallini G, Cecca F, Cresta S, Santantonio M (eds) Atti II Convegno Internazionale Fossili, Evoluzione, Ambiente. Pergola, Italy, pp 163–201

Coccioni R, Fornaciari E, Montanari A, Rio D, Zaverboom D (1997) Potential integrated stratigraphy of the Aquitanian-Burdigalian Section at Santa Croce di Arcevia (Northeastern Apennines, Italy). In: Montanari A, Odin GS, Coccioni R (eds) Miocene Stratigraphy: An Integrated Approach. Elsevier, Amsterdam, pp 279–296

Colacicchi R, Baldanza A (1986) Carbonate turbidites in a Mesozoic pelagic basin (Scaglia Formation, Apennines). Comparison with siliciclastic depositional models. Sedimentary Geology 48: 81–105

Colacicchi R, Passeri L, Pialli G (1970) Nuovi dati sul Giurese umbro-marchigiano ed ipotesi per un inquadramento regionale. Mem Soc Geol It 9: 839–974

Colodner DC, Boyle EA, Edmond JM, Thomson J (1992) Post-depositional mobility of platinum, iridium and rhenium in marine sediments. Nature 358: 402–404

Coltorti M, Bosellini A (1980) Sedimentazione e tettonica nel Giurassico della dorsale marchigiana. Studi Geologici Camerti 6: 189–216

Compston W, Chapman DR (1969) Sr isotope patterns within the Southeast Australian strewnfield. Geochim Cosmochim Acta 33: 1023–1036

Compston W, Williams IS, Jenkins RJF, Gostin VA, Haines PW (1987) Zircon age evidence for the Late Precambrian Acraman ejecta blanket. Austral J Earth Sci 34: 435–445

Cordier P, Gratz AJ (1995) TEM study of shock metamorphism in quartz from the Sedan nuclear test site. Earth Planet Sci Lett 129, 163–170

Corfield RM, Cartlidge JE, Premoli Silva I, Housley RA (1991) Oxygen and carbon isotope stratigraphy of the Paleogene and Cretaceous limestones in the Bottaccione Gorge and the Contessa Highway sections, Umbria, Italy. Terra Nova 3: 414–422

Corner B, Reimold WU, Brandt D, Koeberl C (1997) Morokweng impact structure, Northwest Province, South Africa: Geophysical imaging and some preliminary shock petrographic studies. Earth Planet Sci Lett 146: 351–364

Courtillot V (1999) Evolutionary Catastrophes - The Science of Mass Extinctions. Cambridge University Press, Cambridge, 173 pp

Courtillot V, Jaeger JJ, Yang Z, Féraud G, Hofmann C (1996) The influence of continental flood basalts on mass extinctions: Where do we stand? In: Ryder G, Fastovsky D, Gartner S (eds) New Developments Regarding the KT Event and Other Catastrophes in Earth History. Geological Society of America, Special Paper 307, pp 513–525

Covey C, Thompson SL, Weissman PR, MacCracken MC (1994) Global climatic effects of atmospheric dust from an asteroid or comet impact on Earth. Global and Planetary Change 9: 263–273

Creaser RA, Papanastassiou DA, Wasserburg GJ (1991) Negative thermal ion mass spectrometry of osmium, rhenium and iridium. Geochim Cosmochim Acta 55: 397–401

Crescenti U, Crostella A, Donzelli G, Raffi G (1969) Stratigrafia della serie calcarea dal Lias al Miocene nella regione Marchigiano-Abruzzese (Parte II, litostratigrafia, biostratigrafia, paleogeografia). Mem Soc Geol It 8: 343–420

Crutzen PJ (1987) Acid rain at the K/T boundary. Nature 330: 108–109

Cygan RT, Boslough MB, Kirkpatrick RJ (1990) NMR spectroscopy of experimentally shocked quartz: Shock wave barometry. Proc Lunar Planet Sci Conf 20th: 127–136

Dachille F, Gigle P, Simons PY (1968) Experimental and analytical studies of crystalline damage useful for the recognition of impact structures. In: French BM, Short NM (eds) Shock Metamorphism of Natural Materials. Mono Book Corp, Baltimore, pp 555–570

Dalrymple GB, Ryder G (1993) ^{40}Ar/^{39}Ar Age spectra of Apollo 15 impact melt rocks by laser step-heating and their bearing on the history of lunar basin formation. J Geophys Res 98: 13085–13095

Dalrymple GB, Ryder G (1996) ^{40}Ar/^{39}Ar age spectra of Apollo 17 highlands breccia samples by laser step-heating and the age of the Serenitatis basin. J Geophys Res 101: 26069–26084

D'Argenio B (1970) Evoluzione geottettonica comparata tra alcune piattaforme carbonatiche dei Mediterranei Europeo ed Americano. Atti Accademia Pontiana, NS 20: 3–34

Davis MP, Hut P, Muller RA (1984) Extinction of species by periodic comet showers. Nature 308: 715–717

de Carli PS, Jamieson JC (1961) Formation of diamond by explosive shock. Science 133: 1821–1822

de Laubenfels MW (1956) Dinosaur extinction: One more hypothesis. J Paleontology 30: 207–212

de Silva SL, Wolff JA, Sharpton VL (1990) Explosive volcanism and associated pressures: Implications for models of endogenic shocked quartz. In: Sharpton VL, Ward PD (eds) Global Catastrophes in Earth History. Geological Society of America, Special Paper 247, pp 139–145

Deino AL, Drake RE, Curtis GH, Montanari A (1988) Preliminary laser-fusion ^{40}Ar/^{39}Ar dating results from Oligocene biotites of Gubbio, Italy. In: Premoli Silva I, Coccioni R, Montanari A (eds) The Eocene-Oligocene Boundary in the Marche-Umbria Basin. IUGS Special Publication, F.lli Aniballi Publishers, Ancona, Italy, pp 229–238

Deino A, Channel J, Coccioni R, De Grandis G, DePaolo DJ, Emmanuel L, Fornaciari S, Laurenzi ML, Montanari M, Renard M, Rio D (1997) Integrated stratigraphy of the upper Burdigalian-lower Langhian section at Moria (northeastern Apennines, Italy). In: Montanari A, Odin GS, Coccioni R (eds) Miocene Stratigraphy: An Integrated Approach. Elsevier, Amsterdam, pp 315–342

Delano JW (1992) Australite flanges as flight data recorders. Lunar Planet Sci 23: 301–302

Delano JW, Lindsley D H (1982) Chemical systematics among the moldavite tektites. Geochim Cosmochim Acta 46: 2447–2452

Delano JW, Bouska V, Randa Z (1987) Geochemically inferred redox state in the source-materials of terrestrial impact glasses. Lunar Planet Sci 18: 233–234

Dence MR (1971) Impact melts. J Geophys Res 76: 5525–5565

DePaolo DJ, Kyte FT, Marchall BD, O'Neil JR, Smit J (1983) Rb-Sr, Sm-Nd, K-Ca, O, and H isotopic study of Cretaceous-Tertiary boundary sediments, Caravaca, Spain: Evidence for an oceanic impact site. Earth Planet Sci Lett 64: 356–373

Dercourt J, Ricou LE, Vrielynck B (1993) Atlas Tethys Palaeoenvironmental Maps. Gauthier Villars, Paris, 307 pp

Deutsch A, Schärer U (1994) Dating terrestrial impact events. Meteoritics 29: 301–322

Deutsch A, Buhl D, Langenhorst F (1992) On the significance of crater ages: New ages for Dellen (Sweden) and Araguainha (Brazil). Tectonophysics 216: 205–218

Deutsch A, Grieve RAF, Avermann M, Bischoff L, Brockmeyer P, Buhl D, Lakomy R, Müller-Mohr V, Ostermann M, Stöffler D (1995) The Sudbury Structure (Ontario, Canada): A tectonically deformed multi-ring impact basin. Geologische Rundschau 84: 697–709

Deutsch A, Ostermann M, Masaitis VL (1997) Geochemistry and neodymium-strontium isotope signature of tektite-like objects from Siberia (urengoites, South Ural glass). Meteoritics and Planetary Science 32: 679–686

Deutsch A, Greshake A, Pesonen LJ, Pihlaja P (1998) Unaltered cosmic spherules in an 1.4-Gyr-old sandstone from Finland. Nature 395: 146–148

D'Hondt S (1998) Theories of terrestrial mass extinction by extraterrestrial objects. Earth Sciences History 17: 157–173

D'Hondt S, Keller G, Stallard RF (1987) Major element compositional variation within and between different Late Eocene microtektite strewn fields. Meteoritics 22: 61–79

D'Hondt S, Pilson MEQ, Sigurdsson H, Hanson AK Jr, Carey S (1994) Surface-water acidification and extinction at the Cretaceous-Tertiary boundary. Geology 22: 983–989

D'Hondt S, King J, Gibson C (1996) Oscillatory marine response to the Cretaceous-Tertiary impact. Geology 24: 611–614

Dickin AP, Richardson JM, Crocket JH, McNutt RH, Peredery WV (1992) Osmium isotope evidence for a crustal origin of platinum group elements in the Sudbury nickel ore, Ontario, Canada. Geochim Cosmochim Acta 56: 3531–3537

Diemer E (1997) Libyan Desert Glass: an impactite State of the art in July 1996. In: Proceedings, Silica 96, Meeting on Libyan Desert Glass and related desert events. Pyramids, Milan, Italy, pp 95–110

Dietz RS (1968) Shatter cones in cryptoexplosion structures. In: French BM, Short, NM (eds) Shock Metamorphism of Natural Materials. Mono Book Corp., Baltimore, pp 267–285

Dietz RS (1977) Elgygytgyn crater, Siberia: Probable source of Australian tektite field. Meteoritics 12: 145–157

Doehne E, Margolis SV (1990) Trace-element geochemistry and mineralogy of the Cretaceous/Tertiary boundary; Identification of extraterrestrial components. In: Sharpton VL, Ward PD (eds) Global Catastrophes in Earth History. Geological Society of America, Special Paper 247, pp 367–382

Douglas RG, Woodruff F (1981) Deep-sea benthic foraminifera. In: Emiliani C (ed) The Sea, vol. 7. J. Wiley & Sons, New York, pp 1233–1327

Dressler BO, Sharpton VL (eds) (1999) Large Meteorite Impacts and Planetary Evolution II. Geological Society of America, Special Paper 339, 464 pp

Dressler BO, Grieve RAF, Sharpton VL (eds) (1994) Large Meteorite Impacts and Planetary Evolution. Geological Society of America, Special Paper 293, 348 pp

Dworak U (1969) Stoßwellenmetamorphose des Anorthosits vom Manicouagan Krater, Québec, Canada. Contrib Mineral Petrol 24: 306–347

Dypvik H, Gudlaugson ST, Tsikalas F, Attrep M Jr, Ferrell RE Jr, Krinsley DH, Mork A, Faleide JI, Nagy J (1996) Mjølnir structure: An impact crater in the Barents Sea. Geology 24: 779–782

Eldholm O, Thomas E (1993) Environmental impact of volcanic margin formation. Earth Planet Sci Lett 117: 319–329

El Goresy A, Fechtig H, Ottemann T (1968) The opaque minerals in impactite glasses. In: French BM, Short NM (eds) Shock Metamorphism of Natural Materials. Mono Book Corp., Baltimore, pp 531–554

El Goresy A, Chen M, Gillet P, Künstler F, Stähle V (1999) In situ discovery of shock-induced graphite-diamond phase transformations in polished thin sections of shocked gneisses from the Ries crater, Germany. Meteoritics and Planetary Science 34: A125–A126

Emanuel KA, Speer K, Rotunno R, Srivastava R, Molina M (1995) Hypercanes: A possible link in global extinction scenarios. J Geophys Res 100: 13755–13765

Emmons RC (1943) The Universal Stage (With Five Axes of Rotation). Geological Society of America, Memoir 8, 205 pp

Engelhardt Wv, Bertsch W (1969) Shock induced planar deformation structures in quartz from the Ries crater, Germany. Contributions to Mineralogy and Petrology 20: 203–234

Engelhardt Wv, Luft E, Arndt J, Schock H, Weiskirchner W (1987) Origin of moldavites. Geochim Cosmochim Acta 51: 1425–1443

Englert P, Pal DK, Tuniz C, Moniot RK, Savin W, Kruse TH, Herzog GF (1984) Manganese-53 and beryllium-10 contents of tektites. Lunar Planet Sci 15: 250–251

Erba E (1988) Aptian-Albian calcareous nannofossil biostratigraphy of the Scisti a Fucoidi cored at Piobbico (Central Italy). Riv It Paleontol Strat 94: 249–284

Erba E (1992) Calcareous nannofossil distribution in pelagic rhythmic sediments (Aptian-Albian Piobbico Core, Central Italy). Riv It Paleontol Strat 97: 455–483

Erwin DH (1994) The Permo-Triassic extinction. Nature 367: 231–236

Esser BK, Turekian KK (1989) Osmium isotopic composition of the Raton Basin Cretaceous-Tertiary boundary interval. EOS Trans Am Geophys Un 70: 717

Esser BK, Turekian KK (1993) The osmium isotopic composition of the continental crust. Geochim Cosmochim Acta 57: 3093–3104

Eugster O, Geiss J, Krähenbühl U (1985) Noble gas isotopic abundances and noble metal concentrations in sediments from the Cretaceous-Tertiary boundary. Earth Planet Sci Lett 74: 27–34

Evans NJ, Gregoire DC, Grieve RAF, Goodfellow WD, Veizer J (1993) Use of platinum-group elements for impactor identification: Terrestrial impact craters and Cretaceous-Tertiary boundary. Geochim Cosmochim Acta 57: 3737–3748

Farinacci A, Mariotti N, Nicosia U, Pallini G, Schiavinotto F (1981) Jurassic sediment in the Umbro-Marchean Apennines: an alternative model. In: Farinacci A, Elmi S (eds) Rosso Ammonitico Symposium, Rome 1980, Tectonoscienza, pp 335–398

Farley K (1995) Cenozoic variations in the flux of interplanetary dust recorded by ^{3}He in a deep sea sediment. Nature 376: 153–156

Farley KA, Montanari A, Shoemaker EM, Shoemaker CS (1998) Geochemical evidence for a comet shower in the late Eocene. Science 280: 1250–1253

Farley KA, Mukhopadhyay S, Montanari A (1999) Extraterrestrial Helium-3 in the geologic record: Results from 74 to 29 Ma and from the P-Tr boundary. GSA Annual Meeting, Abstracts with Programs 31(7): A63

Faure G (1986) Principles of Isotope Geology. 2nd Edition. J. Wiley and Sons, New York, 589 pp

Fehn U, Teng R, Elmore D, Kubik PW (1986) Isotopic composition of osmium in terrestrial samples determined by accelerator mass spectrometry. Nature 323: 707–710

Fiet N (1998) Les black shales, un outil chronostratigraphique haute résolution: Example de l'Albien du bassin the Marches-Ombrie (Italie centrale). Bull Soc Géol France 169: 221–231

Finney SC, Berry WBN, Cooper JD, Ripperdan RL, Sweet WC, Jacobson SR, Soufiane A, Achab A, Noble PJ (1999) Late Ordovician mass extinction: A new perspective from stratigraphic sections in central Nevada. Geology 27: 215–218

Fischer AG, Herbert TD, Napoleone G, Premoli Silva I, Ripepe M (1991) Albian pelagic rhytms (Piobbico Core). J Sed Petrol 61: 1164–1172

Fiske PS, Schnetzler CC, McHone J, Chanthavaichith KK, Homsombath I, Phouthakayalat T, Khenthavong B, Xuan PT (1999) Layered tektites of southeast Asia: Field studies in central Laos and Vietnam. Meteoritics and Planetary Science 34: 757–761

Fitzpatrick MEJ (1996) Recovery of Turonian dinoflagellate cyst assemblages from the effects of the oceanic anoxic event at the end of the Cenomanian in southern England. In: Hart MB (ed) Biotic Recovery from Mass Extinction Events. Geological Society London, Special Publication 102: pp 279–297

Fleischer RL, Price PB (1964) Fission-track evidence for the simultaneous origin of tektites and other natural glasses. Geochim Cosmochim Acta 28: 755–760

Fleischer RL, Price PB, Woods RT (1969) A second tektite fall in Australia. Earth Planet Sci Lett 7: 51–52

Ford RJ (1988) An empirical model for the Australasian tektite field. Australian Journal of Earth Sciences 35: 483–490

French BM (1968) Shock metamorphism as a geological process. In: French BM, Short NM (eds) Shock Metamorphism of Natural Materials. Mono Book Corp., Baltimore, pp 1–17

French BM (1987) Comment on "Early Archean silicate spherules of probable impact origin, South Africa and Western Australia". Geology 15: 178–179

French BM (1990a) 25 years of the impact-volcanic controversy: Is there anything new under the Sun or inside the Earth? EOS, Trans Am Geophys Un 71: 411–414

French BM (1990b) Absence of shock-metamorphic effects in the Bushveld Complex, South Africa: Results of an intensive search. Tectonophysics 171: 287–301

French BM (1998) Traces of catastrophe: A handbook of shock-metamorphic effects in terrestrial meteorite impact structures. LPI Contribution 954, Lunar and Planetary Institute, Houston, 120 pp

French BM, Nielsen RL (1990) Vredefort bronzite granophyre: Chemical evidence for an origin as meteorite impact melt. Tectonophysics 171: 119–138

French BM, Short NM (eds) (1968) Shock metamorphism of natural materials. Mono Book Corp, Baltimore, 644 pp

French BM, Underwood JR Jr, Fisk EP (1974) Shock metamorphic features in two meteorite impact structures, Southeastern Libya. Geol Soc Am Bull 85: 1425–1428

French BM, Orth CJ, Quintana CR (1989) Iridium in the Vredefort bronzite granophyre: Impact melting and limits on a possible extraterrestrial component. In: Proceedings, 19th Lunar and Planetary Science Conference, New York, Cambridge University Press, pp 733–744

French BM, Koeberl C, Gilmour I, Shirey SB, Dons JA, Naterstad J (1997) The Gardnos impact structure, Norway: Petrology and geochemistry of target rocks and impactites. Geochim Cosmochim Acta 61: 873–904

Frey H (1980) Crustal evolution of the early earth: The role of major impacts. Precambrian Research 10: 195–216

Fudali RF (1981) The major element chemistry of Libyan Desert Glass and the mineralogy of its precursor. Meteoritics 16: 247–259

Fudali RF (1993) The stratigraphic age of australites revisited. Meteoritics 28: 114–119

Fudali RF, Dyar MD, Griscom DL, Schreiber HD (1987) The oxidation state of iron in tektite glass. Geochim Cosmochim Acta 51: 2749–2756

Galeotti S, Angori E. Coccioni R, Ferrari G, Monechi S, Morettini E, Premoli Silva I (1998) Integrated biostratigraphy and isotope geochemistry across the Paleocene/Eocene boundary in a classical Tethyan setting: the Contessa-Road section. In: La Limite Paléocène-Eocène en Europe: Evénements et Corrélations. Séance spécialisée SGF/CFS, Paris, Strata 9: pp 49–52

Galer SJG, Macdougall JD, Erickson DJ III (1989) Pb isotopic tracers of the Cretaceous-Tertiary extinction event. Geophys Res Lett 16: 1301–1304

Ganapathy R (1980) A major meteorite impact on the earth 65 million years ago: Evidence from the Cretaceous-Tertiary boundary clay. Science 209: 921–923

Ganapathy R (1982) Evidence for a major meteorite impact on the earth 34 million years ago: Implication on the origin of the North American tektites and Eocene extinction. In: Silver LT, Schultz PH (eds) Geological Implications of Impacts of Large Asteroids and Comets on the Earth. Geological Society of America, Special Paper 190, pp 513–516

Ganapathy R (1983) The Tunguska explosion of 1908: Discovery of meteoritic debris near the explosion site and at the South Pole. Science 220: 1158–1161

Ganapathy R, Larimer JW (1983) Nickel-iron spherules in tektites: non-meteoritic in origin. Earth Planet Sci Lett 65: 225–228

Gardin S, Spezzaferri S, Basso D, Coccioni R (1999) Calcareous nannofossil and planktonic foraminiferal response to a meteorite impact in the Umbria-Marche Massignano section (Northeastern Apennines, Italy). EUG Strasbourg Meeting, Journal of Conference Abstracts 4(1): 271

Garlick CD, Naeser CW, O'Neil JR (1971) A cuban tektite. Geochim Cosmochim Acta 35: 731–734

Garvin JB, Schnetzler CC (1994) The Zhamanshin impact feature: a new class of complex crater? In: Dressler BO, Grieve RAF, Sharpton VL (eds) Large Meteorite Impacts and Planetary Evolution. Geological Society of America, Special Paper 293, pp 249–257

Gash PJS (1971) Dynamic mechanism of the formation of shatter cones. Nature Phys Sci 230: 32–35

Gault DE, Quaide WL, Oberbeck VR (1968) Impact cratering mechanics and structures. In: French BM, Short NM (eds) Shock Metamorphism of Natural Materials. Mono Book Corp., Baltimore, pp 87–99

Gayraud J, Robin E, Rocchia R, Froget L (1996) Formation conditions of oxidized Ni-rich spinel and their relevance to the K/T boundary event. In: Ryder G, Fastovsky D, Gartner S (eds) New Developments Regarding the KT Event and Other Catastrophes in Earth History. Geological Society of America, Special Paper 307, pp 425–443

Gentner W, Storzer D, Wagner GA (1969a) Das Alter von Tektiten und verwandten Gläsern. Naturwissenschaften 56: 255–261

Gentner W, Storzer D, Wagner GA (1969b) New fission track ages of tektites and related glasses. Geochim Cosmochim Acta 33: 1075–1081

Gentner W, Glass BP, Storzer D, Wagner GA (1970) Fission track ages and ages of deposition of deep-sea microtektites. Science 168: 359–361

Gersonde R, Kyte FT, Bleil U, Diekmann B, Flores JA, Gohl K, Grahl G, Hagen R, Kuhn G, Sierro FJ, Völker D, Abelmann A, Bostwick JA (1997) Geological record and reconstruction of the late Pliocene impact of the Eltanin asteroid in the Southern Ocean. Nature 390: 357–363

Gibson RL, Reimold WU (2000) Deeply exhumed impact structures: A case study of the Vredefort Structure, South Africa. In: Gilmour I, Koeberl C (eds) Impacts and the Early Earth. Lecture Notes in Earth Sciences 91, Springer Verlag, Heidelberg-Berlin, pp 249–278

Gifford AC (1924) The mountains of the Moon. New Zealand Journal of Science and Technology 7: 129–142

Gilchrist J, Thorpe AN, Senftle FE (1969) Infrared analysis of water in tektites and other glasses. J Geophys Res 74: 1475–1483

Gill R (ed) (1997) Modern Analytical Geochemistry. Addison Wesley Longman, Harlow (UK), 329 pp

Gilmour I (1998) Geochemistry of carbon in terrestrial impact processes. In: Grady MM, Hutchison R, McCall GJH, Rothery DA (eds) Meteorites: Flux with Time and Impact Effects. Geological Society of London, Special Publication 140, pp 205–216

Gilmour I, Koeberl C (eds) (2000) Impacts and the Early Earth. Lecture Notes in Earth Sciences 91, Springer Verlag, Heidelberg-Berlin.

Gilmour I, Wolbach WS, Anders E (1990) Early environmental effects of the terminal Cretaceous impact. In: Sharpton VL, Ward PD (eds) Global Catastrophes in Earth History. Geological Society of America, Special Paper 247, pp 383–390

Gilmour I, Russell SS, Arden JW, Lee MR, Franchi IA, Pillinger CT (1992) Terrestrial carbon and nitrogen isotopic ratios from Cretaceous-Tertiary boundary nanodiamonds. Science 258: 1624–1626

Ginsburg RN (1997a) An attempt to resolve the contoversy over the end-Cretaceous extinction of planktik foraminifera at El Kef, Tunisia using a blind test; Introduction: background and procedures. Marine Micropaleontology 29: 67–68

Ginsburg RN (1997b) Perspectives on the blind test. Marine Micropaleontology 29: 101–103

Glass BP (1967) Microtektites in deep sea sediments. Nature 214: 372–374

Glass BP (1968) Glassy objects (microtektites?) from deep sea sediments near the Ivory Coast. Science 161: 891–893

Glass BP (1969) Chemical composition of Ivory Coast microtektites. Geochim Cosmochim Acta 33: 1135–1147

Glass BP (1970) Zircon and chromite crystals in a Muong Nong-type tektite. Science 169: 766–769

Glass BP (1972) Bottle green microtektites. J Geophys Res 77: 7057–7064

Glass BP (1978) Australasian microtektites and the stratigraphic age of australites. Geol Soc Am Bull 89: 1455–1458

Glass BP (1979) Zhamanshin crater, a possible source of Australasian tektites. Geology 7: 351–353

Glass BP (1986) No evidence for a 0.8-0.9 m.y. old micro-australite layer in deep sea cores. Earth Planet Sci Lett 77: 428–433

Glass BP (1987) Coesite associated with North American tektite debris in DSDP Site 612 on the continental slope off New Jersey. Lunar Planet Sci 18: 328–329

Glass BP (1989) North American tektite debris and impact ejecta from DSDP Site 612. Meteoritics 24: 209–218

Glass BP (1990) Chronostratigraphy of Upper Eocene microspherules. Comment and Reply. Palaios 5: 387–390

Glass BP, Barlow RA (1979) Mineral inclusions in Muong Nong-type indochinites: Implications concerning parent material and process of formation. Meteoritics 14: 55–67

Glass BP, Burns CA (1987) Late Eocene crystal-bearing spherules: two layers or one? Meteoritics 22: 265–279

Glass BP, Burns CA (1988) Microkrystites: A new term for impact produced glassy spherules containing primary crystallites. Proc Lunar Planet Sci Conf 18th, pp 455–458

Glass BP, Koeberl C (1989) Trace element study of high- and low-refractive index Muong Nong-type tektites from Indochina. Meteoritics 24: 143–146

Glass BP, Koeberl C (1999a) ODP Hole 689B spherules and Upper Eocene microtektite and clinopyroxene-bearing spherule strewn fields. Meteoritics and Planetary Science 34, 197–208

Glass BP, Koeberl C (1999b) North American microtektites in the Indian Ocean? Meteoritics and Planetary Science 34, A43–A44

Glass BP, Pizzuto JE (1994) Geographic variation in Australasian microtektite concentrations: Implications concerning the location and size of the source crater. J Geophys Res 99: 19075–19081

Glass BP, Wu J (1992) Search for microaustralites in deep-sea sediments less than 20000 years old. Meteoritics 27: 605–608

Glass BP, Wu J (1993) Coesite and shocked quartz discovered in the Australasian and North American microtektite layers. Geology 21: 435–438

Glass BP, Zwart MJ (1977) North American microtektites, radiolarian extinctions, and the age of the Eocene/Oligocene boundary. In: Swain FM (ed) Stratigraphic Micropaleontology of Atlantic Basins and Borderlands. Elsevier, Amsterdam, pp 553–568

Glass BP, Zwart PA (1979) The Ivory Coast microtektite strewn field: new data. Earth Planet Sci Lett 43: 336–342

Glass BP, Barker RN, Storzer D, Wagner GA (1973) North American microtektites from the Caribbean Sea and their fission track ages. Earth Planet Sci Lett 19: 184–192

Glass BP, Swincki MB, Zwart PA (1979) Australasian, Ivory Coast and North American tektite strewn field: Size, mass and correlation with geomagnetic reversals and other earth events. Proc 10th Lunar Planet Sci Conf: 2535–2545

Glass BP, Burns CA, Lerner DH, Sanfilippo A (1984) North American tektites and microtektites from Barbados, West Indies. Meteoritics 19: 228–229

Glass BP, Burns CA, Crosbie JR, DuBois DL (1985) Late Eocene North American microtektites and clinopyroxene bearing spherules. Proc Lunar Planet Sci Conf 16th, J Geophys Res 90: D175–D196

Glass BP, Muenow DW, Aggrey KE (1986a) Further evidence for the impact origin of tektites. Meteoritics 21: 369–370

Glass BP, Hall CM, York D (1986b) $^{40}Ar/^{39}Ar$ laser-probe dating of North American tektite fragments from Barbados and the age of the Eocene-Oligocene boundary. Chemical Geology 5: 181–186

Glass BP, Senftle FE, Muenow DW, Aggrey KE, Thorpe AN (1988) Atomic bomb glass beads: Tektite and microtektite analogs. In: Konta J (ed) Proceedings of the 2nd International Conference on Natural Glasses. Charles University, Prague, pp 361–369

Glass BP, Wasson JT, Futrell DS (1990) A layered moldavite containing baddeleyite. Proc Lunar Planet Sci Conf 20th, pp 415–420

Glass BP, Kent DV, Schneider DA, Tauxe L (1991) Ivory Coast microtektite strewn field: Description and relation to the Jaramillo geomagnetic event. Earth Planet Sci Lett 107: 182–196

Glass BP, Koeberl C, Blum JD, Senftle F, Izett GA, Evans BJ, Thorpe AN, Povenmire H, Strange RL (1995) A Muong Nong-type Georgia tektite. Geochim Cosmochim Acta 59: 4071–4082

Glass BP, Koeberl C, Blum JB, McHugh CMG (1998) Upper Eocene tektite and impact ejecta layer on the continental slope off New Jersey. Meteoritics and Planetary Science 33: 229–242

Glen W (1994) The Mass Extinction Debates. Stanford University Press, Stanford, California, 371 pp

Glen W (1998) A manifold current upheaval in science. Earth Sciences History 17: 190–209

Glikson AY (ed) (1996) Australian impact structures. AGSO J Austral Geol Geophys 16: 371–607

Göbel E, Reimold WU, Baddenhausen H, Palme H (1980) The projectile of the Lapajärvi impact crater. Zeitschrift für Naturforschung 35a: 197–203

Goltrant O, Cordier P, Doukhan JC (1991) Planar deformation features in shocked quartz: A transmission electron microscopy investigation. Earth Planet Sci Lett 106: 103–115

Goltrant O, Leroux H, Doukhan J-C, Cordier P (1992) Formation mechanism of planar deformation features in naturally shocked quartz. Phys Earth Planet Int 74: 219–240

Gonzalvo C, Molina E (1992) Estudio cuantitativo de los foraminiferos planctònicos en el estratotipo del limite Eoceno/Oligoceno en Massignano (Apeninos, Italia). Geogaceta Soc Geol Espana 12: 64–67

Goodwin AM (1976) Giant impacting and the development of the continental crust. In: Windley BF (ed) The Early History of the Earth. J. Wiley & Sons, London, pp 77–95

Gorter JD, Glikson AY (2000) Origin of a Late Eocene to pre-Miocene buried crater and breccia lens at Fohn-1, North Bonaparte Basin, Timor Sea: A probable extraterrestrial connection. Meteoritics and Planetary Science 35: in press

Gorter JD, Gostin VA, Plummer PS (1989) The enigmatic subdurface Tookoonooka complex in south-west Queensland: its impact origin and implications for hydrocarbon accumulations. In: O'Neil BJ (ed) The Cooper and Eromanga Basins, Australia. Proceedings of the Petroleum Exploration Society, the Society of Petroleum Engineers, and the Australian Society of Exploration Geophysicists (SA Branches), Adelaide, pp 441–456

Gostin VA, Zbik M (1999) Petrology and microstructure of distal impact ejecta from the Flinders Ranges, Australia. Meteoritics and Planetary Science 34: 587–592

Gostin VA, Haines PW, Jenkins RJE, Compston W, Williams IS (1986) Impact ejecta horizon within late Precambrian shales, Adelaide Geosyncline, south Australia. Science 233: 198–200

Gostin VA, Keays RR, Wallace MW (1989) Iridium anomaly from the Acraman impact ejecta horizon: Impacts can produce sedimentary iridium peaks. Nature 340: 542–544

Gradstein FM, Agterberg FP, Ogg JG, Hardenbol J, van Veen P, Thierry J, Huang Z (1994) A Mesozoic time scale. J Geophys Res 99: 24051–24074

Grant JA, Schultz PH (1993) Erosion of ejecta at Meteor Crater, Arizona. J Geophys Res 98: 15033–15047

Gratz AJ, Tyburczy J, Christie JM, Ahrens T, Pongratz P (1988) Shock deformation of deformed quartz. Phys Chem Minerals 16: 221–233

Gratz AJ, Nellis WJ, Christie JM, Brocious W, Swegle J, Cordier P (1992a) Shock metamorphism of quartz with initial temperatures -170 to +1000°C. Phys Chem Minerals 19: 267–288

Gratz AJ, Nellis WJ, Hinsey NA (1992b) Laboratory simulation of explosive volcanic loading and implications from the cause of the K/T boundary. Geophys Res Lett 19: 1391–1394

Gratz AJ, Fisler DK, Bohor BF (1996) Distinguishing shocked from tectonically deformed quartz by the use of the SEM and chemical etching. Earth Planet Sci Lett 142: 513–521

Graup G (1981) Terrestrial chondrules, glass spherules and accretionary lapilli from the suevite, Ries crater, Germany. Earth Planet Sci Lett 55: 407–418

Graup G (1999) Carbonate-silicate liquid immiscibility upon impact melting: Ries crater, Germany. Meteoritics and Planetary Science 34: 425–438

Green DH (1972) Archaean greenstone belts may include terrestrial equivalents of lunar maria? Earth Planet Sci Lett 15: 263–270

Greene MT (1998) Alfred Wegener and the origin of lunar craters. Earth Sciences History 17: 111–138

Grier JA, Swindle TD, Kring DA, Melosh HJ (1999) Argon-40/argon-39 analyses of samples from the Gardnos impact structure, Norway. Meteoritics and Planetary Science 34: 803–807

Grieve RAF (1980) Impact bombardment and its role in protocontinental growth of the early earth. Precambrian Research 10: 217–248

Grieve RAF (1982) The record of impact on Earth: Implication for a major Cretaceous/Tertiary impact event. In: Silver LT, Schultz PH (eds) Geological Implications of Impacts of Large Asteroids and Comets on the Earth. Geological Society of America, Special Paper 190, pp 25–37

Grieve RAF (1987) Terrestrial impact structures. Annual Reviews of Earth and Planetary Science 15: 245–270

Grieve RAF (1991) Terrestrial impact: The record in the rocks. Meteoritics 26: 175–194

Grieve RAF, Cintala MJ (1992) An analysis of differential impact melt-crater scaling and implications for the terrestrial impact record. Meteoritics 27: 526–538

Grieve RAF, Masaitis VL (1994) The economic potential of terrestrial impact craters. International Geology Review 36: 105–151

Grieve RAF, Pilkington M (1996) The signature of terrestrial impacts. AGSO J Austral Geol Geophys 16: 399–420

Grieve RAF, Shoemaker EM (1994) Terrestrial impact cratering. In: Gehrels T (ed) Hazards due to Comets and Asteroids. University of Arizona Press, Tucson, pp 417–462

Grieve RAF, Therriault AM (1995) Planar deformation features in quartz: Target effects. Lunar Planet Sci 26: 515–516

Grieve RAF, Sharpton VL, Goodacre AK, Garvin JB (1985) A perspective on the evidence for periodic cometary impacts on Earth. Earth Planet Sci Lett 76: 1–9

Grieve RAF, Sharpton VL, Rupert JD, Goodacre AK (1988) Detecting a periodic signal in the terrestrial cratering record. Proc Lunar Planet Sci Conf 18th, pp 375–382

Grieve RAF, Rupert J, Smith J, Therriault AM (1995) The record of terrestrial impact cratering. GSA Today 5: 189 & 194–196

Grieve RAF, Langenhorst F, Stöffler D (1996) Shock metamorphism in nature and experiment: II. Significance in geoscience. Meteoritics and Planetary Science 31: 6–35

Guerrera F, Tonelli G, Veneri F, Domeniconi G (1986) Caratteri lito-sedimentologici e mineralogico-petrografici di vulcanoclastiti mioceniche presenti nella successione umbro-marchigiana. Boll Soc Geol It 105: 307–325

Gurov EP, Gurova EP (1980) Shock metamorphosed rocks from the El'gygytgyn meteorite crater in Chukchi National Okrug (in Russian). Meteoritika 39: 102–109

Habib D, Olsson RK, Liu C, Moshkovitz S (1996) High-resolution biostratigraphy of sea-level low, biotic extinction, and chaotic sedimentation at the Cretaceous-Tertiary boundary in Alabama, north of the Chicxulub crater. In: Ryder G, Fastovsky D, Gartner S (eds) New Developments Regarding the KT Event and Other Catastrophes in Earth History. Geological Society of America, Special Paper 307, pp 243–252

Haines PW, Therriault AM, Kelley SP (1999) Evidence for Mid-Cenozoic(?), low-angle multiple impacts in South Australia. Meteoritics and Planetary Science 34: A49–A50

Hallam A (1988) A revaluation of Jurassic eustasy in the light of new data and the revised Exxon curve. In: Wilgus CK, Hastings BS, Ross CA, Posamentier H, Van Wagoner J, Kendall CGStC (eds) Sea-level changes: an integrated approach. Society of Economic Paleontologists and Mineralogists, Special Publication 42: 261–273

Hallam A, Perch-Nielsen K (1990) The biotic record of events in the marine realm at the end of the Cretaceous: calcareous, siliceous, and organic-walled microfossils, and macroinvertebrates. Tectonophysics 171: 347–357

Hallam A, Wignall PB (1997) Mass Extinction and their Aftermath, Oxford Univ Press, Oxford, UK, 320 pp

Haq BU, Hardenbol J, and Vail PR (1988) Mesozoic and Cenozoic chronostratigraphy and eustatic cycles. In: Sea-level changes: an integrated approach. Soc Ec Paleont Mineral Sp Publ 42: 261–27

Harland WB, Armstrong RL, Cox AV, Craig LE, Smith AG, Smith DG (1990) A geologic time scale 1989. Cambridge University Press, Oxford, 263 pp

Harries PJ, Kauffman EG (1990) Pattern of survival and recovery following the Cenomanian-Turonian (Late Cretaceous) mass extinction in the Western Interior Basin, United States. In: Kauffman EG, Walliser OH (eds) Extinction Events in Earth History. Springer, Heidelberg, pp 277–298

Hart MB (1996) Recovery of the food chain after the Late Cretaceous Cenomanian extinction event. In: Hart MB (ed) Biotic Recovery from Mass Extinction Events. Geological Society London, Special Publication 102, pp 265–277

Hartmann WK (1975) Lunar "cataclysm": A misconception? Icarus 24: 181–187

Hartung JB, Anderson RR (1988) A compilation of information and data on the Manson impact structure, LPI Technical Report 88-08, Lunar and Planetary Institute, Houston, 32 pp

Hartung JB, Koeberl C (1994) In search of the Australasian tektite source crater: The Tonle Sap hypothesis. Meteoritics 29: 411–416

Hartung JB, Rivolo AR (1979) A possible source in Cambodia for Australasian tektites. Meteoritics 14: 153–160

Hartung JB, Kunk MJ, Anderson RR (1990) Geology, geophysics, and geochronology of the Manson impact structure. In: Sharpton VL, Ward PD (eds) Global Catastrophes in Earth History. Geological Society of America, Special Paper 247, pp 207–222

Hassler DR, Peucker-Ehrenbrink B, Ravizza GE (1999) Rapid determination of Os isotopic composition by sparging OsO_4 into a magnetic-sector ICP-MS. Chemical Geology, in press

Hazel JE (1989) Chronostratigraphy of upper Eocene microspherules. Palaios 4: 318–329

Heisler J, Tremaine S (1989) How dating uncertainties affect the detection of periodicity in extinctions and craters. Icarus 77: 213–219

Herbert TD, Fischer AG (1986) Milankovitch climatic origin of black shale rhythms in central Italy. Nature 321: 739–743

Heymann D, Chibante LP, Brooks RR, Wolbach WS, Smalley RE (1994) Fullerenes in the Cretaceous-Tertiary boundary layer. Science 265: 645–647

Heymann D, Chibante LPF, Brooks RR, Wolbach WS, Smit J, Korochantsev A, Nazarov MA, Smalley RE (1996) Fullerenes of possible wildfire origin in Cretaceous-Tertiary boundary sediments. In: Ryder G, Fastovsky D, Gartner S (eds) New Developments Regarding the KT Event and Other Catastrophes in Earth History. Geological Society of America, Special Paper 307, pp 453–464

Hildebrand AR, Penfield GT, Kring DA, Pilkington M, Carmargo ZA, Jacobsen SB, Boynton WV (1991) Chicxulub crater: A possible Cretaceous-Tertiary boundary impact crater on the Yucatan Peninsula, Mexico. Geology 19: 867–871

Hildebrand AR, Moholy-Nagy H, Koeberl C, Senftle F, Thorpe AN, Smith PE, York D (1994) Tektites found in the ruins of the Maya city of Tikal, Guatemala. Lunar Planet Sci 25: 549–550

Hildebrand AR, Pilkington M, Connors M, Cortiz-Aleman C, Chavez RE (1995) Size and structure of the Chicxulub crater revealed by horizontal gravity gradients and cenotes. Nature 376: 415–417

Hills JG, Nemchinov IV, Popov S, Teterev AV (1994) Tsunami generated by small asteroidal impacts. In: Gehrels T (ed) Hazards due to Comets and Asteroids. University of Arizona Press, Tucson, pp 779–789

Hodge P (1994) Meteorite Craters and Impact Structures of the Earth. Cambridge University Press, Cambridge, 124 pp

Hodych JP, Dunning GR (1992) Did the Manicouagan impact trigger end-of-Triassic mass extinction? Geology 20: 51–54

Holser WT, Magaritz M (1992) Cretaceous/Tertiary and Permian/Triassic boundary events compared. Geochim Cosmochim Acta 56: 3297–3309

Holser WT, Schönlaub HP, Attrep M Jr, Boeckelmann K, Klein P, Magaritz M, Orth CJ, Fenninger A, Jenny C, Kralik M, Mauritsch H, Pak E, Schramm JM, Stattegger K, Schmöller R (1989) A unique geochemical record at the Permian-Triassic boundary. Nature 337: 39–44

Horn P, Müller-Sohnius D, Köhler H, Graup G (1985) Rb-Sr systematics of rocks related to the Ries crater, Germany. Geochim Cosmochim Acta 49: 384–392

Horn P, Müller-Sohnius D, Schaaf P, Kleinmann B, Storzer D (1997) Potassium-argon and fission-track dating of Libyan Desert Glass, and strontium and neodymium isotope constraints in its source rocks. In: Proceedings, Silica 96, Meeting on Libyan Desert Glass and related desert events. Pyramids, Milan, Italy, pp 59–76

Hörz F (1968) Statistical measurements of deformation structures and refractive indices in experimentally shock loaded quartz. In: French BM, Short NM (eds) Shock Metamorphism of Natural Materials. Mono Book Corp., Baltimore, pp 243–253

Hörz F (1982) Ejecta of the Ries crater, Germany. In: Silver LT, Schultz PH (eds) Geological Implications of Impacts of Large Asteroids and Comets on the Earth. Geological Society of America, Special Paper 190, pp 39–55

Hörz F, Ahrens TJ (1969) Deformation of experimentally shocked biotite. Am J Sci 267: 1213–1229

Hörz F, Quaide WL (1972) Debye-Scherrer investigations of experimentally shocked silicates. The Moon 6: 45–82

Hörz F, Ostertag R, Rainey DA (1983) Bunte Breccia of the Ries: Continuous deposits of large impact craters. Reviews of Geophysics and Space Physics 21: 1667–1725

Hörz F, See TH, Murali AV, Blanchard DP (1989) Heterogeneous dissemination of projectile materials in the impact melts from Wabar crater. Proc Lunar Planet Sci Conf 19th, pp 697–710

Hough RM, Gilmour I, Pillinger CT, Arden JW, Gilkes KWR, Yuan J, Milledge HJ (1995) Diamond and silicon carbide in suevite from the Nördlinger Ries impact crater. Nature 378: 41–44

Hough RM, Gilmour I, Pillinger CT, Langenhorst F, Montanari A (1997) Diamonds from the iridium-rich K-T boundary layer at Arroyo el Mimbral, Tamaulipas, Mexico. Geology 25: 1019–1022

Hough RM, Gilmour I, Pillinger CT (1998) Impact nanodiamonds in Cretaceous-Tertiary boundary fireball and ejecta layers: Comparison with shock-produced diamond and a search for lonsdaleite. Meteoritics and Planetary Science 33: A70–A71

Howard KT, Bunopas S, Burrett CF, Haines PW, Norman MD (2000) The 770ka tektite producing impact event: Evidence for distal environmental effects in NE Thailand. Lunar Planet Sci 31: abs. #1308

Hoyt WG (1987) Coon Mountain Controversies. University of Arizona Press, Tucson, 442 pp

Huber BT (1996) Evidence for planktonic foraminifer reworking versus survivorship across the Cretaceous-Tertiary boundary at high latitudes. In: Ryder G, Fastovsky D, Gartner S (eds) New Developments Regarding the KT Event and Other Catastrophes in Earth History. Geological Society of America, Special Paper 307, pp 319–334

Huber H, Koeberl C, King DT Jr, Petruny LW, Montanari A (1999) High resolution iridium stratigraphy across a distal impactoclastic layer in the Late Eocene type section of Massignano, Italy. Meteoritics and Planetary Science 34: A56–A57

Huber H, Koeberl C, McDonald I, Reimold WU (2000) Use of γ-γ coincidence spectrometry in the geochemical study of diamictites from South Africa. J Radioanalyt Nucl Chem: in press

Huffman AR, Reimold WU (1996) Experimental constraints on shock-induced microstructures in naturally deformed silicates. Tectonophysics 256: 165–217

Huffman AR, Brown JM, Carter NL, Reimold WU (1993) The microstructural response of quartz and feldspar under shock loading at variable temperatures. J Geophys Res 98: 22171–22197

Hut P, Alvarez W, Elder W, Hansen TA, Kauffman EG, Keller G, Shoemaker EM, Weissman P (1987) Comet showers as a cause of mass extinctions. Nature 329: 118–126

Iturralde-Vinent MA (1992) A short note on the Cuban Late Maastrichtian megaturbidite (an impact-derived deposit?). Earth Planet Sci Lett 109: 225–228

Ivanov BA, Badjukov DD, Yakovlev OI, Gerasimov MV, Dikov YP, Pope KO, Ocampo AC (1996) Degassing of sedimentary rocks due to Chicxulub impact: Hydrocode and physical simulations. In: Ryder G, Fastovsky D, Gartner S (eds) New Developments Regarding the KT Event and Other Catastrophes in Earth History. Geological Society of America, Special Paper 307, pp 125–139

Izett GA (1990) The Cretaceous-Tertiary boundary interval, Raton Basin, Colorado and New Mexico, and its content of shock-metamorphosed minerals; Evidence relevant to the K/T boundary impact-extinction hypothesis. Geological Society of America, Special Paper 249, 100 pp

Izett GA (1991) Tektites in Cretaceous-Tertiary boundary rocks on Haiti and their bearing on the Alvarez impact extinction hypothesis. J Geophys Res 96: 20879–20905

Izett GA, Obradovich JD (1992) Laser-fusion $^{40}Ar/^{39}Ar$ ages of Australasian tektites. Lunar Planet Sci 23: 593–594

Izett GA, Cobban WA, Obradovich JD, Kunk MJ (1993) The Manson impact structure: ^{40}Ar-^{39}Ar age and its distal impact ejecta in the Pierre shale in southeastern South Dakota. Science 262: 729–732

Izett GA, Masaitis VL, Shoemaker EM, Dalrymple GB, Steiner MB (1994) Eocene age of the Kamensk buried crater of Russia. Lunar and Planetary Institute, Houston, Contribution 825, p 55

Izett GA, Cobban WA, Dalrymple GB, Obradovich JD (1998) $^{40}Ar/^{39}Ar$ age of the Manson impact structure, Iowa, and correlative impact ejecta in the Crow Creek Member of the Pierre Shale (Upper Cretaceous), South Dakota and Nebraska. Geol Soc Am Bull 110: 361–376

Jackson SE, Fryer BJ, Gosse W, Healey DC, Longerich HP, Strong DF (1990) Determination of the precious metals in geological materials by inductively coupled plasma-mass spectrometry (ICP-MS) with nickel sulphide fire-assay collection and tellurium coprecipitation. Chemical Geology 83: 119–132

Jakes P, Sen S, Matsuishi K (1991) Tektites, experimental equivalents and properties of super-heated (impact) melts. Lunar Planet Sci 22: 633

Jakes P, Sen S, Matsuishi K, Reid AM, King EA, Casanova I (1992) Silicate melts at super liquidus temperatures: Reduction and volatilization. Lunar Planet Sci 23: 599–600

Jansa LF (1993) Cometary impacts into ocean: their recognition and the threshold constraint for biological extinctions. Palaeogeography Palaeoclimatology Palaeoecology 104: 271–286

Jansa LF, Pe-Piper G (1987) Identification of an underwater extra-terrestrial impact crater. Nature 327: 612–614

Jansa LF, Aubry MP, Gradstein FM (1990) Comets and extinctions; Cause and effects? In: Sharpton VL, Ward PD (eds) Global Catastrophes in Earth History. Geological Society of America, Specical Paper 247, pp 223–232

Jéhanno C, Boclet D, Froget L, Lambert B, Robin R, Rocchia R, Turpin L (1992) The Cretaceous-Tertiary boundary at Beloc, Haiti: No evidence for an impact in the Caribbean area. Earth Planet Sci Lett 109: 229–241

Jessberger EK (1988) ^{40}Ar-^{39}Ar dating of the Haughton impact structure. Meteoritics 23: 233–234

Jessberger EK, Gentner W (1972) Mass spectrometric analysis of gas inclusions in Muong Nong glass and Libyan Desert Glass. Earth Planet Sci Lett 14: 221–225

Jessberger EK, Reimold WU (1980) A late Cretaceous ^{40}Ar-^{39}Ar age for the Lappajärvi impact crater, Finland. J Geophys 48: 57–59

Johnsson MJ, Reynolds RC (1986) Clay mineralogy of shale-limestone rhythmites in the Scaglia Rossa (Turonian-Eocene), Italian Apennines. J Sed Petrol 56: 501–509

Jones WB (1985) Chemical analyses of Bosumtwi crater target rocks compared with Ivory Coast tektites. Geochim Cosmochim Acta 49: 2569–2576

Jones WB, Bacon M, Hastings DA (1981) The Lake Bosumtwi impact crater, Ghana. Geol Soc Am Bull 92: 342–349

Kaiho K (1991) Global changes of paleogene aerobic-anaerobic benthic foraminifera and deep-sea circulation. Palaeogeography Palaeoclimatology Palaeoecology 83: 65–85

Kajiwara Y, Kaiho K (1992) Oceanic anoxia at the Cretaceous/Tertiary boundary supported by the sulfur isotope record. Palaeogeography Palaeoclimatology Palaeoecology 99: 151–162

Kamo SL, Krogh TE (1995) Chicxulub crater source for shocked zircons from the Cretaceous-Tertiary boundary layer, Saskatchewan: Evidence from new U-Pb data. Geology 23: 281–284

Kamo SL, Reimold WU, Krogh TE, Colliston WP (1996) A 2.023 Ga age for the Vredefort impact event and a first report of shock metamorphosed zircons in pseudotachylitic breccias and Granophyre. Earth Planet Sci Lett 144: 369–388

Katz ME, Pak DK, Dickens GR, Miller KG (1999) The source and fate of massive carbon input during the latest Paleocene thermal maximum. Science 286: 1531–1533

Kauffman EG (1984) The fabric of Cretaceous marine extinctions. In: Berggren WA, van Couvering JA (eds) Catastrophes and Earth History. Princeton University Press, pp 151–246

Keller G (1986) Stepwise mass extinction and impact events: Late Eocene to Early Oligocene. Marine Micropaleontology 10: 267–293

Keller G (1997) Analysis of El Kef blind test I. Marine Micropaleontology 29: 89–93

Keller G, D'Hondt SL, Vallier TL (1983) Multiple microtektite horizons in Upper Eocene marine sediments: No evidence for mass extinctions. Science 221: 150–152

Keller G, D'Hondt SL, Orth CJ, Gilmore JS, Oliver PQ, Shoemaker EM, Molina E (1987) Late Eocene impact microspherules: Stratigraphy, age, and geochemistry. Meteoritics 22: 25–60

Keller G, Barrera E, Schmitz B, Mattson E (1993) Gradual mass extinction, species survivorship, and long-term environmental changes across the Cretaceous-Tertiary boundary in high latitudes. Geol Soc Am Bull 101: 1408–1419

Kelley SP, Spray JG (1997) A Late Triassic age for the Rochechouart impact structure, France. Meteoritics and Planetary Science 32: 629–636

Kennett JP, Stott LD (1991) Abrupt deep-sea warming, paleoceanographic changes, and benthic extinctions at the end of the Paleocene. Nature 353: 225–229

Kent DV (1998) Impacts on Earth in the Late Triassic. Nature 395: 126

Kieffer SW (1971) Shock metamorphism of the Coconino Sandstone at Meteor Crater, Arizona. J Geophys Res 76: 5449–5473

Kieffer SW, Simonds CH (1980) The role of volatiles and lithology in the impact cratering process. Reviews of Geophysics and Space Physics 18: 143–181

Kieffer SW, Phakey PP, Christie JM (1976a) Shock processes in porous quartzite: transmission electron microscope observations and theory. Contrib Mineral Petrol 59: 41–93

Kieffer SW, Schaal RB, Gibbons R, Hörz F, Milton DJ, Dube A (1976b) Shocked basalt from Lonar impact crater, India, and experimental analogues. Proc 7th Lunar Sci Conf, pp 1391–1412

Kilesev NP, Korotushenko YG (1986) The Bigach astrobleme in eastern Kazakhstan (in Russian). Meteoritika 45: 119–121

King DT Jr, Petruny LW (1999) Do we need stratigraphic nomenclature for impact-derived materials? GSA Annual Meeting, Abstracts with Programs 31(7): A65

King EA (1977) The origin of tektites: A brief review. American Scientist 65: 212–218

King EA (1979) Comment on "Remnants of a probable Tertiary impact crater in south Texas". Geology 7: 328

Kingma KJ, Meade C, Hemley RJ, Mao H-k, Veblen DR (1993) Microstructural observations of α-quartz amorphization. Science 259: 666–669

Kirschner CE, Grantz A, Mullen MW (1992) Impact origin of the Avak Structure, Arctic Alaska, and genesis of the Barrow gas fields. American Association of Petroleum Geologists Bulletin 76: 651–679

Kleinmann B (1969) The breakdown of zircon observed in the Libyan Desert Glass as evidence of its impact origin. Earth Planet Sci Lett 5: 497–501

Knie K, Korschinek G, Faesterman T, Wallner C, Scholten J, Hillebrandt W (1999) Indication of supernova produced [60]Fe activity on Earth. Physical Review Letters 83: 1–21

Koeberl C (1986) Geochemistry of tektites and impact glasses. Annual Reviews of Earth and Planetary Science 14: 323–350

Koeberl C (1988) The Cuban tektite revisited. Meteoritics 23: 161–165

Koeberl C (1989a) New estimates of area and mass for the North American tektite strewn field. Proc Lunar Planet Sci Conf 19th: 745–751

Koeberl C (1989b) Iridium enrichment in volcanic dust from blue ice fields, Antarctica, and possible relevance to the K/T boundary event. Earth Planet Sci Lett 91: 317–322

Koeberl C (1990) The geochemistry of tektites: An overview. Tectonophysics 171: 405–422

Koeberl C (1992a) Geochemistry and origin of Muong Nong-type tektites. Geochim Cosmochim Acta 56: 1033–1064

Koeberl C (1992b) Water content of glasses from the K/T boundary, Haiti: Indicative of impact origin. Geochim Cosmochim Acta 56: 4329–4332

Koeberl C (1993a) Extraterrestrial component associated with Australasian microtektites in a core from ODP Site 758B. Earth Planet Sci Lett 119: 453–458

Koeberl C (1993b) Chicxulub crater, Yucatan: Tektites, impact glasses, and the geochemistry of target rocks and breccias. Geology 21: 211–214

Koeberl C (1993c) Instrumental neutron activation analysis of geochemical and cosmochemical samples: A fast and reliable method for small sample analysis. J Radioanalyt Nucl Chem 168: 47–60

Koeberl C (1994a) African meteorite impact craters: Characteristics and geological importance. J African Earth Sci 18: 263–295

Koeberl C (1994b) Tektite origin by hypervelocity asteroidal or cometary impact: Target rocks, source craters, and mechanisms. In: Dressler BO, Grieve RAF, Sharpton VL (eds) Large meteorite impacts and planetary evolution. Geological Society of America, Special Paper 293, pp 133–152

Koeberl C (1996) Chicxulub - The K-T boundary impact crater: A review of the evidence, and an introduction to impact crater studies. Abhandlungen der Geologischen Bundesanstalt (Wien) 53: 23–50

Koeberl C (1997a) Impact cratering: The mineralogical and geochemical evidence. In: Johnson KS and Campbell JA (eds) Proceedings, "The Ames Structure and Similar Features", Oklahoma Geological Survey Circular 100, pp 30–54

Koeberl C (1997b) Libyan Desert Glass: geochemical composition and origin. In: Proceedings, Silica 96, Meeting on Libyan Desert Glass and related desert events. Pyramids, Milan, Italy, pp 121–132

Koeberl C (1998) Identification of meteoritical components in impactites. In: Grady MM, Hutchison R, McCall GJH, Rothery DA (eds) Meteorites: Flux with Time and Impact Effects. Geological Society of London, Special Publication 140, pp 133–152

Koeberl C, Anderson RR (eds) (1996a) The Manson Impact Structure, Iowa: Anatomy of an Impact Crater. Geological Society of America, Special Paper 302, 484 pp

Koeberl C, Anderson RR (1996b) Manson and company: Impact structures in the United States. In: Koeberl C, Anderson RR (eds) The Manson Impact Structure, Iowa: Anatomy of an Impact Crater. Geological Society of America, Special Paper 302, pp 1–29

Koeberl C, Beran A (1988) Water content of tektites and impact glasses and related chemical studies. Proc Lunar Planet Sci Conf 18th, pp 403–408

Koeberl C, Fredriksson K (1986) Impact glasses from the Zhamanshin crater (USSR): Chemical composition and discussion of origin. Earth Planet Sci Lett 78: 80–88

Koeberl C, Glass BP (1988) Chemical composition of North American microtektites and tektite fragments from Barbados and DSDP Site 612 on the continental slope off New Jersey. Earth Planet Sci Lett 87: 286–292

Koeberl C, Huber H (2000) Optimization of the multiparameter γ-γ coincidence spectrometry for the determination of iridium in geological materials. J Radioanalyt Nucl Chem (Proc. MTAA-10), in press

Koeberl C, Reimold WU (1995a) The Newporte impact structure, North Dakota. Geochim Cosmochim Acta 59: 4747–4767

Koeberl C, Reimold WU (1995b) Early Archaean spherule beds in the Barberton Mountain Land, South Africa: no evidence for impact origin. Precambrian Research 74: 1–33

Koeberl C, Reimold WU (1999) Comment on "Argument supporting explosive igneous activity for the origin of "cryptoexplosion" structures in the midcontinent, United States". Geology 27: 283–284

Koeberl C, Sharpton VL (1988) Giant impacts and their influence on the early earth. In: Papers presented to the Conference on "Origin of the Earth". Lunar and Planetary Institute, Houston, pp 47–48

Koeberl C, Shirey SB (1993) Detection of a meteoritic component in Ivory Coast tektites with rhenium-osmium isotopes. Science 261: 595–598

Koeberl C, Shirey SB (1996) Re-Os isotope study of rocks from the Manson impact structure. In: Koeberl C, Anderson RR (eds) The Manson Impact Structure, Iowa: Anatomy of an Impact Crater. Geological Society of America, Special Paper 302, pp 331–339

Koeberl C, Shirey SB (1997) Re-Os isotope systematics as a diagnostic tool for the study of impact craters and ejecta. Paleogeography Paleoclimatology Paleoecology 132: 25–46

Koeberl C, Sigurdsson H (1992) Geochemistry of impact glasses from the K/T boundary in Haiti: Relation to smectites, and a new type of glass. Geochim Cosmochim Acta 56: 2113–2129

Koeberl C, Storzer D (1988) Chemical composition and fission track age of Zhamanshin crater glass. In: Konta J (ed) Proc 2nd Intern Conf Natural Glasses. Charles University, Praha, pp 207–213

Koeberl C, Brandstätter F, Niedermayr G, Kurat G (1988) Moldavites from Austria. Meteoritics 23: 325–332

Koeberl C, Sharpton VL, Murali AV, Burke K (1990) Kara and Ust-Kara impact structures (USSR) and their relevance to the K/T boundary event. Geology 18: 50–53

Koeberl C, Reimold WU, Boer RH (1993) Geochemistry and mineralogy of Early Archean spherule beds, Barberton Mountain Land, South Africa: Evidence for origin by impact doubtful. Earth Planet Sci Lett 119: 441–452

Koeberl C, Reimold WU, Shirey SB (1994a) Saltpan impact crater, South Africa: Geochemistry of target rocks, breccias, and impact glasses, and osmium isotope systematics. Geochim Cosmochim Acta 58: 2893–2910

Koeberl C, Reimold WU, Shirey SB, Le Roux FG (1994b) Kalkkop crater, Cape Province, South Africa: Confirmation of impact origin using osmium isotope systematics. Geochim Cosmochim Acta 58: 1229–1234

Koeberl C, Sharpton VL, Schuraytz BC, Shirey SB, Blum JD, Marin LE (1994c) Evidence for a meteoritic component in impact melt rock from the Chicxulub structure. Geochim Cosmochim Acta 58: 1679–1684

Koeberl C, Reimold WU, Kracher A, Träxler B, Vormaier A, Körner W (1996a) Mineralogical, petrological, and geochemical studies of drill cores from the Manson impact structure, Iowa. In: Koeberl C, Anderson RR (eds) The Manson Impact Structure, Iowa: Anatomy of an Impact Crater. Geological Society of America, Special Paper 302, pp 145–219

Koeberl C, Reimold WU, Brandt D (1996b) Red Wing Creek structure, North Dakota: Petrographical and geochemical studies, and confirmation of impact origin. Meteoritics and Planetary Science 31, 335–342.

Koeberl C, Poag CW, Reimold WU, Brandt D (1996c) Impact origin of Chesapeake Bay structure and the source of North American tektites. Science 271: 1263–1266

Koeberl C, Reimold WU, Shirey SB (1996d) A Re-Os isotope and geochemical study of the Vredefort Granophyre: Clues to the origin of the Vredefort structure, South Africa. Geology 24: 913–916

Koeberl C, Masaitis VL, Shafranovsky GI, Gilmour I, Langenhorst F, Schrauder M (1997a) Diamonds from the Popigai impact structure, Russia. Geology 25: 967–970

Koeberl C, Bottomley R, Glass BP, Storzer D (1997b) Geochemistry and age of Ivory Coast tektites and microtektites. Geochim Cosmochim Acta 61: 1745–1772

Koeberl C, Armstrong RA, Reimold WU (1997c) Morokweng, South Africa: A large impact structure of Jurassic-Cretaceous boundary age. Geology 25: 731–734

Koeberl C, Reimold WU, Shirey SB (1998a) The Aouelloul Crater, Mauritania: On the problem of confirming the impact origin of a small crater. Meteoritics and Planetary Science 33: 513–517

Koeberl C, Reimold WU, Blum JB, Chamberlain CP (1998b) Petrology and geochemistry of target rocks from the Bosumtwi impact structure, Ghana, and comparison with Ivory Coast tektites. Geochim Cosmochim Acta 62: 2179–2196

Koeberl C, Denison C, Ketcham R, Reimold WU (1998c) High-resolution X-ray computed tomography of impactites: Suevite from the Bosumtwi crater, Ghana. Meteoritics and Planetary Science 33: A84–A85

Koeberl C, Glass BP, Chaussidon M (1999a) Bottle-green microtektites from the Australasian tektite strewn field: Did they form by jetting from the top layer of the target surface? GSA Annual Meeting, Abstracts with Programs 31(7): A63–A64

Koeberl C, Simonson BM, Reimold WU (1999b) Geochemistry and petrography of a Late Archean spherule layer in the Griqualand West Basin, South Africa. Lunar Planet Sci 30: abs. #1755

Koeberl C, Reimold WU, McDonald I, Rosing M (2000a) Search for petrographical and geochemical evidence for the late heavy bombardment on Earth in Early Archean rocks from Isua, Greenland. In: Gilmour I, Koeberl C (eds) Impacts and the Early Earth. Lecture Notes in Earth Sciences 91, Springer, Heidelberg, pp 73–98

Koeberl C, Blum JD, Bottomley R, Chaussidon M, Glass BP, Horn P, Müller-Sohnius D, Storzer D (2000b) Geochemistry and age of high sodium-potassium australites: A separate Australian tektite event. Geochim Cosmochim Acta, submitted

Kolbe P, Pinson WH, Saul JM, Miller EW (1967) Rb-Sr study on country rocks of the Bosumtwi crater, Ghana. Geochim Cosmochim Acta 31: 869–875

Koutsoukos EAM (1998) An extraterrestrial impact in the early Danian: a secondary K/T boundary event? Terra Nova 10: 68–73

Kouwenhoven TJ (1997) The El Kef blind test: How blind was it? - Comment. Marine Micropaleontology 32: 397–398

Krähenbühl U, Geissbühler M, Bühler F, Eberhardt P (1988) The measurement of osmium isotopes in samples from a Cretaceous/Tertiary (K/T) section of the Raton Basin, USA. Meteoritics 23: 282

Kring DA (1993) The Chicxulub impact event and possible causes of K/T boundary extinctions. In: Boaz D, Dornan M (eds) Proceedings of the First Annual Symposium of Fossils of Arizona. Mesa Southwest Museum and Southwest Paleontological Society, Mesa (Arizona), pp 63–79

Kring DA (1997) Air blast produced by the Meteor Crater impact event and a reconstruction of the affected environment. Meteoritics and Planetary Science 32: 517–530

Kring DA, Boynton WV (1991) Altered spherules of impact melt and associated relic glass from the K/T boundary sediments in Haiti. Geochim Cosmochim Acta 55: 1737–1742

Kring DA, Melosh HJ, Hunten DM (1996) Impact-induced perturbations of atmospheric sulfur. Earth Planet Sci Lett 140: 201–212

Krogh TE, Kamo SL, Sharpton VL, Marin LE, Hildebrand AR (1993a) U-Pb ages of single shocked zircons linking distal K/T ejecta to the Chicxulub crater. Nature 366: 731–734

Krogh TE, Kamo SL, Bohor BF (1993b) Fingerprinting the K/T impact site and determining the time of impact by U-Pb dating of single shocked zircons from distal ejecta. Earth Planet Sci Lett 119: 425–429

Kruge MA, Stankiewics BA, Crelling JA, Montanari A, Bensley DF (1994) Fossil charcoal in Cretaceous-Tertiary boundary strata: Evidence for catastrophic firestorm and megawave. Geochim Cosmochim Acta 58: 1393–1397

Kudielka G, Koeberl C, Montanari A, Newton J, Reimold WU (1999) Stable isotope stratigraphy of the J-K boundary, Bosso River Gorge, Italy. In: Buffetaut E, Le Loeuff J (eds) Abstracts, Workshop on Geological and Biological Evidence for Global Catastrophes, Quillan, France. ESF Impact Programme, pp 49–50

Kunk MJ, Izett GA, Haugerud RA, Sutter JF (1989) ^{40}Ar-^{39}Ar dating of the Manson impact structure: A Cretaceous-Tertiary boundary crater candidate. Science 244: 1565–1568

Kyte FT (1998) A meteorite from the Cretaceous/Tertiary boundary. Nature 396: 237–239

Kyte FT, Bohor BF (1995) Ni-rich magnesiowüstite in Cretaceous/Tertiary boundary crystallized from ultramafic, refractory silicate liquid. Geochim Cosmochim Acta 59: 4967–4974

Kyte FT, Bostwick JA (1995) Magnesioferrite spinel in Cretaceous-Tertiary boundary sediments of the Pacific basin: Remnants of hot, early ejecta from the Chicxulub impact? Earth Planet Sci Lett 132: 113–127

Kyte FT, Brownlee DE (1985) Unmelted meteoritic debris in the Late Pliocene iridium anomaly: Evidence for the ocean impact of a nonchondritic asteroid. Geochim Cosmochim Acta 49: 1095–1108

Kyte FT, Smit J (1986) Regional variations in spinel compositions: An important key to the Cretaceous/Tertiary event. Geology 14: 485–487

Kyte FT, Wasson JT (1986) Accretion rate of extraterrestrial matter: Iridium deposited 33 to 67 million years ago. Science 232: 1225–1229

Kyte FT, Zhou Z, Wasson JT (1980) Siderophile-enriched sediments from the Cretaceous-Tertiary boundary. Nature 288: 651–656

Kyte FT, Smit J, Wasson JT (1985) Siderophile interelement variations in the Cretaceous-Tertiary boundary sediments from Caravaca, Spain. Earth Planet Sci Lett 73: 185–195

Kyte FT, Zhou L, Wasson JT (1988) New evidence on the size and possible effects of a Late Pliocene oceanic asteroid impact. Science 241: 63–65

Kyte FT, Zhou L, Lowe DR (1992) Noble metal abundances in an Early Archean impact deposit. Geochim Cosmochim Acta 56: 1365–1372

Kyte FT, Heath GR, Leinen M, Zhou L (1993) Cenozoic sedimentation history of the central North Pacific: Inferences from the elemental geochemistry of core LL44-GPC3. Geochim Cosmochim Acta 57: 1719–1740

Kyte FT, Bostwick JA, Zhou L (1996) The Cretaceous-Tertiary boundary on the Pacific plate: Composition and distribution of impact debris. In: Ryder G, Fastovsky D, Gartner S (eds) New Developments Regarding the KT Event and Other Catastrophes in Earth History. Geological Society of America, Special Paper 307, pp 389–401

Kyte FT, Shukolyukov A, Lugmair GW, Lowe DR, Byerly GR (1999) Early Archean spherule beds - Confirmation of impact origin. GSA Annual Meeting, Abstracts with Programs 31(7): A64–A65

Lacroix A (1935) Les tectites sans formes figurées de l'Indochine. Compt Rend Acad Sci Paris 200: 2129–2132

Lambert P (1981) Breccia dikes: Geological constraints on the formation of complex craters. In: Schultz PH, Merrill RB (eds) Multi-Ring Basins: Formation and Evolution. Pergamon Press, New York, pp 59–78

Lanci L, Lowrie W, Montanari A (1996) Magnetostratigraphy of the Eocene/Oligocene boundary in a short drill-core. Earth Planet Sci Lett 143: 37–48

Langenhorst F (1993) Hochtemperatur-Stoßwellenexperimente an Quarz-Einkristallen. Ph.D. thesis, University of Münster, Germany, 126 pp

Langenhorst F (1994) Shock experiments on α- and ß-quartz: II. Modelling of lattice expansion and amorphization. Earth Planet Sci Lett 128: 683–698

Langenhorst F (1996) Characteristics of shocked quartz in late Eocene impact ejecta from Massignano (Ancona, Italy): clues to shock conditions and source crater. Geology 24: 487–490

Langenhorst F, Deutsch A (1994) Shock experiments on pre-heated α- and ß-quartz: I. Optical and density data. Earth Planet Sci Lett 125: 407–420

Langenhorst F, Deutsch A (1998) Minerals in terrestrial impact structures and their characteristic features. In: Marfunin AS (ed) Advanced mineralogy, Volume 3, Mineral matter in space, mantle, ocean floor, biosphere, environmental management, and jewelry. Springer Verlag, Berlin-Heidelberg, pp 95–119

Langenhorst F, Deutsch A, Hornemann U, Stöffler D (1992) Effect of temperature on shock metamorphism of single crystal quartz. Nature 356: 507–509

Langenhorst F, Shafranovsky GI, Masaitis VL, Koivisto M (1999) Discovery of impact diamonds in a Fennoscandian crater and evidence for their genesis by solid-state transformation. Geology 27: 747–750

Lavecchia G, Brozzetti F, Barchi M, Menichetti M, Keller JVA (1994) Seismotectonic zoning in east-central Italy deduced from an analysis of the Neogene to Present deformations and related stress field. Geol Soc Am Bull 196: 1107–1120

Leckie RM (1989) An oceanographic model for the early evolutionary history of planktonic foraminifera. Palaeogeography Palaeoclimatology Palaeoecology 73: 107–138

Lee M-Y, Wei K-Y (2000) New occurrences of Australasian microtektites in the South China Sea and the West Philippine Sea: Implications for age, size and location of the impact crater. Meteoritics and Planetary Science 35: in press

Leroux H, Reimold WU, Doukhan JC (1994) A T.E.M. investigation of shock metamorphism in quartz from the Vredefort Dome, South Africa. Tectonophysics 230: 223–239

Leroux H, Warme JE, Doukhan J-C (1995) Shocked quartz in the Alamo breccia, southern Nevada: Evidence for a Devonian impact event. Geology 23: 1003–1006

Leroux H, Reimold WU, Koeberl C, Hornemann U, Doukhan J-C (1999) Experimental shock deformation in zircon: A transmission electron microscopic study. Earth Planet Sci Lett 169: 291–301

LeRousseau J, Chauris H, Montanari A, Beaudoin B (1996) Modifications environnementales à la limite K/T dans le bassin profond des Apennins du NE (Italie). In: The Cretaceous-Tertiary boundary: biological and geological aspects. CRNS Meeting, Maison de la Géologie, Paris, p 35

Li C, Ouyang Z, Liu D, An Z (1993) Microtektites and glassy microspherules in loess: Their discoveries and implications. Science in China (Scientia Sinica), Series B 36: 1141–1152

Li C, Ouyang Z, Lin W (1995) Geochemistry of 0.7 Ma B.P. microtektite bearing loess layers; IV, Trace elements. Chinese Science Bulletin 40: 395–399

Lichte FE, Wilson SM, Brooks RR, Reeves RD, Holzbecher J, Ryan DE (1986) New method for the measurement of osmium isotopes applied to a New Zealand Cretaceous/Tertiary boundary shale. Nature 322: 816–817

Liu S, Glass BP (2000) Comparison of the stratigraphic relationship of the upper Eocene couplet of microtektites-microkrystites between holes 689D and 689B. Lunar Planet Sci 31: abs. #1847

Lovering JF, Mason B, Williams GE, McColl DH (1972) Stratigraphical evidence for the terrestrial age of australites. Journal of the Geological Society of Australia 18: 409–418

Lowe DR, Byerly GR (1986) Early Archean silicate spherules of probable impact origin, South Africa and Western Australia. Geology 14: 83–86

Lowe DR, Byerly GR, Asaro F, Kyte FT (1989) Geological and geochemical record of 3400-million-year-old terrestrial meteorite impacts. Science 245: 959–962

Lowrie W, Alvarez W (1984) Lower Cretaceous magnetic stratigraphy in Umbrian pelagic limestones. Earth Planet Sci Lett 71: 315–328

Lowrie W, Lanci L (1994) Magnetostratigraphy of Eocene-Oligocene boundary sections in Italy: No evidence for short subchrons within chrons 12R and 13R. Earth Planet Sci Lett 126: 247–258

Lowrie W, Alvarez W, Asaro F (1990) The origin of the white beds below the Cretaceous-Tertiary boundary in the Gubbio section, Italy. Earth Plan Sci Lett 98: 303–331

Lowrie W, Alvarez W, Napoleone G, Perch-Nielsen K, Premoli Silva I, Toumarkine M (1982) Paleogene magnetic stratigraphy in Umbrian pelagic carbonate rocks: The Contessa sections, Gubbio. Geol Soc Am Bull 93: 414–432

Luck JM, Turekian KK (1983) Osmium-187/Osmium-186 in manganese nodules and the Cretaceous-Tertiary boundary. Science 222: 613–615

Luczaj J (1998) Argument supporting explosive igneous activity for the origin of "cryptoexplosion" structures in the midcontinent, United States". Geology 26: 295–298

Luterbacher HP, Premoli Silva I (1964) Stratigrafia del limite Cretacico-Terziario nell'Appennino centrale. Riv It Paleont Strat 70: 68–128

Lyons JB, Officer CB, Borella PE, Lahodynsky R (1993) Planar lamellar substructures in quartz. Earth Planet Sci Lett 119: 431–440

Lyons JR, Ahrens TJ (1996) Chicxulub impact-induced vaporization: S and C species and their effect on global climate. Lunar Planet Sci 27: 787–788

MacLeod KG (1994) Bioturbation, inoceramid extinction, and mid-Maastrichtian ecological change. Geology 22: 139–142

MacLeod KG, Huber BT (1996) Reorganization of deep ocean circulation accompanying a Late Cretaceous extinction event. Nature 380: 422–425

MacLeod KG, Ward PD (1990) Extinction pattern of Inoceramus (Bivalvia) based on shell fragment biostratigraphy. In: Sharpton VL, Ward PD (eds) Global Catastrophes in Earth History. Geological Society of America, Special Paper 247, pp 509–518

Mak EK, York D, Grieve RAF, Dence MR (1976) The age of the Mistastin Lake Crater, Labrador, Canada. Earth Planet Sci Lett 31: 345–357

Marcucci Passerini M, Bettini P, Dainelli J, Sirugo A (1991) The "Bonarelli Horizon" in the central Apennines (Italy): radiolarian biostratigraphy. Cretaceous Research 12: 321–331

Margolis SV, Claeys P, Kyte FT (1991) Microtektites, microkrystites, and spinels from a Late Pliocene impact in the Southern Ocean. Science 251: 1594–1597

Mark K (1987) Meteorite Craters. University of Arizona Press, Tucson, 288 pp

Marsh SP (1980) LASL Shock Hugoniot Data. University of California Press, Berkeley, 658 pp

Martinez I, Agrinier P (1998) Meteorite impact craters on Earth: major shock-induced effects in rocks and minerals. Compt Rend Acad Sci Paris 327: 75–86

Martinez-Ruiz F, Ortega-Huertas M, Palomo I, Acquafredda P (1997) Quench textures in altered spherules from the Cretaceous-Tertiary boundary layer at Agost and Caravanca, SE Spain. Sedimentary Geology 113: 137–147

Martini E (1971) Standard Tertiary and Quaternary calcareous nannoplankton zonation. Proc 2nd Plankt Conf, Roma, pp 739–786

Martinis B, Pieri G (1962) Alcune notizie sulla formazione evaporitica del Triassico superiore nell'Italia centrale e meridionale. Riv It Paleont Strat 4: 649–678

Martino MD, Farinella P, Longo G (eds) (1998) Special Issue: International Workshop Tunguska 96, Bologna (Italy), 15-17 July 1996. Planetary and Space Science 46(2/3): 125–340

Marvin UB (1990) Impact and its revolutionary implications for geology. In: Sharpton VL, Ward PD (eds) Global Catastrophes in Earth History, Geological Society of America, Special Paper 247, pp 147–154

Masaitis VL (1994) Impactites from Popigai crater. In: Dressler BO, Grieve RAF, Sharpton VL (eds) Large Meteorite Impacts and Planetary Evolution. Geological Society of America, Special Paper 293, pp 153–162

Masaitis VL (1999) Impact structures of northeastern Eurasia: The territories of Russia and adjacent countries. Meteoritics and Planetary Science 34: 691–711

Masaitis VL, Danilin AN, Maschak MS, Raykhlin AI, Selivanovskaya TV, Shadenkov EM (1980) Geology of Astroblemes (in Russian). Nedra Press, Leningrad, 231 pp

Masters BA (1997) El Kef blind test II results. Marine Micropaleontology 29: 77–79

Matsubara K, Matsuda J-I (1991) Anomalous Ne enrichments in tektites. Meteoritics 26: 217–220

Matsubara K, Matsuda J-I, Koeberl C (1991) Noble gases and K-Ar ages in Aouelloul, Zhamanshin, and Libyan Desert impact glasses. Geochim Cosmochim Acta 55: 2951–2955

Matsuda J-I, Matsubara K, Yajima H, Yamamoto K (1989) Anomalous Ne enrichment in obsidians and Darwin Glass: Diffusion of noble gases in silica-rich glasses. Geochim Cosmochim Acta 53: 3025–3033

Matsuda J-I, Matsubara K, Koeberl C (1993) Origin of tektites: Constraints from heavy noble gas concentrations. Meteoritics 28: 586–589

Matsuda J-I, Maruoka T, Pinti DL, Koeberl C (1996) Noble gas study of a philippinite with an unusually large bubble. Meteoritics and Planetary Science 31: 273–277

Mattias P, Crocetti G, Barrese E, Montanari A, Coccioni R, Farabollini P, Parisi E (1992) Caratteristiche mineralogiche e litostratigrafiche della sezione eo-oligocenica di Massignano (Ancona, Italia) comprendente il limite Scaglia Variegata-Scaglia Cinerea. Studi Geologici Camerti 12: 93–103

Matthies D, Koeberl C (1991) Fluorine and boron geochemistry of tektites, impact glasses, and target rocks. Meteoritics 26: 41–45

Maurrasse FJMR, Glass BP (1976) Radiolarian stratigraphy and North American microtektites in Caribbean core RC9-58: Implications concerning the age of the Eocene-Oligocene boundary. Proceedings 7th Caribbean Geological Conference, pp 205–212

Maurrasse FJMR, Sen G (1991) Impacts, tsunamis, and the Haitian Cretaceous-Tertiary boundary. Science 252: 1690–1693

Max MD, Dillon WP, Nishimura C, Hurdle BG (1999) Sea-floor methane blow-out and global firestorm at the K-T boundary. Geo-Marine Letters 18: 285–291

Mazzanti ML (1988) Micropaleontologia, microbiostratigrafia (foraminiferi) e litostratigrafia della Scaglia cretacica del Monte Cònero (Ancona). Laurea Thesis, Univ Urbino, 140 pp

M'Ban EP (1994) Evolution de la sédimentation pélagique au Crétacé supérieur del Marches et de l'Ombrie (Italie Centrale). Litho-biostratigraphie, Géochemie, Paléoenvironment. PhD Thesis, Univ Paris VI, 255 pp

McGetchin TR, Settle M, Head JW (1973) Radial thickness variation in impact crater ejecta: Implications for lunar basin deposits. Earth Planet Sci Lett 20: 226–236

McHone JF, Sorkhabi RB (1994) Apatite fission-track age of Marquez Dome impact structure, Texas. Lunar Planet Sci 25: 881–882

McHone JF, Nieman RA, Lewis CF, Yates AM (1989) Stishovite at the Cretaceous-Tertiary boundary, Raton, New Mexico. Science 243: 1182–1184

McLaren DJ, Goodfellow WD (1990) Geological and biological consequences of giant impacts. Annual Reviews of Earth and Planetary Science 18: 123–171

McLennan SM, Taylor SR (1980) Th and U in sedimentary rocks: Crustal evolution and sedimentary recycling. Nature 285: 621–624

Medenbach O (1985) A new microrefractometer spindle stage and its application. Fortschritte der Mineralogie 63: 111–133

Meisel T, Koeberl C, Ford RJ (1990) Geochemistry of Darwin impact glass and target rocks. Geochim Cosmochim Acta 54: 1463–1474

Meisel T, Krähenbühl U, Nazarov MA (1995) Combined osmium and strontium isotopic study of the Cretaceous-Tertiary boundary at Sumbar, Turkmenistan: A test for an impact vs. volcanic hypothesis. Geology 23: 313–316

Melosh HJ (1989) Impact cratering: A geologic process. Oxford University Press, New York, 245 pp

Melosh HJ (1998) Impact physics constraints on the origin of tektites. Meteoritics Planet Sci 33: A104

Melosh HJ, Ivanov BA (1999) Impact crater collapse. Annual Reviews of Earth and Planetary Science 27: 385–415

Melosh HJ, Schenk P (1993) Split comets and the origin of crater chains on Ganymede and Callisto. Nature 365: 731–733

Melosh HJ, Whittaker EA (1994) Lunar crater chains. Nature 369: 713–714

Melosh HJ, Schneider NM, Zahnle KJ, Latham D (1990) Ignition of global wildfires at the Cretaceous/Tertiary boundary. Nature 343: 251–254

Meyer G (1987) Multiparameter coincidence spectrometry applied to the instrumental neutron activation analysis of rocks and minerals. J Radioanalyt Nuclear Chem 114: 223–230

Middleton R, Klein J (1987) [26]Al: Measurement and applications. Phil Trans Royal Soc London A323: 121–143

Miller KG, Berggren WA, Zhang J, Palmer-Julson AA (1991) Biostratigraphy and isotope stratigraphy of Upper Eocene microtektites at Site 612: How many impacts? Palaios 6: 17–38

Milton DJ (1977) Shatter cones - An outstanding problem in shock mechanics. In: Roddy DJ, Pepin RO, Merrill RB (eds) Impact and Explosion Cratering. Pergamon Press, New York, pp 703–714

Milton DJ, Sutter JF (1987) Revised age for the Gosses Bluff impact structure, Northern Territory, Australia, based on $^{40}Ar/^{39}Ar$ dating. Meteoritics 22: 281–289

Mita H, Shimoyama A (1999) Distribution of polycyclic aromatic hydrocarbons in the K/T boundary sediments at Kawaruppu, Hokkaido, Japan. Geochemical Journal 33: 305–315

Mittlefehldt DW, See TH, Hörz F (1992a) Projectile dissemination in impact melts from Meteor crater, Arizona. Lunar Planet Sci 23: 919–920

Mittlefehldt DW, See TH, Hörz F (1992b) Dissemination and fractionation of projectile materials in the impact melts from Wabar crater, Saudi Arabia. Meteoritics 27: 361–370

Molina E, Gonzalvo C, Keller G (1993) The Eocene–Oligocene planktic foraminiferal transition: Extinctions, impacts and hiatuses. Geol Mag 130: 483–499

Molini-Vesko C, Mayeda TK, Clayton RN (1982) Silicon isotopes: Experimental vapor fractionation and tektites. Meteoritics 17: 225–226

Monaco P, Nocchi M, Ortega-Huertas M, Palomo I, Martinez F, Chiavini G (1994) Depositional trends in the valdorbia section (Central Italy) during the Early Jurassic as revealed by micropaleontology, sedimentology, and geochemistry. Eclogae Geol Helv 87: 1–61

Monaco P, Nocchi M, Parisi G (1987) Analisi stratigrafica e sedimentologica di alcune sequenze pelagiche dell'Umbria sud-orientale durante il Paleogene. Mem Soc Geol It 196: 71–91

Monechi S (1977) Upper Cretaceous and Early Tertiary nannoplankton from the Scaglia Umbra formation. Riv It Paleont Strat 83: 759–802

Monechi S, Thierstein HR (1985) Late Cretaceous-Eocene nannofossil and magnetostratigraphic correlations near Gubbio, Italy. Marine Micropaleontology 9: 419–440

Montanari A (1979) Lineamenti sedimentologici della "Scaglia Bianca" e della "Scaglia Rossa" nelle Marche settentrionali. Laurea Thesis, Univ. Urbino, 289 pp

Montanari A (1983) K-Ar dating of volcanic biotites in pelagic limestones bracketing the Eocene-Oligocene boundary in the Gubbio sequence. In: Baldi T (ed) Report of the TEE Meeting in Visegrad (Hungary), TEE News, Project 174, Newsletter 6, p 3

Montanari A (1986) Spherules from the Cretaceous-Tertiary boundary clay at Gubbio: The problem of outcrop contamination. Geology 14: 1024–1026

Montanari A (1988a) Tectonic implications of hydrothermal mineralization in the Late Cretaceous-Early Tertiary pelagic basin of the Northern Apennines. Boll Soc Geol It 107: 399–411

Montanari A (1988b) Geochemical characterization of volcanic biotites from the Upper Eocene-Upper Miocene pelagic sequence of the Northern Apennines. In: Premoli Silva I, Coccioni R, Montanari A (eds) The Eocene-Oligocene Boundary in the Marche-Umbria Basin (Italy). IUGS Special Publication, F.lli Aniballi Publishers, Ancona, pp 209–228

Montanari A (1990) Geochronology of the terminal Eocene impacts; An update. In: Sharpton VL, Ward PD (eds) Global Catastrophes in Earth History, Geological Society of America, Special Paper 247, pp 607–616

Montanari A (1991) Authigenesis of impact spheroids in the K/T boundary clay from Italy: new constraints for high-resolution stratigraphy of terminal Cretaceous events: J Sed Petrol 61: 315–339

Montanari A, Odin GS (1989) Eocene-Oligocene boundary at Massignano (Ancona, Italy): potential boundary stratotype. 28th International Geological Congress, Washington D.C., Abstracts 2, p 453

Montanari A, Swisher C III (1994) Radioisotopic calibration of the Paleocene-Eocene biomagnetostratigreaphic sequence at Gubbio (Italy). ICOG-8, US Geological Survey, Circular 1107, p 224

Montanari A, Hay RL, Alvarez W, Asaro F, Michel HV, Alvarez LW (1983) Spheroids at the Cretaceous-Tertiary boundary are altered droplets of basaltic composition. Geology 11: 668–671

Montanari A, Drake R, Bice MD, Alvarez W, Curtis GH, Turrin BD, DePaolo DJ (1985) Radiometric time scale for the upper Eocene and Oligocene based on K/Ar and Rb/Sr dating of volcanic biotites from the pelagic sequence of Gubbio, Italy. Geology 13: 596–599

Montanari A, Langenheim VE, Coccioni R (1988a) Stratigraphy and geochronologic potential of the pelagic and hemipelagic sequence of the northeastern Apennines: A research note. Bull Liais Inform, Project No 196, 7: 17–23

Montanari A, Deino A, Drake R, Turrin BD, De Paolo DJ, Odin SG, Curtis GH, Alvarez W, Bice DM (1988b) Radioisotopic dating of the Eocene-Oligocene boundary in the pelagic sequence of the Northern Apennines. In: Premoli Silva I, Coccioni R, Montanari A (eds) The Eocene-Oligocene Boundary in the Marche-Umbria Basin (Italy). IUGS Special Publication, F.lli Aniballi Publishers, Ancona, pp 195–208

Montanari A, Chan LS, Alvarez W (1989) Synsedimentary tectonics in the Late Cretaceous-Early Tertiary pelagic basin of the Northern Apennines. In: Crevello PD, Wilson JL, Sarg JF, Read JF (eds) Controls on Carbonate Platform and Basin Development. SEPM Special Publication 44: pp 379–399

Montanari A, Deino A, Coccioni R, Langenheim VE, Capo R, Monechi S (1991) Geochronology, Sr isotope analysis, magnetostratigraphy, and plankton stratigraphy across the Oligocene-Miocene boundary in the Contessa section (Gubbio, Italy). Newsletters on Stratigraphy 23: 151–180

Montanari A, Asaro F, Kennett JP, Michel E (1993) Iridium anomalies of Late Eocene age at Massignano (Italy), and in ODP Site 689B (Maud Rise, Antarctica). Palaios 8: 420–437

Montanari A, Sandroni P, Clymer A, Collins J, Lanci L, Lowrie W (1994) Preliminary report on a core drilled across the Eocene-Oligocene boundary in the type locality of Massignano (Italy): The Massicore. IUGS-SOG Bull Liais Info, Paris 12: 13–16

Montanari A, Beaudoin B, Chan L, Coccioni R, Deino A, DePaolo DJ, Emmanuel L, Fornaciari E, Kruge M, Lundblad S, Mozzato C, Portier E, Renard M, Rio D, Sandroni P, Stankiewicz A (1997a) Integrated stratigraphy of the middle to upper Miocene pelagic sequence of the Cònero Riviera (Ancona, Italy). In: Montanari A, Odin GS, Coccioni R (eds) Miocene Stratigraphy: An Integrated Approach. Elsevier, Amsterdam, pp 409–450

Montanari A, Bice DM, Coccioni R, Deino A, DePaolo DJ, Emmanuel L, Renard M, Monechi S, and Zevenboom D (1997b) Integrated stratigraphy of the Chattian to mid Burdigalian pelagic sequence of the Contessa Valley (Gubbio, Italy). In: Montanari A, Odin GS, Coccioni R (eds) Miocene Stratigraphy: An Integrated Approach, Eslevier, Amsterdam, pp 249–278

Montanari A, Coccioni R, Fornaciari E, Rio D (1997c) Potential integrated stratigraphy in the Langhian l'Annunziata section near Apiro (Marche region, Italy). In: Montanari A, Odin GS, Coccioni R (eds) Miocene Stratigraphy: An Integrated Approach. Elsevier, Amsterdam, pp 343–349

Montanari A, Campo Bagatin A, Farinella P (1998) Earth cratering record and impact energy flux in the last 150 Ma. Planet Space Sci 46: 271–281

Monteiro JF, Munha J, Ribeiro A (1998) Impact ejecta horizon near the Cenomanian–Turonian boundary, north of Nazaré, Portugal. Meteoritics and Planetary Science 33: A112–A113

Morgan J, Warner M (1999) Chicxulub: The third dimension of a multi-ring impact basin. Geology 27: 407–410

Morgan J, Warner M, Brittan J, Buffler R, Camargo A, Christeson G, Denton P, Hildebrand A, Hobbs R, Macintyre H, MacKenzie G, Maguire P, Marin L, Nakamura Y, Pilkington M, Sharpton VL, Snyder D, Suarez G, Trejo A (1997) Size and morphology of the Chicxulub impact crater. Nature 390: 472–476

Morgan JW (1978) Lonar crater glasses and high-magnesium australites: Trace element volatilization and meteoritic contamination. Proc Lunar Planet Sci Conf 9th: 2713–2730

Morgan JW, Wandless GA (1983) Strangways Crater, Northern Territory, Australia: Siderophile element enrichment and lithophile element fractionation. J Geophys Res 88: A819–A829

Morgan JW, Higuchi H, Ganapathy R, Anders E (1975) Meteoritic material in four terrestrial meteorite craters. Proc 6th Lunar Sci Conf, pp 1609–1623

Morgan JW, Janssens M-J, Hertogen H, Gros H, Takahashi H (1979) Ries impact crater: Search for meteoritic material. Geochim Cosmochim Acta 43: 803–815

Morgan P (1989) Heat flow in the earth. In: James DE (ed) The Encyclopedia of Solid Earth Geophysics. Van Nostrand Reinhold Comp., New York, pp 634–646

Morrow JR, Sandberg CA, Warme JE, Kuehner H-C (1998) Regional and possible global effects of sub-critical Devonian Alamo impact event, southern Nevada, USA. J British Interplanetary Society 51: 451–460

Mossman DJ, Grantham RG, Langenhorst L (1998) A search for shocked quartz at the Triassic-Jurassic boundary in the Fundy and Newark basins of the Newark Supergroup. Can J Earth Sci 35: 101–109

Mukhopadhyay S, Farley KA, Montanari A (1998) ^{3}He evidence for a brief elevation in the interplanetary dust particle flux in the Late Cretaceous. EOS, Trans Am Geophys Un Suppl 79(45): F50

Müller O, Gentner W (1968) Gas content in bubbles of tektites and other natural glasses. Earth Planet Sci Lett 4: 406–410

Müller O, Gentner W (1973) Enrichment of volatile elements in Muong Nong-type tektites: Clue to their formation history? Meteoritics 8: 414–415

Müller WF (1969) Elektronenmikroskopischer Nachweis amorpher Bereiche in stoßwellenbeanspruchtem Quarz. Naturwissenschaften 56, 279

Murali AV, Linstrom EJ, Zolensky ME, Underwood JR Jr, Giegengack RF (1989) Evidence of extraterrestrial component in the Libyan Desert Glass. EOS, Trans Am Geophys Un 70: 1178

Murali AV, Zolensky ME, Underwood JR Jr, Giegengack RF (1997) Chondritic debris in Libyan Desert Glass. In: Proceedings, Silica 96, Meeting on Libyan Desert Glass and related desert events. Pyramids, Milan, Italy, pp 133–142

Murray JB (1999) Arguments for the presence of a distant large undiscovered solar system planet. Monthly Notices Royal Astron Soc 308: 31–34

Nazarov MA, Kolesnikov EM, Badjukov DD, Masaitis VL (1989) Potassium-argon age of the Kara impact event. Lunar Planet Sci 20: 766–767

Nazarov MA, Badjukov DD, Barsukova LD, Alekseev AS (1991) Reconstruction of original morphology of the Kara impact structure and its relevance to the K/T boundary event. Lunar Planet Sci 22: 959–960

Ngo H, Wasserburg GJ, Glass BP (1985) Nd and Sr isotopic compositions of tektite material from Barbardos and their relationship to North American tektites. Geochim Cosmochim Acta 49: 1479–1485

Norris RD, Firth J (1999) Mass failure of the North American margin triggered by the K/T impact. GSA Annual Meeting, Abstracts with Programs 31(7): A123

Norris RD, Röhl U (1999) Carbon cycling and chronology of climate warming during the Paleocene/Eocene transition. Nature 401: 775–778

Norris RD, Huber BT, Self-Trail J (1999) Synchroneity of the K-T oceanic mass extinction and meteorite impact: Blake Nose, western North Atlantic. Geology 27: 419–422

Oberbeck VR (1975) The role of ballistic erosion and sedimentation in lunar stratigraphy. Rev Geophys Space Phys 13: 337–362

Oberbeck VR, Marshall JR, Aggarwal H (1993) Impacts, tillites, and the breakup of Gondwanaland. J Geol 101: 1–19

Oberli F, Meier M (1991) Age of Eocene-Oligocene boundary in the Marche-Umbria basin, Italy, by high resolution U-Th-Pb dating. Terra Abstracts 3: 286

Obradovich JD, Snee LW, Izett GA (1989) Is there more than one glassy impact layer in the Late Eocene? GSA Annual Meeting, Abstracts with Programs 21(6): A134

Ocampo AC, Pope KO, Fischer AG (1996) Ejecta blanket deposits of the Chicxulub crater from Albion Island, Belize. In: Ryder G, Fastovsky D, Gartner S (eds) New Developments Regarding the KT Event and Other Catastrophes in Earth History. Geological Society of America, Special Paper 307, pp 75–88

Odin GS, Dodson MH (1982) Zero isotopic age of glauconies. In: Odin GS (ed) Numerical Dating in Stratigraphy. Wiley, Chichester, pp 277–305

Odin GS, Fullagar (1988) Geologic significance of the glaucony facies. In: Odin GS (ed) Green Marine Clays, Developments in Sedimentology 45, Elsevier, Amsterdam, pp 295–332

Odin GS, Montanari A (1988) Stratigraphy of the Massignano section (Ancona, Italy), and the identification of the Eocene-Oligocene boundary: In: Premoli Silva I, Coccioni R, Montanari A (eds) The Eocene-Oligocene Boundary in the Marche-Umbria Basin (Italy), IUGS Special Publication, F.lli Aniballi Publishers, Ancona, pp 253–263

Odin GS, Montanari A (1989) Age radiométrique et stratotype de la limite Éocène Oligocène. Comptes Rendus Acad Sci Paris 309: 1939–1945

Odin GS, Guise P, Rex DC, Hunziger H (1988) K/Ar and ^{39}Ar/^{40}Ar geochronology of Late Eocene biotites from the northeastern Apennines. In: Premoli Silva I, Coccioni R, Montanari A (eds) The Eocene-Oligocene Boundary in the Marche-Umbria Basin. IUGS Special Publication, F.lli Aniballi Publisher, Ancona, pp 239–248

Odin GS, Montanari A, Deino A, Drake R, Guise P, Kreuzer H, Rex DC (1991) Reliability of volcano-sedimentary biotite ages around the Eocene/Oligocene boundary. Chemical Geology (Isot Geosci) 86: 203–224

Odin GS, Hurford AJ, Montanari A (1992) Study of a presumably volcano-sedimentary layer near the Cretaceous-Paleogene boundary in central Apennines (Italy). IUGS-SOG Bulletin of Liaison and Information 11: 26–28

Odin GS, Montanari A, Coccioni R (1997) Chronostratigraphy of Miocene Stages: a proposal for the definition of precise boundaries. In: Montanari A, Odin GS, Coccioni R (eds) Miocene Stratigraphy: An Integrated Approach. Elsevier, Amsterdam, pp 597–630

Ogg JG, Lowrie W (1986) Magnetostratigraphy of the Jurassic/Cretaceous boundary. Geology 14: 347–550

Okada H, Bukry D (1980) Supplementary modification and introduction of code numbers to the low latitude coccolith biostratigraphic zonation. Marine Micropaleontology 5: 321–325

O'Keefe JA (ed) (1963) Tektites. University of Chicago Press, Chicago, 228 pp

O'Keefe JA (1976) Tektites and their Origin. Elsevier, New York-Amsterdam, 254 pp

Olsson RK (1997) El Kef blind test III results. Marine Micropaleontology 29: 80–84

Öpik EJ (1916) Remarque sur le théorie météorique des cirques lunaires. Bulletin de la Société Russe des Amis de l'Etude de l'Univers No. 3, 21: 125–134

Orth CJ, Attrep M, Mao X, Kauffman EG, Diner R, Elder WP (1988) Iridium abundance maxima in the Upper Cenomanian extinction interval. Geophys Res Lett 15: 346–349

Orth CJ, Attrep M Jr, Quintana LR (1990) Iridium abundance patterns across bio-event horizons in the fossil record. In: Sharpton VL, Ward PD (eds) Global Catastrophes in Earth History. Geological Society of America, Special Paper 247, pp 45–59

Ortiz N (1994) Mass extinction of benthic foraminifera at the Paleocene/Eocene boundary. In: Molina E (ed) Extinction and the fossil record. Seminario Interdisciplinario, Univ Zaragoza 5: pp 201–218

Orue-extebarria X (1997) The El Kef blind test IV results. Marine Micropaleontology 29: 85–88

Oskarsson N, Helgason O, Sigurdsson H (1996) Oxidation state of iron in tektite glasses from the Cretaceous/Tertiary boundary. In: Ryder G, Fastovsky D, Gartner S (eds) New Developments Regarding the KT Event and Other Catastrophes in Earth History. Geological Society of America, Special Paper 307, pp 445–452

Owen MR, Anders MH (1988) Evidence from cathodoluminescence for non-volcanic origin of shocked quartz at the Cretaceous-Tertiary boundary. Nature 334: 145–147

Pal DK, Tuniz C, Moniot RK, Kruse TH, Herzog GF (1982) Beryllium-10 in Australasian tektites: Evidence for a sedimentary precursor. Science 218: 787–789

Palma RL, Rao MN, Rowe MW, Koeberl C (1997) Krypton and xenon fractionation in North American tektites. Meteoritics and Planetary Science 32: 9–14

Palme H (1980) The meteoritic contamination of terrestrial and lunar impact melts and the problem of indigeneous siderophiles in the lunar highland. Proc Lunar Planet Sci Conf 11th, 481–506

Palme H (1982) Identification of projectiles of large terrestrial impact craters and some implications for the interpretation of Ir-rich Cretaceous/Tertiary boundary layers. In: Silver LT, Schultz PH (eds) Geological Implications of Impacts of Large Asteroids and Comets on Earth. Geological Society of America, Special Paper 190, pp 223–233

Palme H, Janssens M-J, Takahasi H, Anders E, Hertogen J (1978) Meteorite material at five large impact craters. Geochim Cosmochim Acta 42: 313–323

Palme H, Göbel E, Grieve RAF (1979) The distribution of volatile and siderophile elements in the impact melt of East Clearwater (Quebec). Proc Lunar Planet Sci Conf 10th: pp 2465–2492

Palme H, Grieve RAF, Wolf R (1981) Identification of the projectile at the Brent crater, and further considerations of projectile types at terrestrial craters. Geochim Cosmochim Acta 45: 2417–2424

Parisi G, Ortega-Huertas M, Nocchi M, Palomo I, Monaco P, Martinez F (1996) Stratigraphy and geochemical anomalies of the early Toarcian oxygen poor interval in the Umbria-Marche Apennines (Italy). Géobios 29: 469–484

Pavic MJ, Brown L, Klein J, Middleton R, Tera F, Valette-Silver N (1983) Beryllium-10: Soils. Yearbook of the Carnegie Institution '82, pp 460–461

Pearson DG, Sigurdsson H, Shirey SB, Murray RW, Lyons TW, Schmitz B (1999) Os isotopes and platinum group elements in K-T boundary clays: Impact signatures versus post-impact processes. GSA Annual Meeting, Abstracts with Programs 31(7): A123–A124

Penfield GT, Camargo ZA (1981) Definition of a major igneous zone in the central Yucatan platform with aeromagnetics and gravity. Techn Program, Abstracts and Biographies, 51st Ann Mtg Soc Exploration Geophysicists, Tulsa, OK, p 37

Pernicka E, Kaether D, Koeberl C (1996) Siderophile element concentrations in drill core samples from the Manson crater. In: Koeberl C, Anderson RR (eds) The Manson Impact Structure, Iowa: Anatomy of an Impact Crater. Geological Society of America, Special Paper 302, pp 325–330

Perry E, Marin L, McClain J, Velàsquez G (1995) Ring of cenotes (sinkholes), northwest Yucatan, Mexico: Its hydrologic characteristics and possible association with the Chicxulub impact crater. Geology 23: 17–20

Peryt D, Lamolda M (1996) Benthic foraminiferal mass extinction and survival assemblages from the Cenomanian-Turonian Boundary Event in the Manoyo section, northern Spain. In: Hart MB (ed) Biotic Recovery from Mass Extinction Events. Geological Society London, Special Publication 102, pp 245–258

Pesonen LJ (1996) The impact cratering record of Fennoscandia. Earth, Moon and Planets 72: 377–393

Pesonen LJ, Henkel H (eds) (1992) Terrestrial impact craters and craterform structures, with a special focus on Fennoscandia. Tectonophysics 216: 1–234

Pettinelli R, Nocchi M, Parisi G (1995) Late Pliensbachian-Toarcian biostratigraphy and environmental interpretations in the Ionian basin (Lefkas Island, western Greece) as compared to the Umbria-Marchean basin (central Italy). Boll Serv Geol Ital 64: 97–158

Peucker-Ehrenbrink B, Ravizza G, Hofmann AW (1995) The marine $^{187}Os/^{186}Os$ record of the past 80 million years. Earth Planet Sci Lett 130: 155–167

Pierazzo E, Chyba CF (1999) Amino acid survival in large cometary impacts. Meteoritics and Planetary Science 34: 909–918

Pierazzo E, Melosh HJ (1999) Hydrocode modeling of Chicxulub as an oblique impact event. Earth Planet Sci Lett 165: 163–176

Pierazzo E, Melosh HJ (2000) Hydrocode modeling of oblique impacts: The fate of the projectile. Meteoritics and Planetary Science 35: 117–130

Pierazzo E, Vickery AM, Melosh HJ (1997) A reevaluation of impact melt production. Icarus 127: 408–423

Pierazzo E, Kring DA, Melosh HJ (1998) Hydrocode simulation of the Chicxulub impact event and the production of climatically active gases. J Geophys Res 103: 28607–28625

Piergiovanni F (1989) Eventi lito-biostratigrafici nella Scaglia Bianca umbro-marchigiana in connessione con l'episodio anossico del "Livello Bonarelli". Boll Soc Geol It 108: 289–314

Pierrard O, Robin E, Rocchia R, Montanari A (1998) Extraterrestrial Ni-rich spinel in upper Eocene sediments from Massignano, Italy. Geology 26: 307–310

Pilkington M, Grieve RAF (1992) The geophysical signature of terrestrial impact craters. Reviews of Geophysics 30: 161–181

Pilkington M, Hildebrand AR, Ortiz-Aleman C (1994) Gravity and magnetic field modeling and structure of the Chicxulub crater, Mexico. J Geophys Res 99: 13147–13162

Pitakpaivan K, Byerly GR, Hazel JE (1994) Pseudomorphs of impact spherules from a Cretaceous Tertiary boundary section at Shell Creek, Alabama. Earth Planet Sci Lett 124: 49–56

Playford PE, McLaren DJ, Orth CJ, Gilmore JS, Goodfellow WD (1984) Iridium anomaly in the Upper Devonian of the Canning Basin, Western Australia. Science 226: 437–439

Poag CW (1997a) Roadblocks on the kill curve: Testing the Raup hypothesis. Palaios 12: 582–590

Poag CW (1997b) The Chesapeake Bay bolide impact: A convulsive event in Atlantic Coastal Plain evolution. Sedimentary Geology 108: 45–90

Poag CW (1999) Chesapeake Invader; Discovering America's Giant Meteorite Crater. Princeton University Press, Princeton, New Jersey, 183 pp

Poag CW, Aubry M-P (1995) Upper Eocene impactites of the U.S. East Coast: Depositional origins, biostratigraphic framework, and correlation. Palaios 10: 16–43

Poag CW, Foster D (2000) Chesapeake Bay impact crater: New seismic evidence of a central peak. Lunar Planet Sci 31: abs. #1358

Poag CW, Powars DS, Poppe LJ, Mixon RB, Edwards LE, Folger DW, Bruce S (1992) Deep Sea Drilling Project Site 612 bolide event: New evidence of a late Eocene impact-wave deposit and a possible impact site, U.S. east coast. Geology 20: 771–774

Poag CW, Powars DS, Poppe LJ, Mixon RB (1994) Meteoroid mayhem in Ole Virginny: Source of the North American tektite strewn field. Geology 22: 691–694

Pope KO, Ocampo AC, Duller CE (1993) Surficial geology of the Chicxulub impact crater, Yucatan, Mexico. Earth, Moon and Planets 63: 93–104

Pope KO, Baines KH, Ocampo AC, Ivanov BA (1994) Impact winter and the Cretaceous/Tertiary extinctions: Results of a Chicxulub asteroid impact model. Earth Planet Sci Lett 128: 719–725

Pope KO, Ocampo AC, Kinsland GL, Smith R (1996) Surface expression of the Chicxulub crater. Geology 24: 527–530

Pomerol C, Premoli Silva I (eds) (1986) Terminal Eocene Events, Elsevier, Amsterdam, 153 pp

Pospichal JJ (1994) Calcareous nannofossils at the K/T boundary, El Kef: No evidence for stepwise, gradual of sequential extinctions. Geology 22: 99–102

Pospichal JJ (1996) Calcareous nannoplankton mass extinction at the Cretaceous/Tertiary boundary: An update. In: Ryder G, Fastovsky D, Gartner S (eds) New Developments Regarding the KT Event and Other Catastrophes in Earth History. Geological Society of America, Special Paper 307, pp 335–360

Potts PJ (1995) A Handbook of Silicate Rock Analysis, paperback edition, Blackie, Glasgow, 622 pp

Preisinger A, Zobetz E, Gratz AJ, Lahodynsky R, Becke M, Mauritsch HJ, Eder G, Grass F, Rögl F, Stradner H, Surenian R (1986) The Cretaceous/Tertiary boundary in the Gosau Basin, Austria. Nature 322: 794–799

Premo WR, Izett GA (1992) Isotopic signature of black tektites from the K-T boundary on Haiti: Implications for the age and type of source material. Meteoritics 27: 413–423

Premoli Silva I (1977) Upper Cretaceous-Paleogene magnetic stratigraphy at Gubbio, Italy: II Biostratigraphy. Geol Soc Am Bull 88: 371–374

Premoli Silva I, Jenkins DJ (1993) Decision on the Eocene-Oligocene boundary stratotype: Episodes 16: 379–381

Premoli Silva I, Montanari A (1989) Intercorrelation of Time Scales: Example of multidisciplinary study from Umbria-Marche pelagic sequences (Italy) at the Eocene-Oligocene transition. 28th International Geological Congress, Washington D.C., Abstracts 2, pp 639

Premoli Silva I, Sliter WV (1995) Cretaceous planktonic foraminiferal biostratigraphy and evolutionary trends from the Bottaccione section, Gubbio, Italy. Palaeontographia Italica 82: 1–26

Premoli Silva I, Coccioni R, Montanari A (eds) (1988) The Eocene-Oligocene Boundary in the Marche-Umbria Basin. IUGS Special Publication, F.lli Aniballi Publishers, Ancona, 268 pp

Premoli Silva I, Tornaghi ME, Ripepe M (1989) Planktonic foraminiferal distributions record productivity cycles: evidence from the Aptian-Albian Piobbico Core (central Italy). Terra Nova 1: 443–448

Prinn RG, Fegley B (1987) Bolide impacts, acid rain, and biospheric traumas at the Cretaceous-Tertiary boundary. Earth Planet Sci Lett 83: 1–15

Racki G (1999) The Frasnian-Famennian biotic crisis: How many (if any) bolide impacts? Geol Rundsch 87: 617–632

Raisbeck GM, Yiou F, Zhou SZ, Koeberl C (1988) [10]Be in irghizite tektites and zhamanshinite impact glasses. Chemical Geology 70: 120

Rampino MR (1994) Tillites, diamictites, and ballistic ejecta of large impacts. J Geol 102: 439–456

Rampino MR, Adler AC (1998) Evidence for abrupt latest Permian mass extinction of foraminifera: Results of tests for the Signor-Lipps effect. Geology 26: 415–418

Rampino MR, Haggerty BM (1996) Impact crises and mass extinctions: A working hypothesis. In: Ryder G, Fastovsky D, Gartner S (eds) New Developments Regarding the KT Event and Other Catastrophes in Earth History. Geological Society of America, Special Paper 307, pp 11–30

Rampino MR, Stothers RB (1984) Terrestrial mass extinctions, cometary impacts and the Sun's motion perpendicular to the galactic plane. Nature 308: 709–712

Rampino MR, Volk T (1996) Multiple impact event in the Paleozoic: Collision with a string of comets or asteroids. Geophys Res Lett 23: 49–52

Rasmussen KL, Clausen HB, Kallemeyn GW (1995) No iridium anomaly after the 1908 Tunguska impact: Evidence from a Greenland ice core. Meteoritics 30: 634–638

Rasmussen KL, Olsen HJF, Gwozdz R, Kolesnikov EM (1999) Evidence for a very high carbon/iridium ratio in the Tunguska impactor. Meteoritics and Planetary Science 34: 891–895

Raup DM (1992) Large-body impact and extinction in the Phanerozoic. Paleobiology 18: 80–88

Raup DM, Sepkoski JJ (1982) Mass extinctions in the marine fossil record. Science 215: 1501–1503

Raup DM, Sepkoski JJ (1984) Periodicity of extinctions in the geologic past. Proceedings of the National Academy of Sciences of the U.S. 81: 801–805

Rehkämper M, Halliday AN, Wentz RF (1998) Low-blank digestion of geological samples for platinum-group element analysis using a modified Carius Tube design. Fresenius J Analyt Chem 361: 217–219

Reid AM, Cohen AJ (1962) Coesite in Darwin Glass. J Geophys Res 67: 1654

Reinhard M (1931) Universaldrehtischmethoden. Birkhäuser Verlag, Basel, 118 pp

Reimold WU (1988) Shock experiments with preheated Witwatersrand quartzite and the Vredefort microdeformation controversy. Lunar Planet Sci 19: 970–971

Reimold WU (1994). Comment on 'Planar lamellar substructures in quartz' by J.B. Lyons, C.B. Officer, P.E. Borella and R. Lahodynsky. Earth Planet Sci Lett 125: 473–477

Reimold WU (1995) Pseudotachylite in impact structures - generation by friction melting and shock brecciation?: A review and discussion. Earth-Science Reviews 39: 247–265

Reimold WU (1998) Exogenic and endogenic breccias: a discussion of major problematics. Earth-Science Reviews 43: 25–47

Reimold WU, Gibson RL (1996) Geology and evolution of the Vredefort impact structure, South Africa. J African Earth Sci 23: 125–162

Reimold WU, Minnitt RCA (1996) Impact-induced shatter cones or sedimentary percussion features in the Klapperkop Formation of the Eastern Transvaal? South African Journal of Geology 99: 299–308

Reimold WU, Barr JM, Grieve RAF, Durrheim RJ (1990) Geochemistry of the melt and country rocks of the Lake St. Martin impact structure, Manitoba, Canada. Geochim Cosmochim Acta 54: 2093–2111

Reimold WU, von Brunn V, Koeberl C (1997) Are diamictites impact ejecta? - No evidence from South African occurrences. J Geol 105: 517–530

Reimold WU, Brandt D, Koeberl C (1998a) Detailed structural analysis of the rim of a large, complex impact crater: Bosumtwi crater, Ghana. Geology 26: 543–546

Reimold WU, Koeberl C, Reddering JSV (1998b) The 1992 drill core from the Kalkkop impact crater, Eastern Cape Province, South Africa: stratigraphy, petrography, geochemistry and age. J African Earth Sci 26: 573–592

Reimold WU, Koeberl C, Brandstätter F, Kruger FJ, Armstrong RA, Bootsman C (1999) The Morokweng impact structure, South Africa: Geological, petrological, and isotopic results, and implications for the size of the structure. In: Dressler BO, Sharpton VL (eds) Large Meteorite Impacts and Planetary Evolution II. Geological Society of America, Special Paper 339, pp 61–90

Reimold WU, Armstrong RA, Koeberl C (2000) New results from the deep borehole at Morokweng, North West Province, South Africa: Constraints on the size of the J/K boundary age impact structure. Lunar Planet Sci 31: abs. #1074

Remane J (1991) The Jurassic-Cretaceous boundary: Problems of definition and procedure. Cretaceous Research 12: 447–453

Renard M (1985) Géochimie des carbonates pélagiques. Thèse d'Etat, Académie de Paris, Université Pierre et Marie Curie, 650 pp

Renon R (1997) Etude environmentale à la limite Crétacé-Tertiaire. Option Sciences de la Terre et Environment, Ecole des Mines de Paris (unpublished), 54 pp

Retallack GJ, Seyedolali A, Krull ES, Holser WT, Ambers CP, Kyte FT (1998) Search for evidence of impact at the Permian-Triassic boundary in Antarctica and Australia. Geology 26: 979–982

Ridenour GS (1986) Evidence for selective volatilization and imperfect mixing in indochinites. Meteoritics 21: 271–281

Rhodes RC (1975) New evidence for impact origin of the Bushveld Complex, South Africa. Geology 3: 549–554

Robertson PB, Dence MR, Vos MA (1968) Deformation in rock-forming minerals from Canadian craters. In: French BM, Short NM (eds) Shock Metamorphism of Natural Materials. Mono Book Corp., Baltimore, pp 433–452

Robin E, Bonté P, Froget L, Jéhanno C, Rocchia R (1992) Formation of spinels in cosmic objects during atmospheric entry: A clue to the Cretaceous-Tertiary boundary event. Earth Planet Sci Lett 108: 181–190

Robin E, Froget L, Jéhanno C, Rocchia R (1993) Evidence for a K/T impact event in the Pacific Ocean. Nature 363: 615–617

Robin E, Pierrard O, Lefevre I, Rocchia R (1999) A search for extraterrestrial spinel in pelagic sediments from the central North Pacific. GSA Annual Meeting, Abstracts with Programs 31(7): A63

Rocchia E, Bonté P, Jéhanno C, Robin E, de Angelis M, Boclet D (1990) Search for the Tunguska event relics in the Antarctic snow and new estimation of the cosmic iridium accretion rate. In: Sharpton VL Ward PD (eds) Global Catastrophes in Earth History. Geological Society of America, Special Paper 247, pp 189–193

Rocchia R, Robin E, Fröhlich F, Meon H, Frogent L, Diemer E (1996a) L'origine des verres du désert libyque: un impact météorique. Comptes Rendus Acad Sci (Paris) 322: 839–845

Rocchia R, Robin E, Froget L, Gayraud J (1996b) Stratigraphic distribution of extraterrestrial markers at the Cretaceous-Tertiary boundary in the Gulf of Mexico area: Implications for the temporal complexity of the event. In: Ryder G, Fastovsky D, Gartner S (eds) New Developments Regarding the KT Event and Other Catastrophes in Earth History. Geological Society of America, Special Paper 307, pp 279–286

Rocchia R, Robin E, Fröhlich F, Ammosse J, Barrat JA, Meon H, Froget L, Diemer E (1997) The impact origin of Libyan Desert Glass. In: Proceedings, Silica 96, Meeting on Libyan Desert Glass and related desert events. Pyramids, Milan, Italy, pp 143–158

Roddy DJ, Pepin RO, Merrill RB (eds) (1977) Impact and Explosion Cratering. Pergamon Press, New York, 1301 pp

Roggenthen WM, Napoleone G (1977) Upper Cretaceous-Paleocene magnetic stratigraphy at Gubbio, Italy, IV. Upper Maastrichtian-Paleogene magnetic stratigraphy. Geol Soc Am Bull 88: 378–382

Ryder G (1990) Lunar samples, lunar accretion, and the early bombardment history of the Moon. EOS Transactions, American Geophysical Union 71: 313–323

Ryder G (1996) The unique significance and origin of the Cretaceous-Tertiary boundary: Historical context and burdens of proof. In: Ryder G, Fastovsky D, Gartner S (eds) New Developments Regarding the KT Event and Other Catastrophes in Earth History. Geological Society of America, Special Paper 307, pp 31–38

Ryder G, Fastovsky D, Gartner S (eds) (1996) The Cretaceous-Tertiary Event and other Catastrophes in Earth History. Geological Society of America, Special Paper 307, 576 pp

Ryder G, Koeberl C, Mojzsis SM (2000) Heavy bombardment of the Earth at ~3.85 Ga: The search for petrographic and geochemical evidence. In: Canup R, Righter K (eds) Origin of the Earth and Moon. University of Arizona Press, Tucson, in press.

Sanfilippo A, Riedel WR, Glass BP, Kyte FT (1985) Late Eocene microtektites and radiolarian extinctions on Barbados. Nature 314: 613–615

Saunders JB, Bernoulli D, Müller-Merz E, Oberhänsli H, Perch-Nielsen K, Riedel WR, Sanfilippo A, Torrini R Jr (1984) Stratigraphy of the late Middle Eocene to Early Oligocene in the Bath Cliff section, Barbados, West Indies. Micropaleontology 30: 390–425

Schmidt G, Pernicka E (1994) The determination of platinum group elements (PGE) in target rocks and fall-back material of the Nördlinger Ries impact crater (Germany). Geochim Cosmochim Acta 58: 5083–5090

Schmidt G, Zhou L, Wasson JT (1993) Iridium anomaly associated with the Australasian-tektite-producing impact: Masses of the impactor and of the Australasian tektites. Geochim Cosmochim Acta 57: 4851–4859

Schmitt RA (1990) A general theory of mass extinctions in the Phanerozoic, I. Observations and constraints. Lunar Planet Sci 21: 1085–1086

Schmitz B, Andersson P, Dahl J (1988) Iridium, sulfur isotopes and rare earth elements in the Cretaceous-Tertiary boundary clays at Stevns Klint, Denmark. Geochim Cosmochim Acta 52: 229–236

Schmitz B, Jeppsson L, Ekvall J (1994) A search for shocked quartz grains and impact ejecta in early Silurian sediments on Gotland, Sweden. Geol Mag 131: 361–367

Schmitz B, Peucker-Ehrenbrink B, Lindström M, Tassinari M (1997) Accretion rates of meteorites and cosmic dust in the Early Ordovician. Science 278: 88–90

Schnabel C, Pierazzo E, Xue S, Herzog GF, Masarik J, Cresswell RG, di Tada ML, Liu K, Fifield LK (1999) Shock melting of the Canyon Diablo impactor: Constraints from nickel-59 contents and numerical modeling. Science 285: 85–88

Schneider DA, Kent DV, Mello GA (1992) A detailed chronology of the Australasian impact event, the Brunhes-Matuyama geomagnetic polarity reversal, and global climate change. Earth Planet Sci Lett 111: 395–405

Schnetzler CC (1970) The lunar origin of tektites R.I.P. Meteoritics 5: 221–222

Schnetzler CC (1992) Mechanism of Muong Nong-type tektite formation and speculation on the source of Australasian tektites. Meteoritics 27: 154–165

Schnetzler CC, Pinson WM (1964) Variation of strontium isotopes in tektites. Geochim Cosmochim Acta 28: 953–969

Schnetzler CC, Pinson WH, Hurley PM (1966) Rubidium-Strontium age of the Bosumtwi crater area, Ghana, compared with the age of the Ivory Coast tektites. Science 151: 817–819

Schnetzler CC, Philpotts JA, Thomas HH (1967) Rare earth and barium abundances in Ivory Coast tektites and rocks from the Bosumtwi crater area, Ghana. Geochim Cosmochim Acta 31: 1987–1993

Schnetzler CC, Walter LS, Marsh JG (1988) Source of the Australasian tektite strewn field: A possible off-shore impact site. Geophys Res Lett 15: 357–360

Schnetzler CC, Fiske PS, Garvin JB, Frawley JJ (1999) Recent developments in the search for the site of the 780,000-year-old southeast Asian impact. Meteoritics and Planetary Science 34: A102–A103

Schopf TJM (1974) Permo-Triassic extinctions: Relation to sea-floor spreading. J Geol 82: 129–143

Schultz PH (1996) Effect of impact angle on vaporization. J Geophys Res 101: 21117–21136

Schultz PH (1998) Shooting the Moon: Understanding the history of lunar impact theories. Earth Sciences History 17: 92–110

Schultz PH, D'Hondt SD (1996) Cretaceous-Tertiary (Chicxulub) impact angle and its consequences. Geology 24: 963–967

Schultz PH, Koeberl C, Bunch T, Grant J, Collins W (1994) Ground truth for oblique impact processes: New insight from the Rio Cuarto, Argentina, crater field. Geology 22: 889–892

Schultz PH, Zarate M, Hames W, Camilión C, King J (1998) A 3.3-Ma impact in Argentina and possible consequences. Science 282: 2061–2063

Schuraytz BC, Sharpton VL, Marin LE (1994) Petrology of impact-melt rocks at the Chicxulub multiring basin, Yucatán, Mexico. Geology 22: 868–872

Schuraytz BC, Lindstrom DJ, Marín LE, Martinez RR, Mittlefehldt DW, Sharpton VL, Wentworth SJ (1996) Iridium metal in Chicxulub impact melt: Forensic chemistry on the K-T smoking gun. Science 271: 1573–1576

Sclater JG, Jaupart C, Galson D (1980) The heat flow through oceanic and continental crust and the heat loss of the earth. Reviews in Geophysics and Space Physics 18: 269–311

Scott ERD (1999) How were tektites formed and ejected? Meteoritics and Planetary Science 34: A103

Seebaugh WR, Strauss AM (1984) A cometary impact model for the source of Libyan Desert Glass. Journal of Non-crystalline Solids 67: 511–519

Sepkoski JJ (1990) The taxonomic structure of periodic extinction. In: Sharpton VL, Ward PD (eds) Global Catastrophes in Earth History. Geological Society of America, Special Paper 247, pp 33–44

Sepkoski JJ Jr (1992) A compendium of fossil marine animal families (second edition). Milwaukee Public Museum Contributions to Biology and Geology 83, 156 pp

Sepkoski JJ (1996) Patterns of Phanerozoic extinctions: a perspective from global data bases: In: Walliser OH (ed) Global events and event stratigraphy. Springer, Berlin, pp 35–52

Seyedolali A, Krinsley DH, Boggs S Jr, O'Hara PF, Dypvik H, Goles GG (1997) Provenance interpretation of quartz by scanning electron microscope-cathodoluminescence fabric analysis. Geology 25: 787–790

Sharp TG, El Goresy A, Wopenka B, Chen M (1999) A post-stishovite SiO_2 polymorph in the meteorite Shergotty: Implications for impact events. Science 284: 1511–1513

Sharpton VL, Grieve RAF (1990) Meteorite impact, cryptoexplosion, and shock metamorphism: A perspective on the evidence at the K/T boundary. In: Sharpton VL, Ward PD (eds) Global Catastrophes in Earth History. Geological Society of America, Special Paper 247, pp 301–318

Sharpton VL, Schuraytz BC (1989) On reported occurrences of shock-deformed clasts in the volcanic ejecta from Toba caldera, Sumatra. Geology 17: 1040–1043

Sharpton VL, Ward PD (eds) (1990) Global Catastrophes in Earth History. Geological Society of America, Special Paper 247, 631 pp

Sharpton VL, Schuraytz BC, Burke K, Murali AV, Ryder G (1990) Detritus in K/T boundary clays of western North America; Evidence against a single oceanic impact. In: Sharpton VL, Ward PD (eds) Global Catastrophes in Earth History. Geological Society of America, Special Paper 247, pp 349–357

Sharpton VL, Dalrymple GB, Marin LE, Ryder G, Schuraytz BC, Urrutia-Fucugauchi J (1992) New links between the Chicxulub impact structure and the Cretaceous/Tertiary boundary. Nature 359: 819–821

Sharpton VL, Burke K, Camargo-Zanoguera A, Hall SA, Lee S, Marín LE, Suárez-Reynoso G, Quezada-Muñeton JM, Spudis PD, Urrutia-Fucugauchi J (1993) Chicxulub multiring impact basin: Size and other characteristics derived from gravity analysis. Science 261: 1564–1567

Sharpton VL, Marin LE, Carney JL, Lee S, Ryder G, Schuraytz BC, Sikora P, Spudis PD (1996) A model of the Chicxulub impact basin based on evaluation of geophysical data, well logs and drill core samples. In: Ryder G, Fastovsky D, Gartner S (eds) New Developments Regarding the KT Event and Other Catastrophes in Earth History. Geological Society of America, Special Paper 307, pp 55–74

Shaw HF, Wasserburg GJ (1982) Age and provenance of the target materials for tektites and possible impactites as inferred from Sm-Nd and Rb-Sr systematics. Earth Planet Sci Lett 60: 155–177

Sheehan PM, Fastovsky DE, Barreto C, Hoffmann RG (1999) Does a "three-meter gap" at the top of the Hell Creek Formation indicate dinosaurs were declining prior to the bolide impact? GSA Annual Meeting, Abstracts with Programs 31(7): A473

Shirey SB (1991) The Rb-Sr, Sm-Nd, and Re-Os isotope systems: A summary and comparison of their applications to the cosmochronology and geochronology of igneous rocks. In: Heman L, Ludden J (eds) Applications of Radiogenic Isotope Systems to Problems in Geology. Mineralogical Association of Canada, Short Course Handbook 19, pp 103–166

Shirey SB, Walker RJ (1995) Carius tube digestion for low-blank rhenium-osmium analysis. Analytical Chemistry 67: 2136–2141

Shoemaker EM (1977) Why study impact craters? In: Roddy DJ, Pepin RO, Merrill RB (eds) Impact and Explosion Cratering. Pergamon Press, New York, pp 1–10

Shoemaker EM, Izett GA (1992) Stratigraphic evidence from western North America for multiple impacts at the K/T boundary. Lunar Planet Sci 23: 1293–1294

Shoemaker EM, Shoemaker CS (1996) The Proterozoic impact record of Australia. AGSO J Austral Geol Geophys 16: 379–398

Shoemaker EM, Uhlherr HR (1999) Stratigraphic relations of australites in the Port Campbell Embayment, Victoria. Meteoritics and Planetary Science 34: 369–384

Shoemaker EM, Wolfe RF, Shoemaker CS (1990) Asteroid and comet flux in the neighborhood of Earth. In: Sharpton VL, Ward PD (eds) Global Catastrophes in Earth History. Geological Society of America, Special Paper 247, pp 155–170

Short NM (1966) Shock-lithification of unconsolidated rock materials. Science 154: 382–384

Short NM (1970) Progressive shock metamorphism of quartzite ejecta from the Sedan nuclear explosion crater. J Geol 78: 705–732

Shukolyukov A, Lugmair GW (1998) Isotopic evidence for the Cretaceous-Tertiary impactor and its type. Science 282: 927–929

Shukolyukov A, Lugmair GW, Koeberl C, Reimold WU (1999) Chromium in the Morokweng impact melt: Isotopic evidence for extraterrestrial components and type of impactor. Meteoritics and Planetary Science 34: A107–A108

Shukolyukov A, Kyte FT, Lugmair GW, Lowe DR, Byerly GR (2000) The oldest impact deposits on earth – first confirmation of an extraterrestrial component. In: Gilmour I, Koeberl C (eds) Impacts and the Early Earth. Lecture Notes in Earth Sciences 91. Springer, Heidelberg-Berlin, pp 99–116

Sigurdsson H, D'Hondt S, Arthur MA, Bralower TJ, Zachos JC, van Fossen M, Channell ET (1991a) Glass from the Cretaceous/Tertiary boundary in Haiti. Nature 349: 482–487

Sigurdsson H, Bonte P, Turpin L, Chaussidon M, Metrich N, Steinberg M, Pradel P, D'Hondt S (1991b) Geochemical constraints on source regions of Cretaceous/Tertiary impact glasses. Nature 353: 839–842

Sigurdsson H, D'Hondt S, Carey S (1992) The impact of the Cretaceous/Tertiary bolide on evaporite terrane and generation of major sulfuric acid aerosol. Earth Planet Sci Lett 109: 543–559

Silver LT, Schultz PH (eds) (1982) Geological Implications of Impacts of Large Asteroids and Comets on the Earth. Geological Society of America, Special Paper 190, 528 pp

Simonson BM (1992) Geological evidence for a strewn field of impact spherules in the early Precambrian Hamersley Basin of Western Australia. Geol Soc Am Bull 104: 829–839

Simonson BM, Davies D (1996) PGEs and quartz grains in a resedimented late Archean impact horizon in the Hamersley Group of Western Australia. Lunar Planet Sci 27: 1203–1204

Simonson BM, Beukes NJ, Hassler S (1997) Discovery of a Neoarchean impact spherule horizon in the Transvaal Supergroup of South Africa and possible correlations to the Hamersley Basin of Western Australia. Lunar Planet Sci 28: 1323–1324

Simonson BM, Davies D, Wallace M, Reeves S, Hassler SW (1998) Iridium anomaly but no shocked quartz from Late Archean microkrystite layer: Oceanic impact ejecta? Geology 26: 195–198

Sinha A, Stott LD (1994) New atmospheric pCO_2 estimates from paleosols during the late Paleocene/early Eocene global warming interval. Global and Planetary Change 9: 297–307

Skala R, Rohovec J (1998) Magic-angle spinning nuclear magnetic resonance spectroscopy of shocked limestones from the Steinheim crater. Meteoritics and Planetary Science 33: A146–A147

Sleep NH, Zahnle KJ, Kasting JF, Morowitz HJ (1989) Annihilation of ecosystems by large asteroid impacts on the early Earth. Nature 342: 139–142

Smit J (1982) Extinction and evolution of planktonic Foraminifera after a major impact at the Cretaceous/Tertiary boundary. In Silver LT, Schultz PH (eds) Geological Implications of Impacts of Large Asteroids and Comets on the Earth. Geological Society of America, Special Paper 190, pp 329–352

Smit J (1999) The global stratigraphy of the Cretaceous-Tertiary boundary impact ejecta. Annual Reviews of Earth and Planetary Science 27: 75–113

Smit J, Hertogen J (1980) An extraterrestrial event at the Cretaceous-Tertiary boundary. Nature 285: 198–200

Smit J, Klaver G (1981) Sanidine spherules at the Cretaceous-Tertiary boundary indicate a large impact event. Nature 292: 47–49

Smit J, Kyte FT (1984) Siderophile-rich magnetic spheroids from the Cretaceous-Tertiary boundary in Umbria, Italy. Nature 310: 403–405

Smit J, Nederbragt AJ (1997) Analysis of El Kef blind test II. Marine Micropaleontology 29: 94–100

Smit J, Groot H, De Jonge R, Smit P (1988) Impact and extinction signatures in complete Cretaceous-Tertiary (K-T) boundary sections. Lunar and Planetary Institute, Contribution 673, pp 182–183

Smit J, Alvarez W, Montanari A, Swinburne N, Van Kempen TM, Klaver GT, Lustenhouwer WJ (1992a) "Tektites" and mikrokrystites at the Cretaceous-Tertiary boundary: Two strewn fields, one crater? Proc Lunar Planet Sci 22: 87–100

Smit J, Montanari A, Swinburne N, Alvarez W, Hildebrand AR, Margolis SV, Claeys P, Lowrie W, Asaro F (1992b) Tektite-bearing, deep-water clastic unit at the Cretaceous-Tertiary boundary in northeastern Mexico. Geology 20: 99–103

Smit J, Roep TB, Alvarez W, Montanari A, Claeys P, Grajales-Nishimura JM, Bermudez J (1996) Coarse-grained, clastic sandstone complex at the K/T boundary around the Gulf of Mexico: Deposition by tsunami waves induced by the Chicxulub impact? In: Ryder G, Fastovsky D, Gartner S (eds) New Developments Regarding the KT Event and Other Catastrophes in Earth History. Geological Society of America, Special Paper 307, pp 151–182

Smit J, Keller G, Zargouni F, Razgallah S, Shimi M, Abdelkader OB, Haj Ali NB, Salem HB (1997) The El Kef sections and sampling procedures. Marine Micropaleontology 29: 65–79

Speijer RP (1994) Extinction and recovery pattern in benthic foraminiferal paleocommunities across the Cretaceous/Paleogene and Paleocene/Eocene boundaries. Mededelingen van Facultait Advertenshappen Universiteit Utrecht 124, pp 1–191

Spicer RA (1989) Plants at the Cretaceous-Tertiary boundary. In: Chaloner WG, Hallam A (eds) Evolution and Extinction. Phil Trans Roy Soc London Series B 325: 291–305

Spicer RA, Corfield RM (1992) A review of terrestrial and marine climates in the Cretaceous with implications for modelling the 'Greenhouse Earth'. Geol Mag 129: 169–180

Spray JG (1995) Pseudotachylyte controversy: Fact or friction? Geology 23: 1119–1122

Spray JG (1997) Superfaults. Geology 25: 579–582

Spray JG (1999) Shocking rocks by cavitation and bubble implosion. Geology 27: 695–698

Spray JG, Thompson LM (1995) Friction melt distribution in a multi-ring impact basin. Nature 373: 130–132

Spray JG, Kelley SP, Rowley DB (1998) Evidence for a late Triassic multiple impact event on Earth. Nature 392: 171–173

Spudis PD (1993) The geology of multi-ring impact basins. Cambridge University Press, New York, 263 pp

Staudacher T, Jessberger EK, Dominik B, Kirsten T, Schaeffer OA (1982) ^{40}Ar-^{39}Ar ages of rocks and glasses from the Nördlinger Ries and the temperature history of impact breccias. J Geophys 51: 1–11

Stauffer PH (1978) Anatomy of the Australasian tektite strewnfield and the probable site of its source crater. 3rd Regional Conference on Geology and Mineral Resources of Southeast Asia, Bangkok, pp 285–289

Stecher O, Ngo HH, Papanastassiou DA, Wasserburg GJ (1989) Nd and Sr isotopic evidence for the origin of tektite material from DSDP Site 612 off the New Jersey Coast. Meteoritics 24: 89–98

Stinnesbeck W, Keller G (1996) K/T boundary coarse-grained siliciclastic deposits in northeastern Mexico and northeastern Brazil: Evidence for mega-tsunami or sea-level changes? In: Ryder G, Fastovsky D, Gartner S (eds) New Developments Regarding the KT Event and Other Catastrophes in Earth History. Geological Society of America, Special Paper 307, pp 197–209

Stinnesbeck W, Barbarin JM, Keller G, Lopez-Oliva JG, Pivnik DA, Lyons JB, Officer CB, Adatte T, Graup G, Rocchia R, Robin E (1993) Deposition of channel deposits near the Cretaceous-Tertiary boundary in northeastern Mexico: Catastrophic or "normal" sedimentary deposits? Geology 21: 797–800

Stinnesbeck W, Keller G, de la Cruz J, de León C, MacLeod N, Whittaker JE (1997) The Cretaceous-Tertiary transition in Guatemala: Limestone breccia deposits from the South Petén basin. Geologische Rundschau 86: 686–709

Stöffler D (1972) Deformation and transformation of rock-forming minerals by natural and experimental shock processes: 1. Behaviour of minerals under shock compression. Fortschritte der Mineralogie 49: 50–113

Stöffler D (1974) Deformation and transformation of rock-forming minerals by natural and experimental processes: 2. Physical properties of shocked minerals. Fortschritte der Mineralogie 51: 256–289

Stöffler D (1984) Glasses formed by hypervelocity impact. Journal of Non-crystalline Solids 67: 465–502

Stöffler D, Grieve RAF (1994a) Classification and nomenclature of impact metamorphic rocks: A proposal to the IUGS subcommission on the systematics of metamorphic rocks. Lunar Planet Sci 25: 1347–1348

Stöffler D, Grieve RAF (1994b) Classification and nomenclature of impact metamorphic rocks: A proposal to the IUGS subcommission on the systematics of metamorphic rocks. In: Montanari A, Smit J (eds) Post-Östersund Newsletter, European Science Foundation (ESF) Scientific Network on Impact Cratering and Evolution of Planet Earth, Strasbourg, pp 9–15

Stöffler D, Hornemann U (1972) Quartz and feldspar glasses produced by natural and experimental shock. Meteoritics 7: 371–394

Stöffler D, Langenhorst F (1994) Shock metamorphism of quartz in nature and experiment: I. Basic observations and theory. Meteoritics 29: 155–181

Stöffler D, Gault DE, Wedekind J, Polkowski G (1975) Experimental hypervelocity impact into quartz sand: Distribution and shock metamorphism of ejecta. J Geophys Res 80: 4062–4077

Stöffler D, Knöll HD, Maerz U (1979) Terrestrial and lunar impact breccias and the classification of lunar highland rocks. Proceedings of the 10th Lunar and Planetary Science Conference, pp 639–675

Stöffler D, Deutsch A, Avermann M, Bischoff L, Brockmeyer P, Buhl D, Lakomy R, Müller-Mohr V (1994) The formation of the Sudbury Structure, Canada: Toward a unified impact model. In: Dressler BO, Grieve RAF, Sharpton VL (eds) Large Meteorite Impacts and Planetary Evolution. Geological Society of America, Special Paper 293, pp 303–318

Storzer D (1985) The fission track age of high sodium/potassium australites revisited. Meteoritics 20: 765–766

Storzer D, Koeberl C (1991) Uranium and zirconium enrichments in Libyan Desert Glass. Lunar Planet Sci 22: 1345–1346

Storzer D, Müller-Sohnius D (1986) The K/Ar age of high sodium/potassium australites. Meteoritics 21: 518–519

Storzer D, Wagner GA (1969) Correction of thermally lowered fission track ages of tektites. Earth Planet Sci Lett 5: 463–468

Storzer D, Wagner GA (1971) Fission-track ages of North American tektites. Earth Planet Sci Lett 10: 435–440

Storzer D, Wagner GA (1977) Fission track dating of meteorite impacts. Meteoritics 12: 368–369

Storzer D, Wagner GA (1979) Fission track dating of Elgygytgyn, Popigai and Zhamanshin craters: No source for Australasian or North American tektites. Meteoritics 14: 541–542

Storzer D, Wagner GA (1980a) Australites older than indochinites - Evidence from fission-track plateau dating. Naturwissenschaften 67: 90–91

Storzer D, Wagner GA (1980b) Two discrete tektite-forming events 140 thousand years apart in the Australian-Southeast Asian area. Meteoritics 15: 372

Storzer D, Wagner GA (1982) The application of fission track dating in stratigraphy: a critical review. In: Odin GS (ed) Numerical Dating in Stratigraphy. J. Wiley & Sons, pp 199–221

Storzer D, Wagner GA, King EA (1973) Fission-track ages and stratigraphic occurrence of Georgia tektites. J Geophys Res 78: 4915–4919

Stothers RB (1993) Impact cratering at geological stage boundaries. Geophysical Research Letters 20: 887–890

Sugita S, Schultz PH, Adams MA (1998) Spectroscopic measurements of vapor clouds due to oblique impacts. J Geophys Res 103: 19427–19441

Svetsov VV (1996) Total ablation of the debris from the 1908 Tunguska explosion. Nature 383: 697–699

Swinburne NHM (1990) The extinction of the rudist bivalves. PhD Thesis, Open University, Milton Keynes, UK, 175 pp

Swinburne NHM (1992) Tethyan extinctions, sea level changes, and the Sr-isotope curve in the 10 Ma preceding the K/T boundary. EOS, Trans Am Geophys Un 72: 267

Swinburne NHM, Naocco A (1992) The platform carbonates of Monte Jouf, Maniago, and the Cretaceous stratigraphy of the Italian Carnian Alps. Geol Croat 46: 25–40

Swisher CC, Grajales-Nishimura JM, Montanari A, Margolis SV, Claeys P, Alvarez W, Renne P, Cedillo-Pardo E, Maurrasse FJMR, Curtis GH, Smit J, McWilliams MO (1992) Coeval $^{40}Ar/^{39}Ar$ ages of 65.0 million years ago from Chicxulub crater melt rock and Cretaceous-Tertiary boundary tektites. Science 257: 954–958

Takayama H, Tada R, Matsui T, Iturralde-Vinent MA, Oji T, Tajika E, Kiyokawa S, Garcia D, Okada H, Hasegawa T, Toyoda K (1999) Origin of a giant event deposit in northwestern Cuba and its relation to K/T boundary impact. Lunar Planet Sci 30: abs. #1534

Taylor HP, Epstein SE (1966) Oxygen isotope studies of Ivory Coast tektites and impactite glass from the Bosumtwi crater, Ghana. Science 153: 173–175

Taylor HP, Epstein S (1969) Correlations between $^{18}O/^{16}O$ ratios and chemical compositions of tektites. J Geophys Res 74: 6834–6844

Taylor SR (1962a) Fusion of soil during meteorite impact and the chemical composition of tektites. Nature 195: 32–33

Taylor SR (1962b) The chemical composition of australites. Geochim Cosmochim Acta 26: 685–722

Taylor SR (1966) Australites, Henbury impact glass, and subgreywacke: A comparison of 51 elements. Geochim Cosmochim Acta 30: 1121–1136

Taylor SR (1973) Tektites: A post-Apollo view. Earth-Science Reviews 9: 101–123

Taylor SR (1982) Planetary Science: A Lunar Perspective. Lunar and Planetary Institute, Houston, 481 pp

Taylor SR (1992) Solar System Evolution. Cambridge University Press, Cambridge, 307 pp

Taylor SR (1993) Early accretion history of the Earth and the Moon-forming event. Lithos 30: 207–221

Taylor SR (1999) The Australasian tektite age paradox. Meteoritics and Planetary Science 34: 311–313

Taylor SR, Kaye M (1969) Genetic significance of the chemical composition of tektites: A review. Geochim Cosmochim Acta 33: 1083–1100

Taylor SR, Koeberl C (1994) The origin of tektites: Comment on a paper by J. A. O'Keefe. Meteoritics 29: 739–742

Taylor SR, McLennan SM (1979) Chemical relationships among irghizites, zhamanshinites, Australasian tektites and Henbury impact glasses. Geochim Cosmochim Acta 43: 1551–1565

Taylor SR, McLennan SM (1985) The Continental Crust: Its Composition and Evolution. Blackwell Scientific Publications, Oxford, 312 pp

Taylor SR, Sachs M (1964) Geochemical evidence for the origin of australites. Geochim Cosmochim Acta 28: 235–264

Taylor SR, Solomon M (1964) The geochemistry of Darwin glass. Geochim Cosmochim Acta 28: 471–494

Tera F, Papanastassiou DA, Wasserburg GJ (1974) Isotopic evidence for a terminal lunar cataclysm. Earth Planet Sci Lett 22: 1–21

Tera F, Middleton R, Klein J, Brown L (1983a) Beryllium-10 in tektites. EOS, Trans Am Geophys Un 64: 284

Tera F, Brown L, Klein J, Middleton R (1983b) Beryllium-10: Tektites. Yearbook of the Carnegie Institution '82, pp 463–465

Thein J (1987) A tektite layer in upper Eocene sediments of the New Jersey continental slope (Site 612, Leg 95). In: Poag CW, Watts A, et al. (eds) Initial Reports of the Deep Sea Drilling Project, Volume 95, U.S. Government Printing Office, Washington, D.C., pp 565–574

Thomas E (1990) Late Cretaceous-early Eocene mass extinctions in the deep sea. In: Sharpton VL, Ward PD (eds) Global Catastrophes in Earth History. Geological Society of America, Special Paper 247, pp 231–247

Tilton GR (1958) Isotopic composition of lead from tektites. Geochim Cosmochim Acta 14: 323–330

Tjalsma RC, Lohman GP (1983) Paleocene-Eocene bathyal and abyssal benthic foraminifera from the Atlantic Ocean. Micropaleontology 4: 1–90

Toon OB, Zahnle K, Morrison D, Turco RP, Covey C (1997) Environmental perturbations caused by the impacts of asteroids and comets. Reviews of Geophysics 35: 41–78

Toutain J-P, Meyer G (1989) Iridium-bearing sublimates at a hot-spot volcano (Piton de la Fournaise, Indian Ocean). Geophys Res Lett 16: 1391–1394

Tredoux M, de Wit MJ, Hart RJ, Lindsay NM, Verhagen B, Sellschop JPF (1989a) Chemostratigraphy across the Cretaceous-Tertiary boundary and a critical assessment of the iridium anomaly. J Geol 97: 585–605

Tredoux M, de Wit MJ, Hart RJ, Armstrong RA, Lindsay NM, Sellschop JPF (1989b) Platinum group elements in a 3.5 Ga nickel-iron occurrence: Possible evidence of a deep mantle origin. J Geophys Res 94: 795–813

Trieloff M, Jessberger, EK (1992) $^{40}Ar/^{39}Ar$ ages of the large impact structures Kara and Manicouagan and their relative relevance to the Cretaceous-Tertiary and the Triassic-Jurassic boundary. Meteoritics 27: 299–300

Trieloff M, Deutsch A, Jessberger EK (1998) The age of the Kara impact structure, Russia. Meteoritics and Planetary Science 33: 361–372

Tschudy RH, Pillmore CL, Orth CJ, Gilmore JS, Knight JD (1984) Disruption of the terrestrial plant ecosystem at the Cretaceous-Tertiary boundary, Western Interior. Science 225: 1030–1032

Tur NA (1996) Planktonic foraminiferal recovery from the Cenomanian-Turonian mass extinction event, northeastern Caucasus. In: Hart MB (ed) Biotic Recovery from Mass Extinction Events. Geological Society London, Special Publication 102, pp 259–264

Turco RP, Toon OB, Park C, Whitten RC, Pollack JB, Noerdlinger P (1982) An analysis of the physical, chemical, optical and historical impacts of the 1908 Tunguska meteor fall. Icarus 50: 1–52

Turekian KK (1982) Potential of ^{187}Os/^{186}Os as a cosmic versus terrestrial indicator in high iridium layers of sedimentary strata. In: Silver LT, Schultz PH (eds) Geological Implications of Impacts of Large Asteroids and Comets on the Earth. Geological Society of America, Special Paper 190, pp 243–249

Urey HC (1973) Cometary collisions and geological periods, Nature 242: 32–33

Valloni R, Zuffa GG (1984) Provenance changes from arenaceous formations of the Northern Apennines, Italy. Geol Soc Am Bull 95: 1035–1039

Valette-Silver N, Brown L, Klein J, Middleton R, Pavic MJ, Tera F (1983) Beryllium-10: Vertical distribution in soils and sediments. Yearbook of the Carnegie Institution '82, pp 462–463

Van Graas G, Viets TC, De Leeuw JW, Schenck PA (1983) A study of the soluble and insoluble organic matter from the Livello Bonarelli, a Cretaceous black shale deposit in the Central Apennines, Italy. Geochim Cosmochim Acta 47: 1051–1059

Velde B, Boyer H (1985) Raman microprobe spectra of naturally shocked microcline feldspars. J Geophys Res 90: 3675–3682

Venkatesan MI, Dahl J (1989) Organic geochemical evidence for global fires at the Cretaceous-Tertiary boundary. Nature 338: 57–60

Vickery AM (1993) The theory of jetting: Application to the origin of tektites. Icarus 105: 441–453

Vickery AM, Browning L (1991) Water depletion in tektites. Meteoritics 26: 403

Vishnevsky SA, Lagutenko VN (1986) The Ragozinka astrobleme: An Eocene crater in the central Urals (in Russian). Dokl Akad Nauk SSSR 14: 1–42

Vishnevsky S, Montanari A (1999) Popigai impact structure (Arctic Siberia, Russia): Geology, petrology, geochemistry, and geochronology of glass-bearing impactites. In: Dressler BO, Sharpton VL (eds) Large Meteorite Impacts and Planetary Evolution II. Geological Society of America, Special Paper 339, pp 19–60

Völkening J, Walczyk T, Heumann KG (1991) Osmium isotope ratio determinations by negative thermal ionization mass spectrometry. International Journal of Mass Spectrometry and Ion Processes 105: 147–159

Vonhof HB, Smit J (1999) Late Eocene microkrystites and microtektites at Maud Rise (Ocean Drilling Project Hole 689B; Southern Ocean) suggest a global extension of the approximately 35.5 Ma Pacific impact ejecta strewn field. Meteoritics and Planetary Science 34: 747–755

Vonhof HB, Wijbrans J, Smit J (1995) The Popigai impact crater: ^{39}Ar/^{40}Ar dating and its expression in the ^{87}Sr/^{86}Sr record of the Massignano section: In Montanari A, Coccioni R (eds) The role of Impacts on the Evolution of Planet Earth, Abstracts and Field Trips, European Science Foundation, Ancona, pp 163–164

Vonhof HB, Smit J, Brinkhuis H, Montanari A (1998) Late Eocene Impacts accelerated global cooling? In: Vonhof HB (PhD Thesis) The Strontium Stratigraphic Record of Selected Geologic Events, Academisch Proefschrift, University of Utrecht, pp 77–90

Walker RJ, Morgan JW, Naldrett AJ, Li C, Fassett JD (1991) Re-Os isotope systematics of Ni-Cu sulfide ores, Sudbury Igneous Complex, Ontario: Evidence for a major crustal component. Earth Planet Sci Lett 105: 416–429

Wallace MW, Gostin VA, Keays RR (1990a) Spherules and shard-like clasts from the late Proterozoic Acraman impact ejecta horizon. Meteoritics 25: 161–165

Wallace MW, Gostin VA, Keays RR (1990b) Acraman impact ejecta and host shales: Evidence for low-temperature mobilization of iridium and other platinoids. Geology 18: 132–135

Wallace MW, Gostin VA, Keays RR (1996) Sedimentology of the Neoproterzoic Acraman impact ejecta horizon, South Australia. AGSO J Austral Geol Geophys 16: 443–451

Walter LS (1965) Coesite discovered in tektites. Science 147: 1029–1032

Walter LS (1967) Tektite compositional trends and experimental vapor fractionation of silicates. Geochim Cosmochim Acta 31: 2043–2063

Walter LS, Carron MK (1964) Vapor pressure and vapor fractionation of silicate melts of tektite composition. Geochim Cosmochim Acta 28: 937–951

Walter LS, Clayton RN (1967) Oxygen isotopes: Experimental fractionation and variations in tektites. Science 156: 1357

Wang K (1992) Glassy microspherules (microtektites) from an Upper Devonian limestone. Science 256: 1547–1550

Wang K, Chatterton BDE (1993) Microspherules in Devonian sediments: origins, geological significance, and contamination problems. Can J Earth Sci 30: 1660–1667

Wang K, Orth CJ, Attrep M Jr, Chatterton BDE, Hou H, Geldsetzer HHJ (1991) Geochemical evidence for a catastrophic biotic event at the Frasnian-Famennian boundary in south China. Geology 19: 776–779

Wang K, Chatterton BDE, Attrep M Jr, Orth CJ (1992) Iridium abundance maxima in the latest Ordovician mass extinction horizon, Yangtze Basin, China: Terrestrial or extraterrestrial? Geology 20: 39–42

Wang K, Attrep M Jr, Orth CJ (1993a) Global iridium anomaly, mass extinction, and redox change at the Devonian-Carboniferous boundary. Geology 21: 1071–1074

Wang K, Chatterton BDE, Attrep M Jr, Orth CJ (1993b) Late Ordovician mass extinction in the Selwyn Basin, northwestern Canada: Geochemical, sedimentological, and paleontological evidence. Canadian J Earth Sci 30: 1870–1880

Wang K, Orth CJ, Attrep M Jr, Chatterton BDE, Wang X, Li JJ (1993c) The last great Ordovician extinction on the South China plate: Chemostratigraphic studies of the Ordovician-Silurian boundary interval on the Yangtze Platform. Palaeogeography Palaeoclimatology Palaeoecology 104: 61–79

Ward PD, Kennedy WJ, MacLeod KG, Mount JF (1991) Ammonite and inoceramid bivalve extinction patterns in the Cretaceous/Tertiary boundary sections of the Biscay region (southwestern France, northern Spain). Geology 19: 1181–1184

Warme JE, Kuehner H-C (1998) Anatomy of an anomaly: The Devonian catastrophic Alamo impact breccia of southern Nevada. International Geology Review 40: 189–216

Warme JE, Sandberg CA (1996) Alamo megabreccia: Record of a late Devonian impact in southern Nevada. GSA Today 6(1): 1–7

Wasson JT (1991) Layered tektites: A multiple impact origin for the Australasian tektites. Earth Planet Sci Lett 102: 95–109

Wasson JT, Moore K (1998) Possible formation of Libyan Desert Glass by a Tunguska-like aerial burst. Meteoritics and Planetary Science 33: A163–A164

Wasson JT, Ouyang X, Zhou L (1990) Uranium volatilization during tektite formation. Meteoritics 25: 419

Weeks RA, Underwood JR Jr, Giegengack R (1984) Libyan Desert Glass: A review. Journal of Non-crystalline Solids 67: 593–619

Wegener A (1921) Die Entstehung der Mondkrater. Friedrich Vieweg & Sohn, Braunschweig, 48 pp

Wei W (1995) How many impact-generated microspherule layers in the upper Eocene? Palaeogeography Palaeoclimatology Palaeoecology 114: 101–110

Weiblen PW, Schultz KJ (1978) Is there any evidence of meteorite impact in the Archean rocks of North America? Proc 9th Lunar Planet Sci Conf, pp 2749–2773

Weihaupt JG (1976) The Wilkes Land anomaly: Evidence for a possible hypervelocity impact crater. J Geophys Res 81: 5651–5663

Weissman PR (1990) The cometary impactor flux at the earth. In: Sharpton VL, Ward PD (eds) Global Catastrophes in Earth History. Geological Society of America, Special Paper 247, pp 171–180

Wetherill GW (1975) Late heavy bombardment of the Moon and terrestrial planets. Proc 6th Lunar Sci Conf, pp 1539–1561

Wezel FC, Vannucci S, Vannucci R (1981) Découvert de divers niveaux riches en iridium dans la "Scaglia rossa" et la "Scaglia bianca" de l'Apennin d'Ombrie-Marches (Italie). Comptes Rendus Acad Sci Paris 293: 837–847

Whitehead J, Spray JG, Grieve RAF, Papanastassiou DA, Ngo HH, Wasserburg GJ (2000) Rb-Sr and Sm-Nd of upper Eocene microtektites: A potential Popigai source. Lunar Planet Sci 31: abs. #1373

Wilding M, Webb S, Dingwell DB (1996) Tektite cooling rates: Calorimetric relaxation geospeedometry applied to a natural glass. Geochim Cosmochim Acta 60: 1099–1103

Williams GE (1986) The Acraman impact structure: Source of ejecta in late Precambrian shales, South Australia. Science 233: 200–203

Williams GE (1994) Acraman, South Australia: Australia's largest meteorite impact structure. Proceedings of the Royal Society of Victoria 106: 105–127

Wilson WF, Wilson DH (1979) Remnants of a probable Tertiary impact crater in south Texas. Geology 7: 144–146

Wittke JH, Barnes VE (1988) Multi-component source for Muong Nong-type bediasite 30775-2. Meteoritics 23: 311

Witzke BJ, Hammond RH, Anderson RR (1996) Deposition of the Crow Creek Member, Campanian, South Dakota and Nebraska. In: Koeberl C, Anderson RR (eds) The Manson Impact Structure, Iowa: Anatomy of an Impact Crater. Geological Society of America, Special Paper 302, pp 433–456

Wolbach WS, Lewis RS, Anders E (1985) Cretaceous extinctions: Evidence for wildfires and search for meteoritic material. Science 230: 167–170

Wolbach WS, Gilmour I, Anders E (1990) Major wildfires at the Cretaceous/Tertiary boundary. In: Sharpton VL, Ward PD (eds) Global Catastrophes in Earth History. Geological Society of America, Special Paper 247, pp 391–400

Wolf R, Woodrow A, Grieve RAF (1980) Meteoritic material at four Canadian impact craters. Geochim Cosmochim Acta 44: 1015–1022

Wolfe JA (1991) Paleobotanical evidence for a June "impact winter" at the Cretaceous-Tertiary boundary. Nature 252: 420–422

Xu D, Yan Z (1993) Carbon isotope and iridium event markers near the Permian/Triassic boundary in the Meishan section, Zhejiang Province, China. Palaeogeography Palaeoclimatology Palaeoecology 104: 171–176

Yakovlev OI, Dikov YP, Gerasimov MV (1993) Oxidation and reduction in impact processes. Geochemistry International 30(7): 1–10

Yin HF, Tong J (1998) Multidisciplinary high-resolution correlation of the Permian-Triassic boundary. Palaeogeography Palaeoclimatology Palaeoecology 143: 199–212

Yiou F, Raisbeck GM, Klein J, Middleton R (1984) ^{26}Al/^{10}Be in terrestrial impact glasses. Journal of Non-crystalline Solids 67: 503–509

Zähringer J (1963) K-Ar measurements of tektites. In: Radioactive Dating: Proceedings Symposium Athens, Nov. 19-23, 1962. IAEA Vienna, pp 289–305

Zahnle KJ (1990) Atmospheric chemistry by large impacts. In: Sharpton VL, Ward PD (eds) Global Catastrophes in Earth History. Geological Society of America, Special Paper 247, pp 271–288

Zahnle K, Grinspoon D (1990) Comet dust as a source of amino acids at the Cretaceous/Tertiary boundary. Nature 348: 157–159

Zappalà V, Cellino A, Gladman BJ, Manley S, Migliorini F (1998) Asteroid showers on Earth after family breakup events. Icarus 134: 176–179

Zhakarov VA, Lapukhov AS, Shenfil OV (1993) Iridium anomaly at the Jurassic-Cretaceous boundary in northern Siberia. Russian Journal of Geology and Geophysics 34: 83–90

Zhao M, Bada JL (1989) Extraterrestrial amino acids in Cretaceous/Tertiary boundary sediments at Stevens Klint, Denmark. Nature 339: 463–465

Zhou L, Kyte FT (1988) The Permian-Triassic boundary event: A geochemical study of three Chinese sections. Earth Planet Sci Lett 90: 411–421

Zhou LP, Shackleton NJ (1999) Misleading positions of geomagnetic reversal boundaries in Eurasian loess and implications for correlation between continental and marine sedimentary sequences. Earth Planet Sci Lett 168: 117–130

Zoller WH, Parrington JR, Kotra JMP (1983) Iridium enrichment in airborne particles from Kilauea volcano. Science 222: 1118–1121

Index

γ-γ coincidence spectrometry 289
^{10}Be 80, 91, 92, 93
^{187}Os/^{188}Os 49, 50, 107, 118, 293, 294
^{188}Os/^{187}Os isotopic ratios 49
29R magnetochron 103, 219
^{3}He 62, 109, 131, 189, 195, 219, 231, 253, 258, 262, 266, 268
^{40}Ar/^{39}Ar 51, 122, 123, 171, 194, 195, 255–257, 274, 275
A. mayaroensis 200
Aalenian 174
Abatomphalus mayaroensis Zone 200
ablation spherules 64, 65, 74, 113, 288
Abruzzo 157, 164, 165
accretion 62, 153, 189, 195, 198, 254, 266, 268
accretionary lapilli 146
accuracy 167, 171, 288
achondrites 44
acid rain 127, 217
acoustic fluidization 21
Acqualagna 204, 228
Adelaide 138
Adria 157, 162
Adriatic Promontory 157
aerodynamically shaped tektites 64
Africa 11, 86, 157
age paradox 90
Agost 106, 253
airblast 12, 61
Alamo Breccia, Nevada 148, 149
Albian 163, 164
Albion Island, Belize 126, 128
alkali feldspar 83, 133
allochthonous 25, 39
Alpine–Himalayan
– orogenesis 57
– orogenic domain 274
Alps 165
amino acids 108
Ammodiscus 210, 248

Anabar Shield 133
Ancona 131, 164, 203, 256–258, 274, 275
anhydrite 125, 162
anoxia 203
anoxic event 164, 174, 189
Antarctica 82, 131, 150, 251, 254
apatite 52, 96, 224
Apennines 158, 162, 165, 172, 173, 185, 186, 190, 197, 200, 201, 203, 206, 239, 247, 250, 256, 278, 280, 283, 284
apographitic diamonds 35
Apollo 8, 153
Appalachian 79, 83
Aptian– 163, 164
– Albian (A/A) 168
Ar–Ar technique 52, 68
Archean 64, 72, 73, 79, 80, 133, 142–144, 148
– spherule layers 142
Arctic
– Russia 194
– Siberia 254
arenaceous
– benthic
– foraminifera 284–286
– foraminiferal tests 207
– foraminifers 227
– benthonic foraminifers 244
Argentina 7, 46, 155
Arizona 5, 15, 34, 43
Arroyo el Mimbral, Mexico 123
Ashanti, Ghana 86
asteroid(s) 1, 3, 6, 14, 63, 153, 198, 209
– impact 3, 14–16, 124, 142
Atlantic Ocean 86, 129, 152, 185, 208, 256, 257
Australasian strewn field(s) 59, 64, 66, 68, 70, 72–80, 84, 88–94
Australia 11, 13, 48, 53, 55, 65, 91, 93, 94, 139, 141, 142, 146, 148, 150, 152, 153

australites 64, 65, 73, 74, 88, 91, 92, 94, 95
Austria 85, 150–152
authigenic 201, 207, 220, 249

B. jacobyi 178
baddeleyite 36, 64, 78, 93, 96
Bahamas 157
Bajocian 163
ballen
– structure 37
– texture 38
Barbados 69, 81, 255–257
Barberton 142–144, 148
Barbetti Quarry 262
Barremian–Aptian boundary 181
Barriasian–Valanginian 163
basaltic melt 207
base surge 39, 58, 126
Barringer, Daniel Moreau 5
Basque Country 192
Bath Cliff section, Barbados 257
bathysiphon 211
Bauma interval 231
Bay of Bengal 91
Bedias, Texas 80
bediasites 74
Belgium 148
Belize 126, 128
Beloc, Haiti 111, 112, 123, 253
benthic
– foraminifera 210, 284, 285
– foraminifers 172, 203, 210, 227, 236, 241, 248
benthos 271
Berriasella jacobi 178
Berriasian 140
Berwind Canyon, Colorado 107, 110, 111
Besednice, Czech Republic 84
bioclasts 201
biogenic calcite 198
biologic turnovers 215
biomicritic limestone 164, 178
biosphere 7, 194
biotite 52, 60, 224, 253, 255, 256, 258, 262, 264, 269–272, 275
bioturbation, 200, 203, 204, 209–213, 215, 218, 222, 231, 235, 237, 239, 244, 271
birefringence 23, 33, 296
Birimian Supergroup 87
Bisciaro 166, 219

bituminous shales 184
bivalves 192
black shale(s) 159, 164, 174, 187
Blake Nose, Florida 107, 129
bleaching 203, 218
blind test 124
Bohemia, Czech Republic 85
Böhm lamellae 28, 295
bolide 15, 106, 108, 109, 112, 117, 141, 217
Bon Accord 145
Bonarelli
– Level 164
– Oceanic Anoxic Event 184
boron isotopes 80
Bosso Gorge 178
Bottaccione Gorge 180, 181, 185, 197, 198, 223–225
bottle-green microtektite 76, 77, 93
bottom current 204, 210
Bouguer gravity anomaly 117
bowl-shaped craters 7, 20
Brazil 28, 29, 55, 129
– twin(s) 28, 29
breccias 24, 25, 36, 39, 40, 42–44, 46–48, 50, 52, 118, 120, 122, 134, 293
Brown Clay Member 181, 182
Brownie Butte 109
Brunhes-Matuyama
– boundary 75
– reversal 88
Bugarone Formation 163
Bunte Breccia 40
Bunyeroo Formation, Adelaide 138
Burano 186
– anhydrite 162
Burdigalian 166
Bushveld Complex 148

Cagli 178, 186, 201
calcarenitic turbidites 165, 201, 203, 211, 231, 241, 247
calcareous
– algae 241
– marl(s) 101, 258, 286
– nannofossils 165, 178, 271
calcic feldspar 204, 220
calciruditic turbidites 239
calderas 3, 7
Callisto 10, 156
Calpionella brevis 179

Cambodia 75, 89
Cambrian 134
Camerino 201
Campanian– 190, 191, 194, 195, 201
– Maastrichtian boundary 194
Campeche platform 217
Canada 7, 26, 46, 54, 55, 82, 112, 117, 131, 151, 153, 184, 274
Candigliano River 228
Caravaca 106, 108
– section 206
carbon isotope stratigraphy 108
carbonaceous chondrites 51
carbonate(s) 27, 146, 154, 157, 262, 264, 269
– apron 239
– dissolution 198
– ooze 198, 200, 218
– platform(s) 157, 162, 164, 165, 192, 201, 239, 247
– reefs 184
Caribbean 81, 130, 207, 254–256, 269
Carius tube 292, 294
cataclasites 39
catastrophism 192
cathodoluminescence 155, 296
CB section 262
Cenomanian–Turonian (C/T) boundary 155, 164, 168, 184, 247
cenotes 117, 119
Cenozoic 72, 133
Central
– Apennines 157, 164
– European strewn field 68, 73, 80, 84, 85
– Pacific Ocean 61
central
– peak(s) 3, 8, 20, 132, 136
– uplift 7, 11, 13, 20, 26
Cetona Formation 162
chalcopyrite 142
charcoal 107, 108
chert(s) 164, 165, 178, 185, 247, 251, 253, 280
China 66, 75, 80, 148, 150, 151
chondrite(s) 44–46, 48, 49, 51, 118, 154
chromite 48, 64, 77, 142
chromium 288
– isotope(s) 43, 107, 50
cinder cones 3
circum-Antarctic current 254

Clear Creek North, Colorado 102
clinopyroxene 81, 91, 130, 204, 207, 220, 256
coccoliths 218
Coccolitophorida 217, 218
coesite 24, 34, 64, 78, 130, 256, 264
collapse 1, 19, 21, 150
Colombacci Formation 159
Colorado 102, 107, 110, 111, 121, 122
comet(s) 3, 14, 15, 63, 88, 109, 131, 142, 150, 156, 189, 195, 198, 253, 258, 266, 268
– impact 3
– shower 156, 159
compaction 219
complex craters 7, 10, 11, 20
compression stage 16, 19, 58
concordia diagram 121
condensed sequence 163
cone-in-cone structure 26
Cònero Riviera 274, 275
conjugate faults 231
contact
– /compression stage 14
– stage 19, 58
contamination 186, 196, 224, 227, 281–283, 287, 294
Contessa 281
– Highway section 180, 224, 225, 253
– Valley 180–182, 224, 250
continental
– crust 74, 76, 79, 153, 157, 171, 220, 291
– drift 4, 5
Coon Butte 5
copious carbonate production 236
corals 241
Corfino 172, 173
Còrniola Formation 163, 174
corundum 64, 78
cosmic spherules 62, 196
cosmogenic radionuclides 91
CQ section 262, 264, 268
Cr isotopic composition 43, 107, 51
Crater Mountain 5
crater
– chains 156
– floor 5, 7, 9, 81
– rays 59, 60
– rim 7, 9, 13, 24, 39, 40, 57, 59, 94, 104, 118, 132

– rocks 1, 52, 87, 88
– fill 39
Cretaceous– 1, 3, 63, 102–104, 107, 109, 114, 120, 124, 125, 127, 129, 130, 132, 134, 137–140, 156, 159, 162, 163, 165, 166, 168, 177, 179–181, 190, 192, 197, 199, 200, 203, 204, 209, 210, 213–219, 222, 223, 229, 231, 233–237, 239–241, 243–245
– Tertiary (K/T) boundary (see K/T boundary) 1, 140, 159
cristobalite 64
Crow Creek Member 137, 138, 152
crushers 283
crustal rocks 3, 17, 21, 38, 43, 44, 49, 64, 72, 79, 80, 98, 141, 266, 291, 295
crystallographic orientations 24, 29, 31, 32, 133
Cuba 80, 128
cyclicity 164, 185
Czech Republic 85

Danian 124, 190
Darwin glass 78
Deccan Traps 125
decompression 19, 47
deep-sea sediments 61, 75, 86, 88, 91
dendritic
– crystals 207
– structures 204, 207, 208, 265
detection limit 49, 91, 132, 154, 288
devitrification 38, 52
Devonian–Carboniferous boundary 155
diagenesis 48, 207
diagenetic alteration 207
diamonds 35, 36, 61, 108, 112, 136, 270
diaplectic glasses 23
Diaspri Formation 163, 178
Dicarinella 184
diffraction pattern 27
dikes 25, 39, 87
dissolution 36, 107, 204, 209, 218, 221, 223, 237, 286, 287, 290–294
distal ejecta 139, 48, 51, 52, 57, 59–61, 63, 69, 75, 94, 99, 101, 104, 113, 115, 122, 123, 132, 137–139, 141, 148, 149, 152, 155, 156, 167, 180, 282, 283, 286
dolomite(s) 17, 134
Dorothia 211
Dresden, Germany 85
DSDP Site

– 216 130, 131
– 292 130, 131
– 462 136
– 465A 206
– 576 106
– 577 106
– 596 31, 114
– 612 69, 78, 82, 131, 257, 269
dumbbell-shaped 204
dysoxia 251

E/O boundary 250, 254–256, 258
Early
– Cretaceous 97, 166
– Eocene 164, 166, 251
– Oligocene 254
Earth's orbit eccentricity cycles 212
earthquakes 6, 16, 237
echinoid 241
eclogites 21
EDX 287, 288
Egypt 95
ejecta 1, 19, 24, 39, 40, 51, 53, 57–61, 85, 92, 99, 101, 104, 112, 113, 122, 123, 126, 128, 129, 131, 133, 136, 138, 139, 141, 144, 146–149, 152, 154, 155, 180, 198, 217, 274
– blanket 59
El Kef, Tunisia 101, 124, 218
electron microprobe 287
Eltanin event 10
endogenic 109, 114, 151
– metamorphism 21, 22
Eocene– 61, 62, 82, 91, 131, 132, 136, 159, 165, 166, 189, 195, 201, 204, 207, 208, 250, 251, 253–259, 261, 264, 266–268, 270, 271
– Oligocene (E/O) boundary 166, 254
equation of state 17, 18
erosion 6, 7, 11, 13, 21, 25, 46, 48, 58, 117, 165, 171, 224, 287
escape velocity 14
Ethiopia 174
eugubina
– layer 215
– limestone 198–200, 214, 215, 219, 221–223, 228, 229, 231, 234–236, 241, 244
Europe 11, 150, 152, 157, 184, 189
Euxinic Shale 166
evaporites 159, 217
excavation 14, 19, 57, 58, 93

– stage 14, 19
Exmore breccia 83, 132, 133

F/F boundary 148, 150
fallback ejecta 7, 9
feldspar(s) 1, 17, 24, 27, 30, 34, 52, 93,
    109, 118, 133, 137, 204, 206, 207, 209,
    284, 295
felsites 72
fibrous structures 204, 208, 267
field documentation 279
Finland 54, 62
fireball 58, 59, 61, 102, 103, 105, 113
fish tooth 236
fission track 46, 51, 52, 54, 94, 224, 255
flanged-button australites 64
Flaxbourne River, New Zealand 108
flexural slip folding 207, 224
flood basalts 125
Florida 129, 157
fluid inclusion 29
flute(s) 241
– casts 241
flysch 159, 165, 166, 224, 239, 274
Fohn-1, Timor Sea 131
fold-and-thrust belt 157, 159
Fonte d'Olio 209, 240, 244, 246
Foraminifera 124, 165, 217, 284–286
foraminiferal turbidites 237
Fornaci 204, 240, 241, 244, 245
– East quarry 245
– Quarry C. section 204
– West quarry 245
fossil meteorite 106
Fossombrone 204
fractive index 23
fragmental impact breccia 40
Frasnian–Famennian (F/F) 148
Frontale 203, 209, 211, 223, 231–236
– outcrop 209
fulgurites 36
fullerenes 61, 108, 109, 150
Furlo 185–188, 201, 203, 204, 210–214,
    217, 218, 222, 223, 236–239, 244
– Gorge 237
– Upper Road section 185–188, 212
– Via Flaminia section 204

G.
– cerroazulensis 256, 257
– cretacea 200, 219, 235

– cretacica 218
– fringa 235
– minutula 235
– semiinvoluta 256, 257
Galilei, Galileo 3
Galileo 2, 3
Ganymede 10, 156
Garrufo 247
gas hydrates 130
Gaspra 2
Gavellinellidae 209
Genga 204
Georgia 80
georgiaites 74
gersdorffite 142
Gessoso–Solfifera Formation 166
Giant Piston Core 253, 266
Gifford, Algernon Charles 5
Gilbert, Grove Karl 4
glacial deposits 155
glass(es) 21, 24, 29, 33–37, 40, 41, 43, 59,
    64, 68, 69, 72, 73, 75–78, 80–82, 86, 88,
    89, 91, 97, 111, 112, 118, 120–123, 139,
    155, 204, 207, 219, 253, 257, 284–287,
    294, 296, 298
glaucony 204, 207, 209, 214, 215, 219–
    222, 224, 239, 244, 249
Global Stratotype Section and Point (see
    GSSP) 101, 159, 258
global wildfires 107, 108
Globigerina eugubina 197, 200
Globigerinateka semiinvoluta 256
globigerinids 197, 223, 231, 236, 239
Globorotalia cerroazulensis 256
Globotruncana 216
Globotruncanidae 197
Globotruncanita 216
goethite 203, 204, 207–209, 214, 215,
    220, 222, 244, 248
Gorgo a Cerbara 204
goyazite 103
grabens 157, 162
gradualism 192, 255
grain flows 239
granite(s) 17, 32, 72, 87, 137
granodiorite 17
graphite 24, 35, 36
gravity 11, 19, 57, 58, 89, 116, 117, 132,
    139, 146
Great Sand Sea, western Egypt 95
greenhouse 217

Greenland   153, 154, 253
GSSP   102, 159, 218, 258, 259, 261, 266
Gruithuisen, Franz von Paula   4
Guangxi, China   66
Guatemala   95, 129
Gubbio, Italy 159, 101, 166, 180, 181,
   185, 190–192, 195, 197, 198, 200, 201,
   203, 210, 212, 213, 216, 217, 219, 223,
   225, 227, 228, 231, 235, 237, 239, 244,
   250, 251, 262, 269
Gulf of Mexico   105, 130, 254–256, 269
Gumbelitria cretacea   200, 218

Hadean Earth   153
Haiti   89, 111, 112, 120–123, 129
halogens   74, 89, 127
Hamersley Basin, Western Australia   146
heat flow   16
heavy noble gases   90
Hedbergella   216
– planispira   181
HEL   17, 18
Helmintoidea   211
Helvetoglobotruncana helvetica   185
hematite   106, 203, 222, 224
hemipelagic   166, 274
Hercynian   162
Herschel, Friedrich Wilhem   3
Heterohelicidae   197
Hettangian Calcare Massiccio   162
Hevelius, Johannes   3
hiatus   198, 200, 215, 274
high pressure
– phases   23
– polymorphs   24, 34
high-sanidine   206, 207
histogram   299, 300
HNa/K australites   74, 94, 95
Hoba iron meteorite   15
Hooke, Robert   3
horst   162
Hospital Hill, Witwatersrand Basin   33
Hugoniot
– elastic limit   17
– equation(s)   17, 18
Hutton, James   4
hydrofluoric acid   29
hydrothermal alteration   48
hypercanes   127
hypervelocity impact 14, 15, 64, 72, 80,
   109

– crater   15

ichnofauna   211
ichnofossil   212, 273
ICP–AES   287
ICP–MS   154, 287, 290, 291, 294
illite   206
Ilyinets   30
impact
– angle   4, 42, 45
– cloud   217
– crater(s)   2, 3, 5, 7–11, 13–16, 19–21,
   24, 26, 34, 37–39, 43, 45, 50–53, 69, 85,
   88, 94, 96, 123, 132, 141, 155, 162, 167,
   171, 173, 250, 251, 255, 268, 274, 277,
   278
   – Acraman, Australia 43, 53, 61, 138,
   139, 146, 152, 154
   – Ames   11
   – Aouelloul, Mauritania 46, 47, 92, 93,
   94
   – Avak   11, 54
   – Bee Bluff, Texas   82
   – Bigach, Kazakhstan   54, 277
   – Bosumtwi, Ghana 37, 41, 42, 46, 54,
   61, 70, 79, 80, 86, 87, 152
   – BP   96–99
   – Brent, Canada   47
   – Chesapeake Bay 11, 54, 62, 70, 79–
   83, 131–133, 136, 152, 168, 195, 268–
   270hicxulub 6, 10, 11, 48, 53, 54, 61,
   63, 101, 104, 108, 113–127, 141, 152,
   168, 190, 195, 268
   – Darwin   78, 93
   – East Clearwater, Canada   46
   – Gardnos, Norway   53
   – Gosses Bluff   13, 21, 55, 141, 177
   – Haughton Dome (Devon Island,
   Canada)   26, 274, 278
   – Henbury   46, 93
   – Isidorus D   8
   – Kalkkop, South Africa   50
   – Kamensk, Russia   54, 171, 251, 253
   – Kara, Russia   54, 115
   – Kara–Ust Kara   171, 190, 194, 195
   – Karla, Russia   277
   – Logoisk, Russia   54, 55, 195, 268
   – Lonar, India   93
   – Manicouagan, Quebec   151, 168
   – Manson 11, 54, 103, 115, 120, 121,
   136–138, 152, 171, 190, 194, 195

– Meteor Crater   5, 14, 34, 43, 46, 47
– Mistastin, Canada   54, 82. 131, 136, 195, 268
– Mjølnir   55, 141, 168, 177
– Montagnais   54, 82, 251, 253
– Morokweng, 46, 51, 55, 139–141, 168, 177, 180
– Newporte   11, 25
– Oasis   96, 97, 98, 99
– Odessa, Texas   46
– Popigai 35, 54, 62, 82, 131, 133–136, 152, 168, 195, 254, 268–270
– Red Wing Creek   11, 27
– Ries 36, 40, 54, 70, 73, 80, 85, 86, 143, 152, 275, 278
– Rio Cuarto   7, 46, 155
– Rochechouart, France   46
– Roter Kamm   13
– Saltpan, South Africa   38, 46
– Steen River, Canada   54, 184
– Sudbury   7, 39
– Theophilus   8
– Tswaing, South Africa   38, 46
– Tycho   59
– Vredefort 7, 10, 21, 29, 39, 52, 117, 143, 148
– Wabar, Saudi Arabia   46, 47, 52, 93
– Wanapitei, Canada 54, 82, 131, 136, 268
– West Clearwater, Canada   48
– Zhamanshin   46, 54, 93
– cratering   1, 3, 4, 5, 16, 17
– diamond   24
– ejecta 1, 6, 10, 14, 48, 52, 53, 139, 180, 279
– excavation   19
– experiments   4, 7
– fallout   203, 204
– glass(es) 24, 36, 40, 47, 63, 69, 71, 89, 90–94, 96, 104, 111, 112, 114, 120, 121, 123, 130, 131, 136, 138, 139, 142, 148
– lithologies   24, 69
– melt(s) 1, 7, 24, 25, 36–40, 43, 44, 46–49, 51, 52, 118, 120–123, 133–137, 140, 152, 286, 293
– breccia(s), 39, 40, 69
– rock(s) 1, 7, 36, 37, 40, 43, 44, 46–48, 51, 269, 286, 293
– microspherules   206
– structures 1, 6, 10–12, 21, 34, 37, 39, 40, 43, 46, 48, 49, 51–54, 57, 61, 63, 81,
83, 87, 88, 93, 96–99, 103, 114–117, 122–124, 131–139, 141, 143, 144, 146, 148, 151–153, 156, 168, 171, 189, 194, 268, 270
– winter   217
impactite(s) 29, 39, 40, 42–44, 46, 47, 48, 49, 52, 53, 134, 144, 147, 282, 291
impactoclastic layer(s) 53, 60, 91, 101, 130, 131, 136, 143, 152, 189, 195, 258, 265–273, 282, 283, 286
impactor(s) 15, 42, 43, 46, 60, 97, 106, 107, 113, 118, 168, 189
INAA   186, 187, 204, 208, 287, 289, 291
India   93, 125
Indian Ocean   82, 131, 136, 152, 256
Indochina 67, 72, 74, 75, 78, 79, 80, 88, 89, 130
indochinite(s)   68, 74, 78, 92, 94, 95
inoceramids   192
interplanetary dust particles 62, 189, 195, 266, 268
inverted stratigraphy   59
ion microprobe   287
Iowa   103, 115, 136–138, 194
irghizites   94
iridium (Ir) 44, 45, 48, 61, 63, 76, 91, 96, 104–107, 113, 118, 128, 129, 131, 140, 142, 143, 146–150, 154, 159, 170, 185–187, 189, 195, 197, 198, 203, 204, 206, 207, 218, 220, 221, 223, 239, 244, 254, 256–258, 262–266, 269, 277, 287, 289, 290–292
iron
– meteorite(s)   5, 15, 43, 44–46, 50, 142
– sulfide   249
isochron   52, 293
isotope
– dilution   292, 293
– methods   51
isotropization   23, 29, 33
Isua (Greenland)   154
Ivory Coast, West Africa 61, 74, 79, 86–88, 91, 92, 152
– strewn fields   68, 75, 80, 86
– tektites   61, 74, 79, 86, 88, 91, 92

J/K boundary   140, 141, 168, 177, 178
jadeite   34
jetting   19, 77
Jurassic– 79, 134, 139, 140–142, 151, 162, 163, 168, 172, 173, 177, 179–181, 219

– Cretaceous (J/K) boundary 140, 168

K/T boundary 3, 6, 30, 31, 36, 48, 49, 51,
  53, 61–63, 89, 101–116, 118, 120–125,
  127–129, 137, 141, 143, 150, 152, 159,
  164, 168, 171, 172, 190, 192, 194–204,
  206–216, 218–232, 234, 236, 237, 239–
  241, 243–249, 254, 267, 271, 282–285,
  289
kaolinite   102, 103, 206
K–Ar method   51, 52, 68
Kara Sea   194
Kathu core   146
Kendelbachgraben, Austria   151
kerogen   107
K-feldspar 52, 83, 133, 146, 203, 204,
  206, 207, 209, 215, 220, 244, 248, 266,
  267
K-feldspars   106
kimberlites   34
Kimmeridgian   163
kinetic energy   14–16
kink bands   23
kinkbanding   133
Kyoto   258

L'Annunziata section, Apiro   275
La Lajilla, Mexico   129
La Vedova section   275
Laga Formation   159, 239
Langhian   166
Langren, Michel Florent van   3
Laos   64
Late
– Devonian   142, 148, 149, 150
– Eocene 82, 91, 101, 109, 130, 131, 136,
  143, 146, 162, 165, 168, 195, 204, 207,
  250, 253, 254, 256, 257, 262, 263, 265,
  266, 268, 272, 273, 283, 284, 286
– Triassic   152, 157, 162
– Triassic–Early Jurassic   157
late heavy bombardment   153, 154
layered tektites   64
LDG   95–99
lechatelierite 24, 34, 36, 38, 69, 74–76, 92,
  93, 96, 111
Lias   162
Libya   95
Libyan Desert Glass (see LDG) 95, 97–99
limestone(s) 27, 112, 129, 137, 149, 152,
  159, 163–166, 172, 173, 178, 185, 195–

  200, 203, 204, 207, 209–211, 213–215,
  217–219, 221–223, 228, 229, 231–233,
  235–237, 239–241, 244, 247, 248, 280–
  284, 286
lithic   36, 39, 40, 57, 99, 227
– (fragmental) breccias   39
– breccia   40
loess   64, 72, 75, 80
lonsdaleite   36
lower bathyal   210
lunar
– crater(s)   3–5, 7, 57, 59, 93
– highlands   153
– maria   10
Luochuan   75
Lyell, Charles   4

maars   3
Maastrichtian 101, 124, 128, 190, 192–
  195, 215, 231, 253
– crisis   190, 195
Macchia di Sole   247
Macigno   165
Madrid   107
magnesioferrite   62, 113, 114, 204, 208
– spinel   204
magnesiowüstite   113
magnetic spherules   196, 208, 266
magnetite   48, 113, 208, 224, 288
Maiolica Formation   163, 178, 180, 239
majorite   34
Malaysia   91
manganese nodules   106
mantle 48, 49, 79, 107, 120, 121, 140,
  153, 291, 293
– rocks   48, 49, 291
Marche   53, 163
marine plankton   203, 217, 220
marly limestone(s) 164–166, 222, 251,
  255, 258, 264, 269, 271, 273, 275, 285,
  286
Marne
– a Fucoidi Formation   163
– a Posidonia Formation   163
– del Serrone   174, 175
Marnoso–Arenacea Formation 159, 224,
  227, 239
Mars   5, 10, 153
Martha's Vineyard   80
maskelynite   24, 34

359

mass extinction(s) 1, 3, 51, 53, 63, 123, 124, 127, 137, 148, 150, 151, 155, 156, 159, 168, 172, 184, 186, 189
Massa Martana 172
MASSICORE 259, 266, 268
Massignano 131, 159, 189, 195, 253, 256–259, 261, 262, 264–273, 283
matrix 83, 208, 282, 286, 291, 295
Mauritania 46, 93, 94
mayaroensis limestone 241
Mediterranean 166
megabreccias 150, 163
Mekong
– River 80
– Valley 88
melting 19, 23, 25, 27, 34, 36, 38, 43, 64, 72, 79, 89, 92, 93, 96, 110, 133, 154, 287
Mercury 5, 10, 60
Mesozoic 133, 134, 159, 167
Messinian Gessoso–Solfifera
– Formation 159
– salinity crisis 166
metamorphic basement 162
meteor 15
meteorite(s) 4–7, 10, 14, 15, 21, 40, 42–47, 49, 50, 105, 107, 108, 139, 142, 144, 156, 291, 293
meteoritic
– component 38, 42–44, 46–48, 51, 91, 93, 96, 99, 107, 112, 118, 132, 139, 140, 142, 144, 279, 286
– matter 198
– projectile(s) 1, 10, 291
meteoroid 15
methane 130
Mexico 54, 63, 104, 105, 107, 112, 115, 116, 123, 127–130, 141
micas 23
microcline 227
micro
– krystites 130, 131, 136, 146, 152, 204, 207, 214, 215, 220, 221, 244, 256–258, 266, 267, 283, 285
– meteorites 198
– spherules 140, 206, 283
– tektites 59–61, 64, 69, 75–79, 81, 82, 85, 86, 88, 90, 91, 93, 130, 136, 146, 148, 152, 204, 207, 255–257, 269, 283, 287
Middle

– Eocene 165, 166, 254
– /Late Miocene 168
Milankovich cyclicity 164, 185, 215
Miller indices 28, 297, 299
mills 283, 288
Mimbral, Mexico 112, 123, 127, 128, 253
mixed layered illite-smectite 206
modification stage 14, 19
molasse 72, 86, 159, 274
Moldau 85
moldavite(s) 68, 73, 79, 84, 85
monazite 77, 96, 269
monomict
– breccias 25, 40
– fragmental impact breccias 39
Montagna dei Fiori 201, 203, 209, 210, 213, 223, 247, 249
Monte
– Cagnéro 256
– Cerchio 172, 173
– Cònero 164, 166, 201, 203, 204, 209, 210, 213, 217, 222, 223, 239, 241, 242, 244, 247, 256, 258
– dei Corvi 275, 277
– Nerone 163
– San Vicino 201, 211, 231, 232
– Serrone 175
Monteville Formation 146
Monti
– Martani 172
– Sibillini 201
Moon 1, 3–5, 7–10, 14, 59, 72, 153, 156
Moravia, Czech Republic 85
mosaicism 27, 83, 118, 133, 296
Mount St. Helens 16, 108
multiple impacts 80, 103, 148, 185, 189
multiring basins 10, 117
Muong Nong-type tektites 64, 73–75, 77, 89, 90, 92, 111

Namibia 13, 15
nannoplankton 103
Nebraska 137, 152
nekton 217
neritic facies 162
neutron activation analysis 154, 186, 287, 291
Nevada 149
New Zealand 5, 107, 108, 110
Ni-rich spinel(s) 61, 102, 131, 258, 265, 267

NMR   24, 34, 154, 296
nodular chert(s)   164, 165, 178
nomenclature   39, 42, 104
Norian   168
North American
– microtektite layer   78, 131
– microtektites   82
– strewn field(s)   68, 73, 75, 78, 80–82,
  91, 130–132, 136, 256, 262
– tektite(s)   74, 79, 82, 131, 133, 136,
  255, 269
North Atlantic   107, 130, 157, 251, 256
Nova Scotia   82, 151
NTIMS   293, 294
Nubian sandstone   97
nuclear
– explosion crater   58
– magnetic resonance (see NMR)   34

O₂ minimum zone   184
oblique impact(s)   7, 46, 99, 109, 117, 155
ocean floor   10
oceanic anoxic event   164, 184, 189
ODP Site
– 536   105
– 588B   277
– 689B   82, 257, 277
– 758B   91
– 904A   82
– 1049   107, 129
olivine   17, 24, 27, 34, 288, 295
Oort cloud   156
Öpik, Ernest   5
optical mosaicism   23
orbital velocity   15
orbitoids   241
ordinary chondrites   51
osmium (Os) isotope(s) 49, 93, 107, 112,
  292–294
Osangulariidae   209
outer rim   7, 132, 134
oxygen isotopes   79

P. eugubina Zone   200, 203
P0 Zone   200, 215, 218
PAHs   108
Paleocene–Eocene boundary   253
Paleodychtion   211
Paleogene   159, 197, 250, 258
paleontology   6, 192
Paleozoic   132, 137, 167

palynomorphs   172
parautochthonous rocks   39
Paris Basin   251
Parvularuglobigerina eugubina Zone   200
PDF(s), 87, 109, 110, 118, 131, 133, 138,
  139, 149, 151, 208, 209, 264, 295–300
peak ring   8, 116, 118, 132
pebbly mudstones   163
pelagic 80, 182, 185, 192, 193, 195, 197,
  198, 201, 203, 204, 207, 210, 211, 213,
  215, 217, 219, 231, 237, 239, 241, 247,
  250, 255, 258, 259, 262, 264, 269, 274
– biomicrites   237
– limestones   159, 164
– ooze   203
Peñalver Formation   128
Pennide–Liguride Ocean   157
perennial fallout   198
Pering mine   146
periodicity   51, 53, 155, 156, 168
Permian–   134, 150, 168, 254
– Triassic boundary (P/TR) 61, 150, 168,
Perugia   162
Petriccio 113, 199, 203, 204, 206, 208–
  210, 213, 214, 220, 223, 227–231, 235
– outcrop   209
– section   228
petrographic microscope   283, 284, 286,
  295
PFs   27–29, 32
PGE(s) 44, 47, 48, 61, 102, 104–107, 113,
  139, 140, 142–144, 147, 150, 154, 282,
  289, 291, 292, 294
Phanerozoic   184, 254
Philippines   74
philippinite   90
photic zone   157, 217
photosynthesis   217
phytoplankton   217
Pian dell'Elmo   231
Pierre Shale, South Dakota   137, 138
piggyback basins   159
pillow basalts   207
Piobbico   181, 201, 256
plagioclase   24, 34, 52, 83, 133, 227, 296
planar
– deformation features (see PDFs), 23, 24,
  27–29, 32, 34, 118, 146, 264
– features   110, 111, 139, 284, 295, 300
– fractures   24, 28, 154
planetesimals   153

plankton   180, 190, 217, 218, 241, 254, 271, 274, 275
planktonic foraminifera   165, 181, 194, 195, 197, 199, 201, 209, 214, 215, 217, 218, 231, 236, 237, 269, 271
Planolites   211–215, 222, 233, 234, 239, 240, 266, 267, 271
plate tectonics   4, 6, 11
platinum group elements (see PGEs)   44, 140, 142, 198, 282
Pleistocene   136, 157, 159
Pliensbachian–   163, 168, 173, 174, 176
– Toarcian boundary (P/T)   168
Pliocene   11, 166
Podrina   228
Poggio   181–183, 195, 231–233, 240
– Le Guaine   181, 182
– San Vicino   195, 231–233
polycyclic aromatic hydrocarbons (see PAHs)   108
polymict breccia(s)   25, 40, 118, 135, 136
Port Campbell, Victoria, Australia   91
power drill   281
Praeglobotruncana   184
Precambrian   79, 82, 138, 148
precision   51–53, 112, 132, 156, 167, 198, 288
pressure-solution   26, 210, 232
– pseudobedding (stylolitization)   210
primary productivity   194, 195, 210
productivity   211, 215, 217, 241, 254
projectile   15, 16, 19, 42, 43, 45–49, 91, 144, 209
proximal ejecta   39, 57, 59–61, 63, 104, 113, 128, 149, 282
pseudomorphic   201
pseudotachylitic breccias   25, 40
Pueblo, Colorado   185
pyrite   118, 142, 146, 203, 209, 210, 213, 220, 249
pyroxene   34

Quagliotti quarry   244, 246
quartz   1, 17, 23, 24, 27–34, 36–38, 61, 71, 77, 78, 83, 88, 93, 96, 102, 109–111, 115, 118, 128, 131, 133, 134, 137–139, 146, 149–151, 172, 195, 203, 204, 207–209, 220, 221, 227, 244, 256, 258, 262–266, 269, 283–285, 287, 294–300
quartzite   32, 33, 149, 283
quenched silicate melt   204

quenching   21, 64, 72, 267
R. cushmani   184
Racemiguemelina fructicosa   216
radiochemical neutron activation analysis (RNAA)   291
radiolarites   184
Raman spectroscopy   296
rare earth element(s) (REE)   37, 64, 72, 73, 94, 96, 98, 106
rare fish   209
rarefaction wave   18, 19
Raton
– Basin   30
– Pass, Colorado   102
– New Mexico   102, 107, 109
Rb–Sr   49, 51, 52, 72, 79, 80–83, 89, 112, 120, 136
– isochron method   52, 293
– isotopic system   79
recrystallization   35, 37, 52, 118, 144
redox   207, 210, 239
reduction   91, 162, 218, 288, 290, 291
REE patterns   64, 72, 73, 94, 96
refractive index   296
release wave   18, 19
relict mineral(s)   75, 77
Reophax   211
retene   108
Rhaetavicula contorta limestones   162
Rhyzammina   211
ringwoodite   24, 34
RNAA   291
Rocca Leonella   201
Rosita contusa   216
Rosso Ammonitico Formation   163, 174
Rotalipora   184, 185
– *cushmani*   185
rudist(s)   192, 241
Rugoglobigenindae   197
Rugoglobigerina   216
Russia   12, 35, 54, 55, 115, 131, 194, 277
rutile   64, 77

sample preparation   279, 282, 287, 290, 295
San Francisco   16
sandstone(s)   27, 62, 93, 96, 98, 127, 128, 132, 134, 137, 139, 224, 227, 280
sanidine   137, 206, 227, 275, 284
Sassoferrato   175

Saudi Arabia 52
Scaglia
- Bianca 164, 165, 181, 182, 184, 185, 187, 188, 247
- Cinerea, 164–166, 250, 255, 274
- Rossa 164, 165, 181, 182, 184, 187, 188, 190, 195, 196, 198, 201, 203, 209–213, 215, 217–219, 222, 224, 231, 237–239, 241, 244, 247, 250, 251, 253
- Variegata 164, 165, 250, 255, 268
scanning electron microscopy (see SEM) 287
Scheggia 175, 223
Schlier Formation 166
schlieren 64, 93, 96
Scolicia 211
secondary craters 58
sediment loading 210
seismic data 11, 117
Selli Level 163, 164, 181, 182
SEM 29, 31, 103, 110, 111, 114, 204, 207, 208, 227, 236, 249, 267, 286–288, 295, 300
shales 27, 64, 72, 137, 138, 174
shallow water environments 184, 192
Shatsky Rise 107
shatter cones 10, 25, 26, 27, 138
shock
- deformation 10
- effects 29, 87, 154, 155, 282
- metamorphic 1, 3, 10, 21, 23, 24, 29, 109, 139, 154, 296
- metamorphism 10, 21–23, 27–29, 38–40, 49, 77, 83, 87, 109, 118, 132, 133, 139, 143, 154, 209
- recovery experiments 22
- wave(s) 1, 16–19, 22, 27, 33, 36, 42, 59
shocked
- minerals 3, 19, 24, 27, 33, 36, 52, 61, 75, 78, 91, 99, 103, 109, 110, 114, 130, 136, 137, 146, 148, 152, 154, 168, 170, 185, 187, 279, 295
- quartz 27, 29, 35, 83, 109, 110, 133, 138, 151, 220, 265, 269, 283, 284, 286, 300
Siberia 6, 12, 15, 35, 61, 82, 94, 131, 133, 135, 140, 254
siderophile element(s) 3, 43, 44, 47, 48, 60, 61, 76, 93, 96, 103, 105, 106, 130, 140, 142, 143, 146, 154, 282. 287
silica glass 29

siliciclastic 159, 165, 166, 198, 209, 219, 274
- turbidites 165, 166
silicon carbide 36
simple craters 7, 11
Sirolo 240, 258
skeletal crystals 204, 207, 208, 267
skeletal structure 204
slope facies 164
slumping 11, 19, 129, 201, 204, 231
slumps 163, 181
smectite 103, 206, 286
Sm–Nd, 49, 51, 72, 79, 81–83, 89, 112, 120, 136
solar system 14, 15, 131, 156
soot 107, 108, 130
Sorosphaera 211
source crater 1, 6, 53, 57, 61, 69–71, 79, 80, 82, 83, 85, 86, 88, 89, 92–95, 99, 104, 110, 115, 132, 138, 139, 142, 148, 149
South
- Africa 7, 29, 38, 46, 50–52, 55, 117, 139, 142, 143, 145–148
- Australia 61, 138
- China Sea 80
- Dakota 137, 152
Southern Ocean 11, 82, 131, 152
spallation 42
spheroids 207, 224, 244, 248, 266
spherules 43, 62, 76, 81, 91, 102, 103, 106, 111, 113, 114, 130, 142–146, 148, 149, 168, 170, 172, 185, 187, 195, 196, 201, 204, 206–209, 214, 219–224, 227, 233, 234, 239, 244, 253, 256, 262, 266, 271
spindle stage 29, 284, 296
spinel 61, 62, 106, 113, 195, 204, 208, 262, 265–267, 286, 287
Spiroplectammina 210, 211, 227
splash-form tektites 64, 74, 89, 90
Spoleto 172
Starkville South, Colorado 107
stepwise
- extinction 124
- heating 53, 123
Stevns Klint 105–107, 109, 110
stishovite 24, 34, 78, 111, 264
stony meteorites 15, 43
stony-iron meteorites 44
stress wave 16

strewn field(s) 59, 64, 66, 68–75, 77, 79–82, 84–86, 88, 89, 91, 93, 130–132, 136, 146, 155, 256
Subphyllochorda 211
subsidence 1, 157, 159, 162, 163
– rate 163
suevite 36, 37, 40, 42, 87, 134, 286
suevitic 36, 38–41
– (fragmental) 25
– breccia(s) 36, 39, 40, 134
sulfur isotopes 108
Sumbar, Turkmenistan 107
superheating 36, 92, 217
supernova explosions 124
supersonic shock wave 16

target 1, 7, 11, 16, 17, 19–21, 24–27, 32, 34, 35, 37, 39, 42–44, 47, 49, 50, 57, 59, 60, 65, 69, 75, 77, 79, 85, 87, 91–93, 97, 98, 108, 127, 133, 139, 141, 146, 154, 155, 265, 269
– rock(s) 87, 92, 98, 108, 282, 286, 293, 294
Tasman Sea 277
Tasmania 93
Teapot Dome, Wyoming 103
tectonic
– deformation 23, 165, 171, 173
– shearing 219, 224, 235
tektite(s) 52, 57, 60, 61, 63–72, 74–95, 99, 111, 112, 131, 133, 136, 152, 255–257, 262, 269
– classification 69
TEM 28, 29, 36, 286, 295, 300
Tertiary 102, 107, 118, 127, 129, 140, 159, 163, 190, 192, 197, 199, 200, 203, 206, 209, 212, 214, 215, 219, 222, 223, 231, 233–237, 239, 241, 243–245
Terzo San Severo 172
Tethyan Ocean 157
Textulariina 209, 210, 211
Thailand 66, 68, 88
thailandites 66
thin section(s) 132, 145, 200, 204, 207, 215, 227, 266, 267, 284–287, 295, 298
thrust-and-fold belt 239
Tikal, Guatemala 95
tillites 155
Timor Sea 131
titano-magnetite 224
Tithonian– 140

– Berriasian boundary 168, 177
Toarcian 168, 173, 174, 176
Tonle Sap, Cambodia 89
Torricella section (Furlo anticline) 214
Tortonian–Messinian boundary 166
trace fossils 210, 212, 213, 229, 233, 244, 271
transient cavity 19, 21, 59
transmission electron microscope (see TEM) 28, 264
Transvaal Supergroup 145, 146
Triassic– 159, 162, 166
– Jurassic (TR/J) boundary 151, 152, 168, 172, 173
Tritaxia 211
tsunami deposits 128, 149
Tunguska, Siberia 6, 12, 15, 61, 97
turbidite 159, 164, 201, 203, 212, 222, 231, 237, 244, 274
Turkmenistan 107
Turonian 164, 184, 237
Tuscan basin 165

Ukraine 30, 54, 55
ultramafic 48, 49
Umbria–Marche 53, 101, 131, 141, 157, 161, 244
– Apennines 157, 161
– sequence 53
uniformitarian(ism) 4, 6, 114
United States of America 12, 131, 132, 137, 185, 189
universal stage 29, 31, 118, 296, 298
U–Pb age 121
Upper Cretaceous–Oligocene 163
upper crust 65, 71, 72, 73, 91
Upper Freshwater-Molasse 86
Ural Glass 94
Uranus 3
urengoites 94
U-stage 284, 296–298, 300
U–Th–Pb dating 51, 52
Valdorbìa 173–176
vapor fractionation 75, 77, 89
vaporization 19, 36, 217
Venezuela Basin 256
Venus 5
Verneulina 211
Vietnam 88
Vispi quarry 181, 182
Vltava 85

volatile elements   74, 89
volatilization fractionation   89
volcanic 3–7, 16, 17, 21, 35–38, 60, 167,
  206, 219, 224, 253, 262, 274
– ashes   165, 166, 253
– eruption(s)   6, 4, 108, 124
– features   3
– hypothesis   3, 4
– structures   7
volcanism   1, 3, 6, 16

W. archeocretacea   185
W. hornestowensis   235
Washington   258
water cooling   215
water–sediment interface   200, 203, 210,
  215, 218, 219, 221, 239, 248
Weddel Sea   277
Wegener, Alfred   4
Western Interior   103, 110, 185
Whiteinella archeocretacea   185
winnowing   204, 218, 239

Wittenoom Formation   146
Witwatersrand Basin   33
Wolf Creek   46
Woodside Creek, New Zealand   107, 108
Wyoming   103

X–ray
– computed tomography   42
– diffraction   27, 36, 296
– fluorescence (XRF)   287

Yucatan   217
– Peninsula, Mexico   63
– Mexico   115–117, 119, 120, 125

zhamanshinites   94
zircon(s) 24, 29, 36, 52, 63, 64, 77, 78, 96,
  110, 111, 121, 122, 137, 139, 140, 153,
  154, 224, 269
Zoophycos 211–213, 215, 222, 223, 231,
  233, 239, 240, 271
zooplankton   198, 217

Lecture Notes in Earth Sciences

For information about Vols. 1–19
please contact your bookseller or Springer-Verlag

Vol. 20: P. Baccini (Ed.), The Landfill. IX, 439 pages. 1989.

Vol. 21: U. Förstner, Contaminated Sediments. V, 157 pages. 1989.

Vol. 22: I. I. Mueller, S. Zerbini (Eds.), The Interdisciplinary Role of Space Geodesy. XV, 300 pages. 1989.

Vol. 23: K. B. Föllmi, Evolution of the Mid-Cretaceous Triad. VII, 153 pages. 1989.

Vol. 24: B. Knipping, Basalt Intrusions in Evaporites. VI, 132 pages. 1989.

Vol. 25: F. Sansò, R. Rummel (Eds.), Theory of Satellite Geodesy and Gravity Field Theory. XII, 491 pages. 1989.

Vol. 26: R. D. Stoll, Sediment Acoustics. V, 155 pages. 1989.

Vol. 27: G.-P. Merkler, H. Militzer, H. Hötzl, H. Armbruster, J. Brauns (Eds.), Detection of Subsurface Flow Phenomena. IX, 514 pages. 1989.

Vol. 28: V. Mosbrugger, The Tree Habit in Land Plants. V, 161 pages. 1990.

Vol. 29: F. K. Brunner, C. Rizos (Eds.), Developments in Four-Dimensional Geodesy. X, 264 pages. 1990.

Vol. 30: E. G. Kauffman, O.H. Walliser (Eds.), Extinction Events in Earth History. VI, 432 pages. 1990.

Vol. 31: K.-R. Koch, Bayesian Inference with Geodetic Applications. IX, 198 pages. 1990.

Vol. 32: B. Lehmann, Metallogeny of Tin. VIII, 211 pages. 1990.

Vol. 33: B. Allard, H. Borén, A. Grimvall (Eds.), Humic Substances in the Aquatic and Terrestrial Environment. VIII, 514 pages. 1991.

Vol. 34: R. Stein, Accumulation of Organic Carbon in Marine Sediments. XIII, 217 pages. 1991.

Vol. 35: L. Håkanson, Ecometric and Dynamic Modelling. VI, 158 pages. 1991.

Vol. 36: D. Shangguan, Cellular Growth of Crystals. XV, 209 pages. 1991.

Vol. 37: A. Armanini, G. Di Silvio (Eds.), Fluvial Hydraulics of Mountain Regions. X, 468 pages. 1991.

Vol. 38: W. Smykatz-Kloss, S. St. J. Warne, Thermal Analysis in the Geosciences. XII, 379 pages. 1991.

Vol. 39: S.-E. Hjelt, Pragmatic Inversion of Geophysical Data. IX, 262 pages. 1992.

Vol. 40: S. W. Petters, Regional Geology of Africa. XXIII, 722 pages. 1991.

Vol. 41: R. Pflug, J. W. Harbaugh (Eds.), Computer Graphics in Geology. XVII, 298 pages. 1992.

Vol. 42: A. Cendrero, G. Lüttig, F. Chr. Wolff (Eds.), Planning the Use of the Earth's Surface. IX, 556 pages. 1992.

Vol. 43: N. Clauer, S. Chaudhuri (Eds.), Isotopic Signatures and Sedimentary Records. VIII, 529 pages. 1992.

Vol. 44: D. A. Edwards, Turbidity Currents: Dynamics, Deposits and Reversals. XIII, 175 pages. 1993.

Vol. 45: A. G. Herrmann, B. Knipping, Waste Disposal and Evaporites. XII, 193 pages. 1993.

Vol. 46: G. Galli, Temporal and Spatial Patterns in Carbonate Platforms. IX, 325 pages. 1993.

Vol. 47: R. L. Littke, Deposition, Diagenesis and Weathering of Organic Matter-Rich Sediments. IX, 216 pages. 1993.

Vol. 48: B. R. Roberts, Water Management in Desert Environments. XVII, 337 pages. 1993.

Vol. 49: J. F. W. Negendank, B. Zolitschka (Eds.), Paleolimnology of European Maar Lakes. IX, 513 pages. 1993.

Vol. 50: R. Rummel, F. Sansò (Eds.), Satellite Altimetry in Geodesy and Oceanography. XII, 479 pages. 1993.

Vol. 51: W. Ricken, Sedimentation as a Three-Component System. XII, 211 pages. 1993.

Vol. 52: P. Ergenzinger, K.-H. Schmidt (Eds.), Dynamics and Geomorphology of Mountain Rivers. VIII, 326 pages. 1994.

Vol. 53: F. Scherbaum, Basic Concepts in Digital Signal Processing for Seismologists. X, 158 pages. 1994.

Vol. 54: J. J. P. Zijlstra, The Sedimentology of Chalk. IX, 194 pages. 1995.

Vol. 55: J. A. Scales, Theory of Seismic Imaging. XV, 291 pages. 1995.

Vol. 56: D. Müller, D. I. Groves, Potassic Igneous Rocks and Associated Gold-Copper Mineralization. 2nd updated and enlarged Edition. XIII, 238 pages. 1997.

Vol. 57: E. Lallier-Vergès, N.-P. Tribovillard, P. Bertrand (Eds.), Organic Matter Accumulation. VIII, 187 pages. 1995.

Vol. 58: G. Sarwar, G. M. Friedman, Post-Devonian Sediment Cover over New York State. VIII, 113 pages. 1995.

Vol. 59: A. C. Kibblewhite, C. Y. Wu, Wave Interactions As a Seismo-acoustic Source. XIX, 313 pages. 1996.

Vol. 60: A. Kleusberg, P. J. G. Teunissen (Eds.), GPS for Geodesy. VII, 407 pages. 1996.

Vol. 61: M. Breunig, Integration of Spatial Information for Geo-Information Systems. XI, 171 pages. 1996.

Vol. 62: H. V. Lyatsky, Continental-Crust Structures on the Continental Margin of Western North America. XIX, 352 pages. 1996.

Vol. 63: B. H. Jacobsen, K. Mosegaard, P. Sibani (Eds.), Inverse Methods. XVI, 341 pages, 1996.

Vol. 64: A. Armanini, M. Michiue (Eds.), Recent Developments on Debris Flows. X, 226 pages. 1997.

Vol. 65: F. Sansò, R. Rummel (Eds.), Geodetic Boundary Value Problems in View of the One Centimeter Geoid. XIX, 592 pages. 1997.

Vol. 66: H. Wilhelm, W. Zürn, H.-G. Wenzel (Eds.), Tidal Phenomena. VII, 398 pages. 1997.

Vol. 67: S. L. Webb, Silicate Melts. VIII. 74 pages. 1997.

Vol. 68: P. Stille, G. Shields, Radiogenetic Isotope Geochemistry of Sedimentary and Aquatic Systems. XI, 217 pages. 1997.

Vol. 69: S. P. Singal (Ed.), Acoustic Remote Sensing Applications. XIII, 585 pages. 1997.

Vol. 70: R. H. Charlier, C. P. De Meyer, Coastal Erosion – Response and Management. XVI, 343 pages. 1998.

Vol. 71: T. M. Will, Phase Equilibria in Metamorphic Rocks. XIV, 315 pages. 1998.

Vol. 72: J. C. Wasserman, E. V. Silva-Filho, R. Villas-Boas (Eds.), Environmental Geochemistry in the Tropics. XIV, 305 pages. 1998.

Vol. 73: Z. Martinec, Boundary-Value Problems for Gravimetric Determination of a Precise Geoid. XII, 223 pages. 1998.

Vol. 74: M. Beniston, J. L. Innes (Eds.), The Impacts of Climate Variability on Forests. XIV, 329 pages. 1998.

Vol. 75: H. Westphal, Carbonate Platform Slopes – A Record of Changing Conditions. XI, 197 pages. 1998.

Vol. 76: J. Trappe, Phanerozoic Phosphorite Depositional Systems. XII, 316 pages. 1998.

Vol. 77: C. Goltz, Fractal and Chaotic Properties of Earthquakes. XIII, 178 pages. 1998.

Vol. 78: S. Hergarten, H. J. Neugebauer (Eds.), Process Modelling and Landform Evolution. X, 305 pages. 1999.

Vol. 79: G. H. Dutton, A Hierarchical Coordinate System for Geoprocessing and Cartography. XVIII, 231 pages. 1999.

Vol. 80: S. A. Shapiro, P. Hubral, Elastic Waves in Random Media. XIV, 191 pages. 1999.

Vol. 81: Y. Song, G. Müller, Sediment-Water Interactions in Anoxic Freshwater Sediments. VI, 111 pages. 1999.

Vol. 82: T. M. Løseth, Submarine Massflow Sedimentation. IX, 156 pages. 1999.

Vol. 83: K. K. Roy, S. K. Verma, K. Mallick (Eds.), Deep Electromagnetic Exploration. X, 652 pages. 1999.

Vol. 84: H. V. Lyatsky, G. M. Friedman, V. B. Lyatsky. Principles of Practical Tectonic Analysis of Cratonic Regions. XX, 369 pages. 1999.

Vol. 85: C. Clauser, Thermal Signatures of Heat Transfer Processes in the Earth's Crust. X, 111 pages. 1999.

Vol. 86: H. V. Lyatsky, V. B. Lyatsky, The Cordilleran Miogeosyncline in North America. XX, 384 pages. 1999.

Vol. 87: M. Tiefelsdorf, Modelling Spatial Processes. XVIII, 167 pages. 2000.

Vol. 88: S. Rodrigues-Filho, G. Müller, A Holocene Sedimentary Record from Lake Silvana, SE Brazil. XII, 96 pages. 1999.

Vol. 90: R. Klees, R. Haagmans (Eds.), Wavelets in the Geosciences. XVIII, 241 pages. 2000.

Vol. 91: I. Gilmour, C. Koeberl (Eds.), Impacts and the Early Earth. XVIII, 445 pages. 2000.

Vol. 93: A. Montanari, Ch. Koeberl, Impact Stratigraphy. XIII, 364 pages. 2000.